Nuffield Co-ordinated Sciences

CHEMISTRY

Published for the Nuffield—Chelsea Curriculum Trust by
Longman Group UK Limited

General Editors,
Nuffield Co-ordinated Sciences
Geoffrey Dorling
Andrew Hunt
Grace Monger

Author and General Editor, Chemistry
Andrew Hunt

Contributors
David Barlex
Sir Lawrence Bragg
John Groves
R. B. Ingle
Roland Jackson
Gordon Van Praagh

The Nuffield–Chelsea Curriculum Trust acknowledge their debt to the many contributors to earlier Nuffield Science schemes.

The Trust would like to thank the team of advisers who guided the development of Nuffield Co-ordinated Sciences Chemistry:

David Barlex, Goldsmiths College
Tim Brosnan, St Mary's College, Twickenham
Tony Dempsey, The Feltham School, Middlesex
John Groves, Banbury School
Roland Jackson, Backwell Comprehensive School, Avon
Roger Norris, Wymondham College, Norfolk
Pam Rivaz, Hertfordshire Country Council
Lindsay Taylor, Groby Community College

The Trust would also like to record their gratitude to the following for advice on the industrial aspects of Nuffield Co-ordinated Sciences Chemistry:

Exxon Chemicals at Southampton
P. N. Bailey
D. Beattie
M. N. Chaplain
J. E. Hackitt
P. E. Moore
J. F. C. Nicholls
R. C. Rivaz
N. J. Tilling
F. P. Vickers
D. Whately

Exxon Chemicals at Mossmorran in Fife
D. W. S. Wright

ICI
R. A. Finch, Schools Liaison Officer, Welwyn Garden City
C. S. Major, Plant Protection Division
D. M. Raw, Agricultural Division
P. Shorrocks, Agricultural Division

Mineral Industry Manpower and Careers Unit
Mr G. A. Cox, Director
Miss E. Barrett, Richmond-upon-Thames College

RTZ
J. Welch, Publications and Publicity Manager
C. Harris, RTZ Metals

Water Research Council
K. V. Bowles

Help was also received from many organizations who contributed to a resource bank of up-to-date information about the chemical industry for the project. These sources of information are listed in the appendix to the teachers' guide.

Finally the Trust would like to thank the Headmaster and Governors of Durrants School for their willing support in releasing the editor from his duties at the school during the development of the project.

Longman Group UK Limited
Longman House, Burnt Mill, Harlow, Essex CM20 2JE, England and
Associated Companies throughout the World

First published 1988
Sixth impression 1990
Copyright © The Nuffield-Chelsea Curriculum Trust 1988

Illustrations by Peter Edwards, Oxford Illustrators and Chris Ryley
Cover illustration by Plus 2 Design

Filmset in Times Roman
Produced by Longman Group (FE) Ltd
Printed in Hong Kong

ISBN 0 582 04256 9

Contents

Introduction   1

Topic C1   Raw materials   9

Chapter C1   The elements of chemistry   10
Chapter C2   Petrochemicals   20
Chapter C3   Chemicals from plants   46
Chapter C4   Chemicals and rocks   57

Topic C2   Materials in use   85

Chapter C5   Materials and structures   86
Chapter C6   Glasses and ceramics   113
Chapter C7   Metals and alloys   129
Chapter C8   Polymers   145

Topic C3   Chemicals in our homes   163

Chapter C9   Foams, emulsions, sols and gels   164
Chapter C10   Keeping clean   178
Chapter C11   Dyes and dyeing   201
Chapter C12   Chemicals in the medicine cupboard   213

Topic C4   Energy changes in chemistry   223

Chapter C13   Fuels and fires   224
Chapter C14   Batteries   248

Topic C5   Soil and agriculture   255

Chapter C15   Soil   256
Chapter C16   Fertilizers   268

Topic C6   The Periodic Table, atoms and bonding   287

Chapter C17   The Periodic Table   288
Chapter C18   Atoms and bonding   298

Data section   315
Index   330

Acknowledgements

AFRC, Institute of Food Research, Reading Laboratory: 9.6, 9.7
Aerofilms Ltd: 9.5
Agricultural Research Council: 3.17
The Astbury Dept of Biophysics, University of Leeds: 3.16
Dr A. R. Bailey: 7.11
BASF: 16.14
BDH Ltd: 12.14a and b
Jit Baran: 0.8, 1.11a, b and c, 2.1, 2.3, 2.4, 2.7, 2.8, 2.9, 2.36, 2.39a, b and c, 2.41a, b and c, 4.19, 4.27, 4.28, 5.8, 5.10, 5.15, 5.18, 5.19, 5.20, 5.31b, 5.37, 5.38, 5.39, 5.40, 5.51, 5.52, 5.53, 8.40, 8.41, 9.15, 9.16, 9.17, 9.20, 9.22, 9.23, 9.24, 9.31, 10.21, 12.2, 13.4, 13.9, 15.6, 15.10
Barnabys Picture Library: page 9, 5.1, 10.24, 13.41
J. Bibby Science Products Ltd: 6.5, 6.6
BOC Ltd: 17.7
The Boots Company: page 163
British Aerospace: 5.6
British Alcan Aluminium: 0.2, 4.34, 4.39, 7.31
British Gas: 2.2g, 13.5, 13.31
British Glass Industry Research Association: 6.16
British Geological Survey: 4.1, 4.3, 4.4, 15.2, 15.5, 15.9
British Launderers Research Association: 10.8
British Library: 1.7
BP Chemicals Ltd: 2.2a, 2.2e, 3.4, 8.3, 8.5
BP Oil Ltd: 0.9b, 0.14, 2.2f, 2.2i
The British Petroleum Company: 2.2h, 13.24, 13.27, 13.36, 13.49
British Railways: 7.36
BSC Stainless: 7.29
British Steel Corporation: 4.31, 7.10, 7.18, 7.22, 7.42, 7.43
Building Research Establishment: 15.8
Camera Press: 3.7, 13.9
Camping Gaz (GB) Ltd: 13.7
The J. Allan Cash Photo Library: 0.1, 0.12, 3.1, 7.12, 7.32, 10.1, 12.7, 15.12
Central Electricity Generating Board: 6.17, 13.37
Ceram Research: 6.11, 6.25, 6.29
Chemical Society: 5.34d
Chloride Ltd: 14.11, 14.13
Corning Ltd: 6.20
Courtaulds Speciality Plastics: 8.22
Courtaulds Acetate: 8.24
Crown House Tableware: 5.3, 6.4
Daily Telegraph Colour Library: 9.33
De Beers Consolidated Mines Ltd: 7.13
Deutches Museum, Munich: 16.12
The Dumfermline Press: 2.35
Dyfed County Council: page 223
ESPI: 8.2a and b, 8.32, 8.34, 8.37, 8.38, 8.39
Equal Opportunities Commission: 12.15
Esso Petroleum Company Ltd: 13.23, 13.26
Ever Ready: 14.8
Exxon Chemical Olefins Inc., Fife Ethylene Plant: 2.28, 2.29, 2.30, 2.31, 2.32, 2.33, 2.34, 2.50
Exxon Chemical Belgium: 2.46, 2.48, 2.49

Farmers Weekly: 2.2d, page 255, 15.21, 16.3
Finncell, 0.9a, 3.2
Fisons Fertilizers Ltd: 16.9
Ford Motor Company Ltd: 7.28
Forestry Commission: 13.43
Forgemasters Steels Ltd: 7.20
Format Photographers Ltd/Roshini Kempadoo: 0.3
Format Photographers Ltd/Jenny Mathews: 0.13
Format Photographers Ltd/Maggie Murray: 0.15
Format Photographers Ltd/Joanne O'Brien: 12.1, 12.4
Format Photographers Ltd/Brenda Prince: 0.11
Geological Museum: 4.5, 4.6, 4.8, 4.9, 4.10, 4.11, 4.32, 5.28b, 7.14
Glass and Glazing Federation, 6.22
Dr L. F. Haber: 16.11
Honeychurch Toys Ltd: 5.9b
Professor Dorothy Hodgkin: 5.47
Houseman (Burnham) Ltd: 10.17, 10.28, 10.29
Howmedica Inc: 7.1
Timothy Hunt: drafts of drawings of molecules
Ilford: 9.9
Imperial Chemical Industries plc, Agricultural Division: 16.16, 16.19, 16.22, 16.23, 16.24
Imperial Chemical Industries plc, Mond Division: 4.20, 4.23, 4.24, 4.30, 4.33
Imperial Chemical Industries plc, Organics Division: 11.16, 11.17, 11.18, 11.19
Imperial Chemical Industries plc, Paints Division: 9.2, 9.8a, b and c
Imperial Chemical Industries plc, Petrochemicals and Plastics Division: 8.31
Imperial Chemical Industries plc, Plant Protection Division: 10.12
Industrial Diamond Information Bureau: 5.21
International Institute for Cotton: 3.5
International Tin Research Institute: 7.33
International Wool Secretariat: 8.16
Johnson Matthey: 7.6
Journal of the Society of Dyers and Colorists: 11.9
Keep Britain Tidy Group: 13.38
King, Taudevin & Gregson Ltd: 6.15
King's College (London) Audio-Visual Service Unit: 5.31a and b, 7.8, 7.9, 8.10, 17.15a and b, 18.11
The Kleenoff Company: photo on page 197
Frank Lane Picture Agency Ltd: 3.3, 9.25
Lead Development Association: 7.3
Lego UK Ltd: 5.9a
Ian Logan Design Co: 5.9c
London Waste Regulation Authority: 6.31
Longman Group (UK) Ltd: 5.30
Malaysian Rubber Producers Association: 3.6
Macaulay Institute for Soil Research: 15.14
Mansell Collection: 8.17, 14.1
Marks, L. O. and Smith, D. T. *J. Microsc.* **130**, p. 249, 1983: 1.3
Masonlite/Louise Holland: 17.9
Medical Research Council: 5.49
Metal Box: 7.23, 7.38
Joseph Metcalf Ltd: 15.22

Meteorological Office, Bracknell/Air Force Cambridge Research Laboratory, Massachusetts: 9.32
Museum of the History of Science: 4.26d
NERC – reproduced by permission of The Director, Institute of Geological Sciences: 9.4
National Coal Board: 13.8, 13.14, 13.15, 13.16, 13.17, 13.18, 13.34
National Coal Board, Restoration and Research Branch: 13.33a and b
National Dairy Council: 9.1
National Film Archive: 8.20
Noblelight: 17.10
Nottinghamshire County Council Fire Brigade: 13.44, 13.46
Novosti Press Agency: page 287, 17.3
Nuffield–Chelsea Curriculum Trust: 3.9, 5.41, 5.42, 5.44, 5.45, 5.48, 7.7, 10.5, 15.7
Nuffield–Chelsea Curriculum Trust/Michael Plomer: 5.22, 5.32
Oxford Scientific Films: 8.15
Pharmaceutical Society of Great Britain: 0.10
Philips Lighting: 17.8
Pilkington: 6.21
Pilkington Glass Museum: 6.1a and b
Eileen Preston: 6.12, 10.23
Punch Publications Ltd: 11.12
RTZ Ltd: 4.37, 4.40, 4.41, 4.42, page 85, 7.21
Rank Taylor Hobson: 6.2
Redland plc: 6.26
Rex Features Ltd: 0.9c, 13.35, 13.42
Rolls Royce plc: 7.5
Rothamsted Experimental Station: 16.7, 16.30
Royal Institution: 4.26b, c, e, f, g, h and i, 5.34a, b, c, e and f
Royal Mint: 5.2, 7.27, 7.34
School of Chemistry, University of Bristol: 6.13
Science Museum: 1.8, 1.9, 3.8, 8.18, 8.19, 11.3, 11.8, 11.10, 11.11, 14.2
Science Photo Library: 5.4
Science Photo Library/Russ Kinne: 5.7
Science Photo Library/NASA: 13.1, 14.15, 15.1
Shell Photographic Library: 2.2b and c, 2.12, 2.14, 2.15, 2.16, 2.17, 2.19, 2.21, 2.23, 2.24, 5.5, 8.6, 8.7, 13.21, 13.22
Sisis Equipment Ltd: 15.7
A. O. Smith Harvestore Products Ltd: 10.34
Stone Manganese Marine Ltd: 7.30
Swedish Institute for Cultural Relations: 0.5
Tate & Lyle: 3.23
Thames Water: 0.4, 10.19, 10.35
Times Newspapers: 0.6
Topham Picture Library: 3.25, 13.47, 13.50, 15.3
Tudor Safety Products Ltd: 13.45
USIS: 15.4
Unilever Educational Publications: 10.3
United Glass Containers: 6.3
Van den Berghs & Jurgens Ltd: 2.38
Vauxhall Motors Ltd: 7.41, 8.4, 9.36, 14.12
Josiah Wedgwood and Sons Ltd: 2.24, 6.23, 6.27, 6.28, 7.17
Whitby Literary and Philosophical Society: 4.14, 4.15, 4.17
ZDA: 7.2, 7.4, 7.19, 7.40

Introduction

What is science?

Before you can join in and be a scientist you need to know something of what scientists have already discovered and how they do their work. All the people pictured in figures 1 to 4 are scientists. What is special about being a scientist? What do scientists do and how do they do it?

Figure 1
A chemist in a plastics firm.

Figure 2
Laboratory technician at work.

Figure 3
A scientist doing research in the science laboratory of the Institute for Marine Affairs, Trinidad.

Figure 4
Biologists sampling the Thames near Pangbourne.

Figure 5
The Nobel medal which is given to the winners of the annual prizes for physics, chemistry, medicine, literature, and the promotion of peace. The awards are made from the bequest of Alfred Nobel, the Swedish inventor of dynamite, who died in 1896. The prize has been awarded to several of the scientists mentioned in this book, including Dorothy Hodgkin and Lawrence Bragg.

Some people compare being a scientist with playing a game. It can be great fun but it is also very competitive. Many scientists want to be first to publish new results and theories so that they will win awards and be remembered for their work. Scientists, like sports professionals, need plenty of training and have to work hard for success.

Often scientists work in teams and enjoy the excitement of working out ideas and making discoveries as a group. Members of a team play the game according to agreed rules. One of the rules in the game of science is that the results of experiments are not accepted until they have been published and checked by other scientists. The editors of science journals send new articles to "referees" to make sure that they are written according to the rules and are fit to print.

Science helps us to make sense of the world we live in. Every day we experience regular patterns: the Sun rises each morning, sugar always dissolves in water, plants wilt without water. This gives us a sense of order which lets us make predictions: the Sun will rise tomorrow morning, the next spoonful of sugar will dissolve in a cup of tea, the rubber plant will die if we forget to water it.

So scientists look for order in what they observe. To see order, they must first make observations and take measurements. In this way they collect much information about the world we live in. They look for patterns in their knowledge.

For example, meteorologists are scientists who study the weather. They record huge numbers of measurements of air pressure, rainfall, temperature and wind speed. The measurements are not taken at random. Meteorologists have theories about which are the important readings to take. Computers are programmed according to these theories to process the mass of data.

We see one way of making a pattern out of all the facts about the climate every time we watch a weather forecast on television. The lines on the weather map have been worked out from all the measurements and satellite observations.

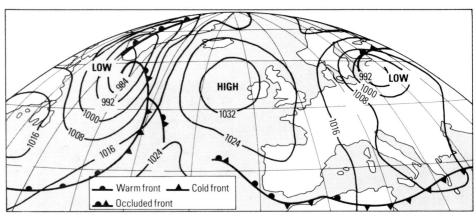

Figure 6
Weather maps like this are published in newspapers.

Once scientists have found patterns it becomes possible to make predictions. Some predictions are very reliable. Astronomers can predict the movement of the planets and comets very accurately. Sometimes the predictions are less certain, as you know if you have ever been misled by a weather forecast.

As they are collecting facts and looking for patterns scientists also seek explanations. Why is the world as it is? Why do things behave in the way that they do? Inventing theories to explain what we know is an important part of science.

One of the things that you will be doing as you study this science course is seeing for yourself some of the things which scientists have already found. We will be taking you on a "conducted tour" to give you firsthand experience of what scientists have discovered.

While on this tour you will be taught some of the practical skills needed to use scientific equipment and methods. Learning skills needs practice, as you will know if you play a musical instrument or are learning to type. The practice may not always be enjoyable but it is necessary.

If you have ever been a tourist on holiday you will know that it can be interesting – at least for a time. However in the end being a tourist is not as interesting as living your own life where you belong. So it is with science.

To understand science you need to have experience of asking questions and planning investigations. You have to be involved in experiments where the answer is not known in advance. We hope that by the end of this course you will have had several opportunities to join in and be a scientist using your knowledge and skills to carry out investigations.

What are the questions and problems which interest chemists?

Chemistry involves finding out what different substances are made of and what their properties are. Chemistry as a subject is partly the knowledge and understanding written down in books such as this; but it is also an activity – chemistry is what chemists do. So one way to understand what chemistry is all about is to study the life and work of famous chemists. Even better is to do some chemistry yourself.

Chemists have invented many ways of making discoveries. In Chapter **C4** you can see how Humphry Davy used electrolysis to discover new elements. In Chapter **C5** you will find an account of the work of Michael Faraday. He studied the theory of electrolysis and showed that a new theory was needed to explain his observations.

Chemistry is about the way things change. Why does toast go black if you leave it too long under the grill? Why do batteries start to leak when they have run down? Why does iron go rusty slowly when a firework can go off with a bang? Why can we keep ourselves warm by burning things? Why can't we live off grass if sheep and cows can? Why are some berries safe to eat while others are poisonous?

Scientists look for patterns in their knowledge. One of the most important patterns in chemistry is the Periodic Table, which helps to make sense of all the chemical elements. The way in which the Periodic Table pattern was discovered is described in Chapter **C17**.

Chemists try to find theories to explain what they know about the properties of materials. These theories often involve ideas about atoms and molecules. We cannot see atoms and molecules so chemists make models to help explain what the theories mean. The use of models is discussed in Chapter **C5**.

Also in Chapter **C5**, Lawrence Bragg describes how he and his father discovered a way of finding out how atoms are arranged in crystals. This story illustrates the links between the sciences. Both Lawrence Bragg and his father were physicists, but their discoveries have proved enormously important for chemistry and biology. Dorothy Hodgkin studied chemistry at university. She

Figure 7
Kekulé's dream.

Figure 8
A model of a benzene ring.

then went on to use the methods discovered by the Braggs to investigate complicated biological molecules such as vitamins.

A very famous discovery in chemistry came about as a result of a dream. In 1865, the German chemist, Kekulé was trying to work out the structure of a compound called benzene (see figure 5.38 on page 101). At that time no-one could explain the unusual properties of this compound.

Kekulé tells the story of how he was sitting in front of the fire and began to doze:

> "The atoms flitted before my eyes. Long rows, variously, more closely, united; all in movement wriggling and turning like snakes. And see, what was that? One of the snakes seized its own tail and the image whirled scornfully before my eyes. As though from a flash of lightning I awoke; I occupied the rest of the night working out the consequences of the theory."

For the first time chemists realized that atoms could be arranged in rings. The discovery of the structure of benzene was a big advance in the understanding of the chemicals used to make dyes and drugs. You can see that there is a ring structure in aspirin if you look at figure 12.10 on page 217.

Kekulé gave this advice to his fellow scientists:

> "Let us learn to dream, and we may perhaps find the truth." He was then careful to add: "But let us beware of publishing our dreams before they have been tested by a discerning mind which is wide awake."

What is the practical importance of chemistry?

Chemistry is not only a theoretical subject, it is also about making things. It is about taking raw materials from the world about us and turning them into products which help to satisfy our everyday needs.

We have to eat and drink. We wear clothes and take shelter in our homes. We travel about on foot or perhaps by bicycle, bus or car. We communicate with others by talking, reading and writing, or using the telephone, radio or television. If we are ill we may take a medicine to make us feel better. The products of chemistry and the chemical industry help us to do all these things more easily and more safely.

The process of taking simple chemicals and joining them together to make new and more complicated ones is called *synthesis*. The products of chemical synthesis include dyes, medicines, fertilizers and plastics. In Chapter C11 you can read about the discovery of the first synthetic dye by William Perkin.

In Chapter C12 you will find an account of the discovery of aspirin. This story started with chemists trying to find out why willow bark could be used to cure pain. It ended, chemically, with the discovery of methods of making aspirin using chemicals from coal or oil. The aspirin story is still an unfinished one for science, because we do not yet fully understand why aspirin helps to relieve pain and reduce inflammation.

Figure 9 shows that the chemical industry converts raw materials found in the earth, sea and air into new substances which we value more and are willing to pay for. On average the products of the chemical industry are worth about one and a half times as much as the raw materials from which they were made. In this way the industry adds to the wealth of the country.

Chemists have had to discover ways of manufacturing substances on a large scale. One of the most famous processes is the Haber process for producing ammonia. Ammonia is needed for making fertilizers. Fritz Haber

Raw materials
from mines, forests, sea, air, farms, oil and gas wells

The chemical industry
converts the raw materials into pure substances such
as acids, alkalis, salts, solvents, compressed gases,
and organic chemicals

The chemical industry itself
uses the chemicals it has made
to produce cosmetics,
detergents and soap,
disinfectants,
drugs and medicines,
dyes and inks,
explosives,
fertilizers and pesticides,
plastics,
fibres and rubbers,
paints and many others

Other industries
use chemicals to produce
products made of metal, glass,
plastic and wood,
drinks and food products,
vehicles and machinery,
paper and paper products,
electrical equipment,
textiles,
hardware for building,
fuels and lubricants,
packaging and many others

The ultimate market:
human needs for food, clothing, shelter, health,
defence, communications, transport, and so on

Figure 9
The work of the chemical industry.

was a theoretical chemist. He needed the engineering skill of Carl Bosch to build a manufacturing plant large enough to make huge amounts of fertilizers. This story is described in Chapter **C**16.

The chemical industry continues to develop. In Chapter **C**2 you can read about one of the newest parts of the petrochemical industry in Scotland based on gas from the North Sea.

Links between biology and chemistry are leading to the introduction of new methods of manufacture based on biotechnology. In this way we may be able to produce fertilizers, food and fuels by methods which do not depend so much on energy resources such as oil and gas.

A knowledge of chemistry is important in many other branches of technology. Figures 10 to 15 show some of the careers open to people today with qualifications in chemistry.

Figure 10
Hospital pharmacist.

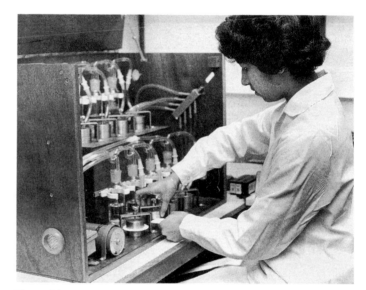

Figure 11
Trainee laboratory technician.

Figure 12
Chemist working for a firm which manufactures video tapes. He is measuring the specific gravity of one of the tape "binder" chemicals.

Figure 13
Doctor interviewing a patient. To be a doctor you need some qualifications in chemistry.

Figure 14
Chemical engineer at work: he is setting an instrument in a research and development department at BP.

Figure 15
Chemist operating an automatic analyser.

Another important aspect of chemistry is *analysis* which means splitting things up to find out what they are made of. Chemical analysis is used to test our food to check that it is fit to eat and not contaminated by harmful substances. Analysis is used to control pollution and to test the purity of the water we drink and the air we breathe.

Co-ordinated Sciences

This is a Co-ordinated Sciences course. You may have seen the word *co-ordinated* used elsewhere. What do you think the following newspaper story means?

"Police carried out a *co-ordinated* raid on ten warehouses in the London area last night. As a result fifteen people were arrested and are being held for questioning."

To be *co-ordinated* the raid must have been carefully organized. Each person in the police team knew exactly what was happening elsewhere and when it was happening. That is what we and your teachers have tried to do with this science course.

You have different books for Biology, Chemistry and Physics, but the separate parts of the course have been carefully co-ordinated so that ideas you meet in one subject can be used in another. For example you will need to use ideas about energy in all your sciences. You will learn about energy first in Physics. When you come to use energy ideas in Biology and Chemistry, your teachers will know and expect to use what you have learned in Physics.

To help you make the most of this co-ordination we have included a large number of cross references in each of the books. For example when you first come to the subject of X-rays in Chemistry you will be reminded that you can find out more about the wavelength of X-rays in your Physics book. You are also told that chemists use X-rays to investigate structure in a way which is quite different from the use of X-rays in Biology.

Sciences are only a part of your school timetable. Your work in this course will also have links with other subjects such as mathematics, CDT, geography and home economics. You should be on the look out for opportunities to take advantage of these links.

How to use this book

This book is divided into six topics. It is possible that you will study them in the order they are printed. But your teacher may decide on another order. You will have to make use of the knowledge and understanding you have gained in Biology and Physics if you are to make sense of all the chapters.

The commentary text points out the main cross-links between the three parts of the course. Commentary text is printed in type like this.

You do not have to know everything in this book. The book contains more than you have to study so that you and your teacher have some choice about the way you will learn.

Your teacher will guide you. The commentary text will also help you to decide which bits to study in detail, which parts to read for interest and which sections to leave out all together.

Some of the more theoretical topics are printed in boxes so that you can read the rest of the text without having to deal with the more advanced theory. There are other boxes which include practical problems for you to solve and results of experiments for you to interpret. In some boxes there are symbols to help you to recognize the skills and processes which you need to apply. The meaning of these symbols is explained on page 6 of your Physics book.

The historical sections which tell you about the work of famous scientists have been included to give you an idea of how scientific ideas develop. You will not be expected to remember the details in written examinations.

You are expected to appreciate the practical importance of chemistry. Several manufacturing processes are described in this book so that you can appreciate the scale and importance of industry. You will not be expected to remember all the details. You should be able to read about a chemical process and show that you can make sense of the information.

You will probably start most topics by gaining practical experience in the laboratory. Some of the experiments are designed to give you experience of the properties of materials. This will help you to understand the theory. For example, you will be able to understand the properties of glass better when you have made some glass for yourself and tried shaping glass.

You will also be able to plan your own experiments. You will find some suggested problems to investigate in this book. You may have better ideas of your own. Successful investigations require careful planning and your teacher will give you guidance about how to set up a scientific inquiry.

Reading a textbook is not the same as reading a story. You have been familiar with stories ever since you started to read. You know the ways in which story books are organized. Textbooks are different. You will have to learn how to make sense of technical writing. You may be given worksheets by your teacher which will help you to make sense of some of the types of text in this book.

Topic C1   Raw materials

Chapter **C1**   **The elements of chemistry**   10
Chapter **C2**   **Petrochemicals**   20
Chapter **C3**   **Chemicals from plants**   46
Chapter **C4**   **Chemicals and rocks**   57

Chemists can conjure metals out of rocks and turn black coal into white nylon, or dirty oil into beautiful dyes. In this first topic you will find out how some of these tricks are done. At the same time you will gain a clearer understanding of atoms and molecules, starting with the simple molecules in air and water but going on to some of the very complicated molecules in living things. In the laboratory you will have the chance to try some industrial processes on a small scale and see for yourself how heating and electrolysis can change one thing into another.

Oil palm in Malaya — a source of vegetable oil.

Chapter C1

The elements of chemistry

You should already be familiar with the chemistry included in this chapter. It reminds you of the difference between elements and compounds. The chapter includes a section which will help you to practise using chemical symbols and writing equations.

This chapter also contains an introduction to the Data section at the end of this book.

C1.1 What are raw materials made of?

You should already know that everything in the universe is made up of about one hundred *elements*. You will find the elements listed in table 2 of the Data section.

Sometimes the elements are found free in nature. Two valuable examples are gold and carbon (as diamond). The air consists mainly of two other elements mixed up but not chemically combined. Usually, however, the elements are mixed up and joined together so that the first job for chemists is to separate them and sort them out.

	Estimated temperature range in °C	Physical state
Core	3000 to 5000	semi-solid
Mantle	650 to 3000	liquid
Crust	0 to 650	solid
Atmosphere	−90 to 60	gas
Hydrosphere	0 to 30	liquid plus polar ice caps

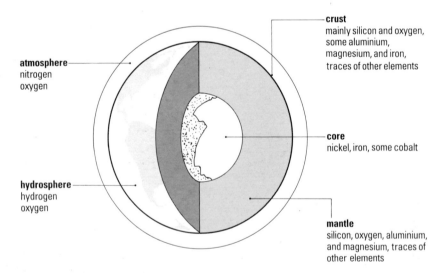

Figure 1.1
The structure of the Earth and the distribution of elements in it. In the mantle, crust and hydrosphere the elements are present as compounds. Which compound makes up most of the hydrosphere?

Substances made of two or more elements joined together are called *compounds*. Table salt is a two-element compound made of sodium and chlorine. The chemical name for salt is sodium chloride. Sugar is a three-element compound made of carbon, hydrogen, and oxygen. The names and formulae of many common compounds are listed in tables 4 and 5 of the Data section.

Element	Percentage by mass in the Earth's crust
Oxygen	50
Silicon	26
Aluminium	7
Iron	4
Calcium	3
Potassium	2.5
Sodium	2.5
Magnesium	2
Hydrogen	1
Other elements	2

Figure 1.2
The abundance of elements in the Earth's crust.

1 Draw a picture of a bicycle, car or house. Label the picture to show what each part is made from. (If you prefer you can cut your picture from a magazine.)

2 Look at the labels on containers for foods, cosmetics or garden chemicals. Find some with detailed lists of ingredients. Choose five examples, name them, state what they are used for, and list the chemicals they contain.

3 The table in figure 1.2 lists the commoner elements in the crust of the Earth. Put this information into the form of a pie chart or bar graph.

4a Which two gases make up most of the air?
b What proportion of the volume of the air is made up of the two gases you have named in **a**?
c Name three other gases which are present in the air.
d Which of the gases in the air are elements and which are compounds?

5 Choose the odd one out in each of these groups of four elements and give the reason for your choice.
a iron, mercury, nickel, zinc
b aluminium, lead, sulphur, tin
c argon, carbon, chlorine, oxygen
d bromine, chlorine, helium, iodine

6 Give a brief description of three methods which are used in chemistry to separate and sort out mixtures. (Possible examples include filtration, centrifuging, evaporation, distillation, and chromatography.)

The difference between an element and its compounds can be explained using the theory that everything is made of atoms. Atoms are minute. They are so very small that you cannot see them; but it is now possible, with the most powerful electron microscopes, to take photographs to show the positions of atoms.

In any one element all the atoms are the same. Each gold atom is the same as any other gold atom. The atoms of carbon are also all the same as each other but they differ from the atoms of gold.

Every element has its own kind of atom. This means that there are as many kinds of atom as there are elements. Chemists use shorthand symbols to label the elements and their atoms. There is a list of these symbols for many of the elements in table 3 of the Data section.

7 What are the chemical symbols for the following elements?
Carbon, silver, nitrogen, magnesium, sodium, and lead.

8 What are the names of the following elements? S, Cu, Ni, K, and Br.

9 Use tables 4 and 5 in the Data section to find out which elements are combined in these compounds: methane (natural gas), ethanol (alcohol), ethanoic acid (the acid in vinegar), propanone (nail varnish remover), calcium carbonate (chalk), ammonia (used in household cleaners), potassium nitrate (used to make gunpowder).

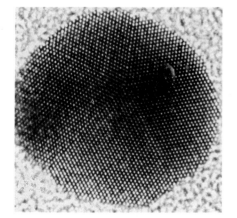

Figure 1.3
A small particle of gold. The rows of atoms are viewed end-on, with each row showing up as a white spot.

In some elements the atoms are joined together in small groups. These small groups of atoms are called molecules. You will find examples of elements which are molecular listed in table 2 of the Data section, and the table also shows the formulae of the molecules. In oxygen, for example, all the atoms are joined in pairs. Each pair is an oxygen molecule.

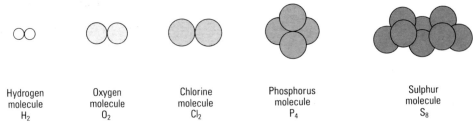

| Hydrogen molecule H_2 | Oxygen molecule O_2 | Chlorine molecule Cl_2 | Phosphorus molecule P_4 | Sulphur molecule S_8 |

Figure 1.4
The molecules of several elements.

The symbol for an oxygen molecule is O_2. The "2" shows that the molecule is made of two atoms joined.

The product of burning hydrogen in oxygen is water. The chemical reaction can be summarized using a *word equation*:

hydrogen + oxygen $\longrightarrow$ water

BOX 1   Experiment
The synthesis of water

INTERPRET

The diagram in figure 1.5 illustrates an experiment which shows two elements joining to form a compound. Hydrogen burns in the oxygen of the air. This is an example of a chemical reaction. The reaction gives out much energy. As the vapour from the flame is drawn into the apparatus it condenses as a liquid in the cold test-tube. The condensed liquid is colourless, it turns cobalt chloride paper from blue to pink and boils at 100 °C.

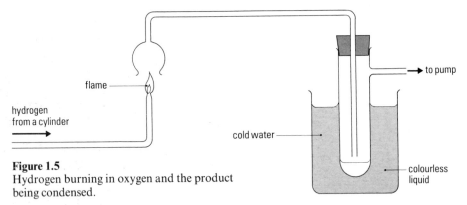

flame

hydrogen from a cylinder

to pump

cold water

colourless liquid

Figure 1.5
Hydrogen burning in oxygen and the product being condensed.

10 How can you tell, when watching this experiment, that the chemical reaction of hydrogen with oxygen gives out energy? What happens to the energy?

11 What is the evidence that the product of the reaction is water?

12 In what ways is the water made in the reaction different from the two elements of which it is composed?

13 What does the word *synthesis* mean? (You will meet the word again in Chapter **B3** of your Biology book when you study photo*synthesis*.)

This word equation can be made more descriptive if *state symbols* are included. State symbols show whether a substance in an equation is solid (s), liquid (l) or gaseous (g).

hydrogen(g) + oxygen(g)⟶water(l)

Figure 1.6 gives a picture of what is happening to the atoms and molecules when hydrogen reacts with oxygen.

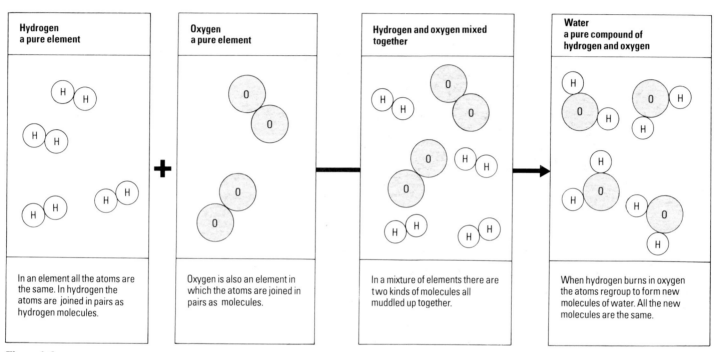

Figure 1.6
What happens to the molecules as hydrogen combines with oxygen.

14 Use Worksheet **C**1A to make a flick book which will give you a moving picture of what happens when the molecules of two elements react to form a compound.

15 Make your own flick book to show what happens when hydrogen and oxygen react to make water.

16 How many atoms are there in each of these molecules? For each substance, explain whether it is an element or compound.
a Cl_2, chlorine
b CH_4, methane
c S_8, sulphur
d P_4, phosphorus
e C_2H_5OH, ethanol.

17 The chemical formula of glucose is: $C_6H_{12}O_6$.
a Name the three elements in glucose.
b How many hydrogen atoms are there in a glucose molecule?
c What is the total number of atoms in a glucose molecule?

18 Hydrogen gas, H_2, burns in chlorine gas, Cl_2, to form hydrogen chloride, HCl, which is also a gas. Write a word equation for the reaction, with state symbols, and then draw a diagram similar to figure 1.6 to show what happens to the molecules.

TABLEAU DES SUBSTANCES SIMPLES.

Figure 1.7
Lavoisier's list of elements from his book
Traité Elémentaire de Chimie.

C1.2   When were the elements discovered?

This section assumes that you have already studied some of the elements in the Periodic Table. You have probably been shown reactive metals such as sodium and potassium and seen what happens when they are added to water. You may also have some knowledge of the halogens, including chlorine, bromine and iodine. The discovery and importance of the Periodic Table is described in Chapter C17.

How many elements are there? Is there a limit? Are there patterns in the properties of the elements? Scientists began to ask these questions at the beginning of the nineteenth century when many new elements were being discovered.

In 1789 Antoine Lavoisier, a famous French chemist, published a book called *Traité Elémentaire de Chimie* which explained his theories about burning and the air. This was the beginning of modern chemistry as we know it. In the book he listed thirty-three substances which he thought were elements. (See figure 1.7.)

Lavoisier included light and caloric (heat) in his list, but we no longer think of these as material substances. He also included eight other substances, including lime and magnesia, which we now know to be compounds. At that time no one was able to break them down into anything simpler.

Lavoisier's ideas started a race to discover new elements. Fourteen were found between 1800 and 1810. By 1830 the list had grown to fifty-five.

This was a very exciting time in the history of chemistry. While all these new elements were being discovered, John Dalton and others were working out ideas about atoms.

Dalton wrote down his ideas about atoms in his notebook. The first entry was made on the 6th September 1803. In his notes he wrote that everything is made up of atoms and that these atoms cannot be split up, created or destroyed. These were not new ideas, but he went on to suggest that each element has its own kind of atoms which are identical. He showed that the reaction of elements to form compounds could be explained using the idea of atoms combining. The way this theory works is explained in section C1.1.

Perhaps the most important new idea put forward by Dalton was the idea that it is useful to know the masses of atoms. He suggested that the atoms of different elements differ in mass and he showed that it is possible to use this idea to work out the formulae of compounds.

---

19 Look at Dalton's list of elements in figure 1.9.
a Which of the substances named in his list are not now thought to be elements?
b Devise and draw your own ''Dalton style'' atoms for three other elements.
c Why do you think that we no longer use Dalton's symbols?

20 Choose any element you like. Find out who discovered it, where it was discovered and how. What is the element used for now and has it any important compounds? (Dictionaries, encyclopedias and text books will help you with this question.)

Figure 1.8
John Dalton.

For nearly one hundred years after John Dalton put forward his atomic theory, people thought of atoms as solid, indestructible particles. They had no reason and no experimental evidence to think otherwise. Now we know much more about atomic structure but even so Dalton's theory is a useful one.

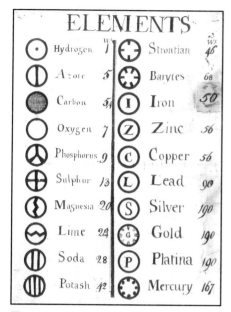

Figure 1.9
Dalton's list of chemical elements with their symbols.

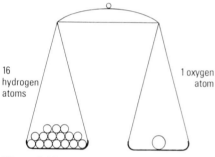

Figure 1.10
An oxygen atom is balanced by sixteen hydrogen atoms.

Lavoisier's ideas about elements and Dalton's ideas about atoms eventually led to the discovery of the Periodic Table. The way in which this came about is explained in Chapter C17. To understand the story you have to know something about atomic masses.

Atoms are so very small that a special scale has to be used to measure their masses. The unit is called the *atomic mass unit* and it is given the symbol u.

Hydrogen is the element with the lightest atoms and so the mass of a hydrogen atom is chosen to be 1 u. Figure 1.10 shows an imaginary balance on which one oxygen atom is balanced by sixteen hydrogen atoms. The mass of an oxygen atom is 16 u. The atomic masses of the elements are listed in table 3 of the Data section.

The Periodic Table was discovered by arranging the elements in order of their atomic masses. A modern form of the table is shown in table 1 of the Data section. The horizontal rows in the table are called *periods*. The vertical columns are called *groups*.

The Periodic Table helps us to make sense of all the elements because it shows us patterns in the behaviour of the elements. Elements in the same group have similar chemical properties. The reactive metals in group 1 are called the alkali metals and they include lithium, sodium and potassium. The halogens in group 7 include chlorine, bromine and iodine; you will find pictures of all these elements in Chapter C5.

21 The information you need for this question is in table 2 of the Data section. You will need Worksheet **C1B** which is an outline of the Periodic Table. Use a different colour of pen or pencil for each set of elements, and give a key to the colours you use. Mark in the elements which:

a have been known since ancient times

b were discovered between 1700 and 1799

c were discovered between 1800 and 1810

d were discovered between 1811 and 1899

e were discovered between 1900 and the present day.

22 You will need the data in table 3 of the Data section. Imagine that you are balancing atoms on a scale such as that shown in figure 1.10. How many hydrogen atoms would you need to balance:

a one carbon atom   **c** two nitrogen atoms   **e** one oxygen molecule

b one magnesium atom   **d** three sulphur atoms   **f** two water molecules?

23 In what ways are lithium, sodium and potassium similar? Make a table or chart to show as many similarities as you can find. (Use reference books and the Data section, and look back to your notes on earlier work that you have done in science to investigate these elements.)

C1.3   How do we write chemical equations?

You may not have had much experience of writing equations before; at first it can be difficult to work out how to use the symbols. It is a bit like learning a new language and the rules seem strange until you are used to them. This section shows you what to do. You will have opportunities to practise the skill in later chapters.

You cannot start to work out a symbol equation until you can write the word equation for the reaction. What this means is that you have to know the results of experiments to investigate what happens during the reaction.

Here we use the reaction of hydrogen with oxygen as the first example. You already know what happens if you have studied the experiment in section **C**1.1.

$$\text{hydrogen(g)} + \text{oxygen(g)} \longrightarrow \text{water(l)}$$

The hydrogen and oxygen on the left are the *reactants* which combine to give water, the *product*, on the right. Figure 1.11 shows how models can give another picture of what is happening to the atoms and molecules during the reaction. Compare this model equation with figure 1.6.

 **+**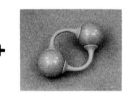

Figure 1.11
Ball-and-spring models to show the reaction of hydrogen with oxygen to form water.

We cannot take photographs of models every time we want to show what happens during a chemical reaction and it takes a long time to draw a picture of the molecules so we write symbol equations instead. Here is the symbol equation for the model reaction in figure 1.11:

$$2H_2(g) + O_2(g) \longrightarrow 2H_2O(l)$$

H_2 means a molecule of hydrogen in which two hydrogen atoms are joined together.

H_2O means a molecule of water in which two hydrogen atoms are joined to one oxygen atom.

$2H_2O$ means two whole molecules of water, as shown in figure 1.11.

If you look carefully at figure 1.11 you will see that the numbers of atoms have not changed during the reaction. There are four hydrogen atoms on the lefthand side of the equation and four on the right. There are two oxygen atoms on the left and two on the right. This is why we say that the equation is balanced. (See figure 1.12.)

The example in box 2 shows the steps to follow when writing chemical equations. The reaction is the one you use every time you heat something by burning gas in a kitchen or in a laboratory. Natural gas is methane, CH_4 and when it burns it reacts with oxygen; the products are carbon dioxide and water. The discussion in figure 1.13 (overleaf) will help you to see how you can think your way through the steps of balancing an equation.

You will find it helpful to refer to box 2 when you are asked to write equations in the other chapters of this book.

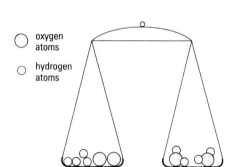

○ oxygen atoms

○ hydrogen atoms

Figure 1.12
The equation for the reaction of hydrogen with oxygen is balanced.

Summary

The passage opposite summarizes some of the main ideas in this chapter. Copy the passage filling in the blanks with the words listed. There is one word to each blank. The words are listed in alphabetical order.

BOX 2   How to write a symbol equation

Example
What is the balanced equation for the reaction of methane with oxygen?

Answer
Step 1: Write down the word equation.

methane + oxygen $\longrightarrow$ carbon dioxide + water

Step 2: Write down the formulae for the reactants and products. (You will find the formulae in tables 1, 4 and 5 of the Data section.)

$CH_4 + O_2 \longrightarrow CO_2 + H_2O$

Step 3: Balance the equation.
You must not change any of the formulae to balance the equation. You balance the equation by writing numbers in front of the formulae. These numbers then refer to the whole formula.

$CH_4 + 2O_2 \longrightarrow CO_2 + 2H_2O$

Lefthand side   **Righthand side**
1 carbon atom   1 carbon atom
4 hydrogen atoms   4 hydrogen atoms
4 oxygen atoms   4 oxygen atoms

Step 4: Add state symbols which normally show the states of the reactants and products at room temperature.

$CH_4(g) + 2O_2(g) \longrightarrow CO_2(g) + 2H_2O(l)$

atomic	conduct	hydrogen	mass	ninety	shiny
brittle	different	insulators	metals	right	silicon
cannot	elements	joined	molecules	same	theory
compounds	groups	left			

An element is a substance which _____ be broken down into simpler substances. Everything in the universe is made of _____. About _____ elements occur in nature. Most of the crust of the earth is made of five elements: oxygen, _____, aluminium, iron and calcium.

Elements are classified as _____ and non-metals. Metals usually _____ electricity; they are _____ and strong. Non-metals are electrical _____, they are dull and _____.

Elements combine to form _____. A compound is a substance made of two or more elements _____. Compounds have _____ properties from the elements from which they are made.

Dalton's atomic _____ suggests that everything is made of minute particles called atoms. All the atoms of an element are the _____. The atoms of one element are not the same as the atoms of another element because they differ in _____.

Small groups of atoms joined together are called _____. One of the elements which consists of molecules is _____.

In the Periodic Table the elements are arranged in order of _____ mass. Elements with similar properties come together in the table as vertical columns called _____. Metal elements appear on the _____ hand side of the table; non-metal elements are on the _____.

Balancing an equation

Figure 1.13
How to balance an equation.

Chapter **C2**   **Petrochemicals**

In this chapter you are introduced to the chemistry of carbon compounds. At the same time you will find out about the importance of the modern chemical industry based on oil and gas. You will discover that oil is the main source of the "chemical building bricks" which are used to make plastics, dyes, drugs and many other useful products. You will learn to use formulae and models to help you to make sense of the chemistry of oil chemicals.

C2.1   What are petrochemicals?

As you will learn in Chapter **C**13, most oil and gas is used for fuel. In 1980, 82 per cent of each barrel of oil used in Britain was burned, in one form or another, to generate electricity, to drive cars and lorries, to heat our homes, and to provide power for industry. Another 8 per cent was used to make lubricants, waxes, and bitumen for surfacing roads. Only 10 per cent was used by the chemical industry.

Giving a chemist some oil is a bit like letting a young child loose in a room full of building bricks. The playroom shown in figure 2.1 is full of models and bits of models, big and small, all mixed up together. Gavin has started to play straight away.

Figure 2.1

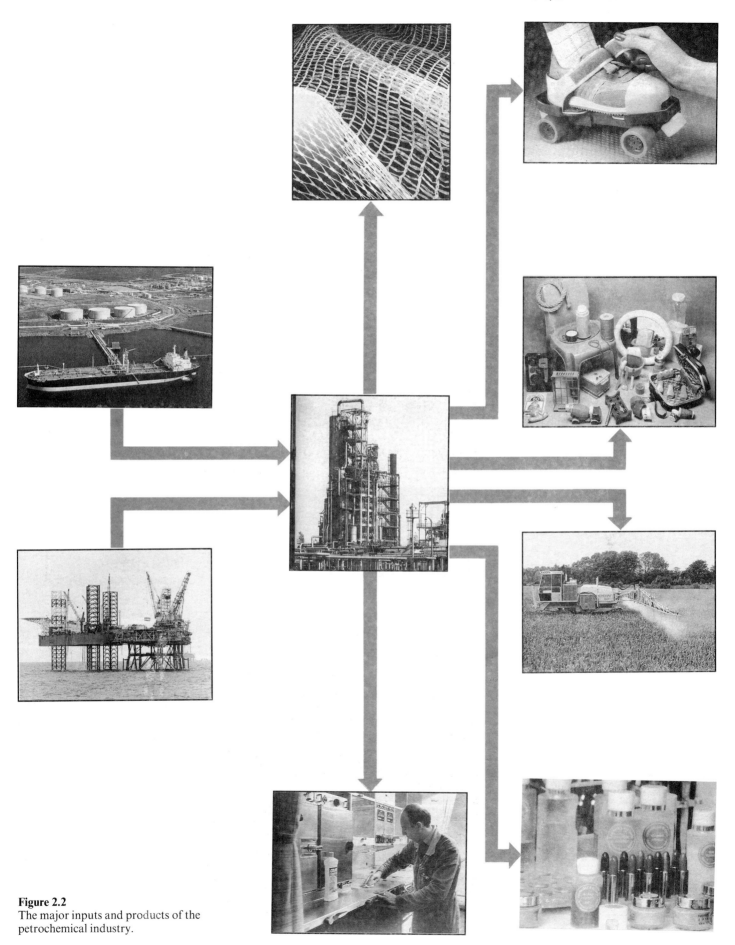

Figure 2.2
The major inputs and products of the petrochemical industry.

You can imagine what Gavin's parents will do when they see the mess: they will want to clear it up and sort it out. Will they settle for the arrangement in figure 2.3? At least it looks tidier and all the models are arranged neatly by size.

Figure 2.3

Figure 2.4

When Gavin asks his parents to help him make some new models, they cannot get started until they have broken up some of the bigger models into single bricks and sorted them out again. Figure 2.4 shows Gavin and his parents all ready to start building.

The oil used in industry is like Gavin's playroom in figure 2.1. It is a messy mixture of lots of different molecules of different sizes. So in one way it is complicated because there are many types of molecule in it; but in another way it is simple because most of the molecules are made of just two elements. Figure 2.5 shows a few of the molecules which might be found mixed up in oil; you can see that they are all made of hydrogen atoms and carbon atoms. Almost all the molecules in crude oil are like this, that is, they are *hydrocarbons*. A hydrocarbon is a compound of only two elements: hydrogen and carbon.

There are hundreds of different hydrocarbons in crude oil. The smallest molecules have 1 to 4 carbon atoms in them and are present in crude oil as dissolved gas. The largest molecules may consist of up to fifty carbon atoms.

Carbon probably forms more compounds than all the other elements put together. The reason for this is that carbon atoms are able to join up in chains, branched chains and rings. No other atoms can do this to the same extent. There are so many carbon compounds that there is a special name for their chemistry: organic chemistry. It is called this because many of the most interesting carbon compounds come from living things.

---

1 Draw an outline of a barrel of oil. Divide up your outline, to scale, to show how the oil in a barrel is split up for different uses, as described in the pictures on the previous page.

2 How can you tell, by looking at figure 2.5, that all the substances in oil are compounds?

3 How can you tell, by looking at figure 2.5, that oil is a mixture of compounds?

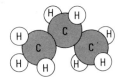

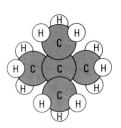

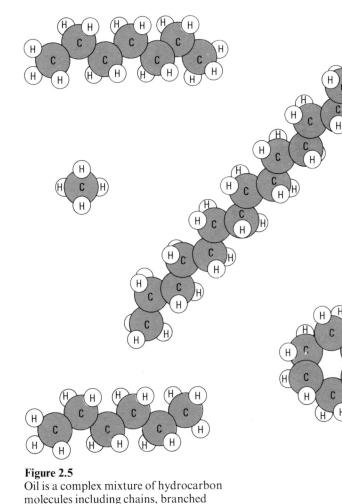

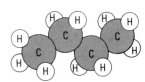

Figure 2.5
Oil is a complex mixture of hydrocarbon
molecules including chains, branched
chains and rings.

4 What is a hydrocarbon?

5 Pick out from figure 2.5 examples of molecules in which the carbon atoms are
arranged **a** in a chain, **b** in a branched chain, and **c** in a ring. Draw one example of
each.

6 How many carbon atoms are there in the smallest molecule in figure 2.5? How many
carbon atoms are there in the largest molecule in figure 2.5?

Chemists and engineers in an oil refinery set out to do to the molecules in
oil what Gavin and his parents did to his models. First they sort them out
roughly according to size and then they break up some of the bigger
molecules into smaller pieces ready to make new molecules.

Methane CH₄

Ethane C₂H₆

Propane C₃H₈

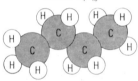

Butane C₄H₁₀

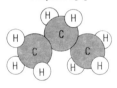

Figure 2.6
Models, formulae and names for the first four alkanes.

C2.2   What are the rules for making molecules?

*You will find it helpful to make models of the molecules discussed while studying this section. Worksheet **C2A** shows how you can do this. The use of models in chemistry is discussed in more detail in Chapter **C5**.*

In this section, ideas about bonding are limited to hydrogen and carbon atoms, but you will find that molecules involving other elements can be dealt with in the same way.

There are so many kinds of molecule in oil that it helps, at least in theory, to group them in families. You can pick out the members of one of these families from figure 2.5 by choosing all the ones in which the carbon atoms are simply in a chain. The compounds in this family of hydrocarbons are called the *alkanes*. Some examples are listed in figure 2.6. You can see that each member of the family has a name and a *molecular formula*.

The molecular formula of a compound tells how many atoms of each element there are in its molecules. The formulae of compounds have had to be discovered by experiment. Nowadays the formulae of most common compounds are already known; there is a list of the names and formulae of the alkanes in table 4 of the Data section.

Figures 2.7, 2.8 and 2.9 show models of three alkane molecules. The same colour code is used as in the earlier pictures: the carbon atoms are black balls and the hydrogen atoms are white balls. The atoms in the molecules are held together by chemical bonds which, in these models, are represented by springs.

Figure 2.7
Ball and spring model of a methane molecule.

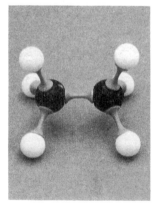

Figure 2.8
Ball and spring model of an ethane molecule.

Figure 2.9
Ball and spring model of a propane molecule.

If you look carefully at the models you will see that they obey these rules:

● each hydrogen atom can only form one bond
● each carbon atom can form four bonds.

Using these two rules, you can work the arrangement of atoms in many hydrocarbons.

It takes time to draw diagrams of molecules, and so a shorthand way of showing the structure of molecules has been invented. Examples of this shorthand are shown in figure 2.10 for the three simplest alkanes. You can see that the atoms are represented by their normal chemical symbols and that lines are drawn to show the chemical bonds between the atoms. There is no attempt in these shorthand formulae to show the shape of the molecules. They are called *graphical formulae*.

Name of the alkane	Molecular formula	Graphical formula
Methane	CH_4	H | H—C—H | H
Ethane	C_2H_6	H H | | H—C—C—H | | H H
Propane	C_3H_8	H H H | | | H—C—C—C—H | | | H H H

Figure 2.10
Names and formulae of the first three alkanes.

Figure 2.11
Ball and spring models of two hydrocarbons with the same molecular formula.

7 Draw a table similar to that in figure 2.10 to show the names, molecular formulae and graphical formulae of the next three alkanes: butane, pentane and hexane. (You will find figure 2.6 and table 4 of the Data section helpful.)

8 Both the molecules in figure 2.11 have the same molecular formula.
a What is the molecular formula of the molecules?
b Draw the graphical formula of each molecule using the shorthand method used in figure 2.10.

9 How many molecules with the formula C_5H_{12} can be made which obey the bonding rules? (You may find it helpful to have a set of molecular models when you answer this question.) Draw the graphical formula of each molecule.

C2.3   How can the molecules in oil be sorted out?

If you look at the list of boiling-points in table 4 of the Data section, you can see that there is a connection between the boiling-point of an alkane and the size of its molecules.

10 On a graph, plot the boiling-points of the alkanes (vertical axis) against the number of carbon atoms in their molecules (horizontal axis).

11 Decane is the alkane with ten carbon atoms in its molecules.
a Draw the graphical formula of a decane molecule.
b If the formula of decane is written C_xH_y, what are the values of x and y?
c Use the graph you have plotted in question **10** to predict the boiling-point of decane.

12 Which of the alkanes listed in table 4 of the Data section are gases at room temperature (20 °C) and pressure? Which are liquids and which are solids? (See box 1 on page 91.)

Table 4 of the Data section shows that the larger the molecules of an alkane, the higher its boiling-point. This is true of all the families of hydrocarbon compounds, not just the alkanes. It means that it is possible to sort out the molecules of oil by *distilling* them at different temperatures. In this way the oil is split into a series of fractions. In each fraction the molecules are about the same size. The process used is called *fractional distillation* and is used to obtain fuels and lubricants from oil as described in Chapter **C**13, section **C**13.3. The use of fractional distillation to separate liquefied hydrocarbon gases is described in this chapter.

One of the richest sources of oil and gas in the North Sea is the Brent Field, just over a hundred miles north east of the Shetland Islands. So important is this field that there are now four platforms standing in deep water to tap its huge reserves of hydrocarbons. The first Brent platform came into operation in 1971. The last and largest began production in June 1981. These platforms were built to survive fierce gales and massive waves.

Figure 2.12
Platform in the Brent field.

The hydrocarbons from the Brent field are unusually rich in a mixture of the small alkane molecules. 80 per cent of this mixture is natural gas, methane. The rest is mainly ethane, propane, and butane together with some of the hydrocarbons found in petrol. Ethane, propane and butane are

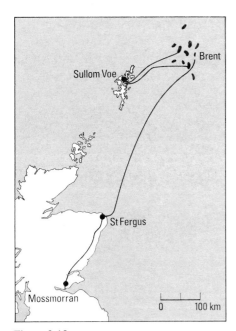

Figure 2.13
The Brent field and the pipelines to Shetland and St Fergus and on to Mossmorran.

normally gases, but they can be liquefied under pressure at room temperature. (Camping gas cylinders contain either liquid propane or liquid butane under pressure.) They can also be liquefied at atmospheric pressure if they are cooled enough.

The mixture of small molecules is very valuable and so it is separated from the oil at one of the Brent platforms. One pipeline takes the oil to Sullom Voe in Shetland, while a second pipeline takes the natural gas and liquefied gases on a two- to three-day journey of 278 miles to St Fergus on the coast near Peterhead.

Arriving at St Fergus, the pipeline delivers liquid hydrocarbons under pressure at about $+4\,^\circ$C. The natural gas is separated from the other hydrocarbons by a process which lowers the pressure and cools the mixture from $+4\,^\circ$C to $-90\,^\circ$C. Methane gas separates from the other hydrocarbons, which stay liquid. The methane is supplied to a nearby British Gas plant to be distributed through the national grid of natural gas pipelines for use as a fuel (see Chapter C13). At the moment 10 to 15 per cent of the natural gas used in the United Kingdom comes from the Brent Field via St Fergus.

Pipelines are the safest and most efficient way of transporting liquids. So once the natural gas has been removed at St Fergus, the remaining 20 per cent of the hydrocarbons continue their journey to Mossmorran by pipeline. Installing the pipeline was a major engineering achievement. Between

Figure 2.14
The St Fergus gas separation plant.

St Fergus and Mossmorran it crosses under seventeen rivers, twenty-four major roads including two motorway crossings, and fourteen railway lines. On its way it passes by Aberdeen, Forfar, Perth and Cowdenbeath, and carries the liquid gases underneath some of Scotland's best agricultural land.

Figure 2.15
Laying the St Fergus to Mossmorran pipeline.

Figure 2.16
The same part of the pipeline route after reinstatement.

13 According to table 4 in the Data section, which of methane, ethane, propane and butane are gases and which are liquids at $-90°C$ and atmospheric pressure?

14 Why is methane from the North Sea called *natural gas*?

15 The St Fergus to Mossmorran pipeline can carry up to 15 000 tonnes of liquefied gas in 24 hours. A road tanker can carry 30 tonnes and would have to make a round trip of 276 miles from St Fergus to Mossmorran and back each day.
a How many tankers would have to make the round trip each day to carry the same amount of liquefied gas as the pipeline?
b If you lived on the road route how often would a tanker pass your house (in either direction) if there were no pipeline?

16 For most of its length, the St Fergus to Mossmorran pipeline is made of steel (9.5 mm thick) coated with coal tar enamel. In places, the wall thickness is increased to 11.1 mm, and some parts of the pipe are bedded in concrete. The pipe is never less than 1.2 m below ground level. Remote controlled isolation valves are installed at 11 km intervals.
a Suggest a reason for coating the steel pipe with coal tar enamel.
b In which areas do you think the engineers chose to use thicker walled pipe?
c For what reason might some lengths of the pipe be bedded in concrete?
d Why must the pipeline be at least 1.2 m below ground?
e what do you imagine is the purpose of the isolation valves?
f How might the measuring instruments which record the flow of liquid through the pipe be used to check that there are no leaks?

When the mixture of liquids arrives at Mossmorran, it is separated into its four main parts by distillation in a series of three columns. Ethane is removed from the top of the first column and is carried by pipeline to the neighbouring cracker unit. (This is described in the next section of this chapter.)

Figure 2.17
The Mossmorran gas separation plant.

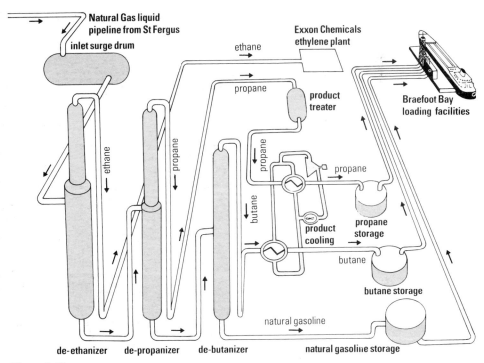

Figure 2.18
The gas separation process.

Propane is removed from the top of the second column and butane from the top of the third. These two products are stored under pressure in refrigerated tanks. Propane and butane are valuable fuels because they give a clean flame which is easy to control. They are used on a large scale in aluminium smelting, brick making and paper making. On a small scale they are supplied in cylinders for domestic heating and cooking. The remaining liquid from the bottom of the de-butanizer is natural gasoline which can be used to make petrol or supplied to the chemical industry.

Figure 2.19
Aerial view of the Mossmorran gas separation plant.

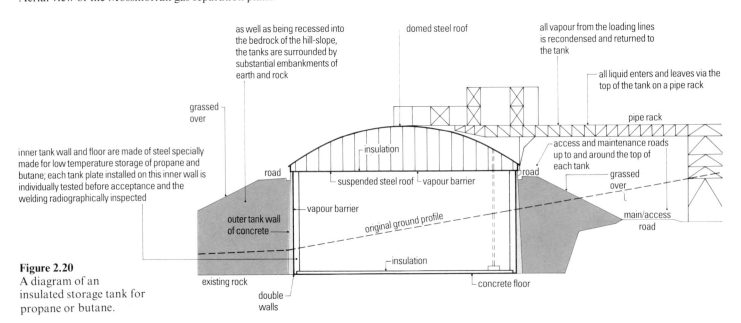

as well as being recessed into the bedrock of the hill-slope, the tanks are surrounded by substantial embankments of earth and rock

domed steel roof

all vapour from the loading lines is recondensed and returned to the tank

all liquid enters and leaves via the top of the tank on a pipe rack

pipe rack

grassed over

inner tank wall and floor are made of steel specially made for low temperature storage of propane and butane; each tank plate installed on this inner wall is individually tested before acceptance and the welding radiographically inspected

road

insulation

road

access and maintenance roads up to and around the top of each tank

grassed over

suspended steel roof   vapour barrier

outer tank wall of concrete

vapour barrier

original ground profile

main/access road

Figure 2.20
A diagram of an insulated storage tank for propane or butane.

existing rock

double walls

insulation

concrete floor

From the storage tanks, the products complete their journey through Scotland by pipeline to the Braefoot Bay terminal where they are loaded onto tankers to be shipped to customers.

Figure 2.21
A tanker loading at the Braefoot Bay terminal.

You could picture the complete process by imagining a butane molecule leaving the rocks 3000 metres below Brent at 6 o'clock on a Monday morning. It will travel by pipeline for 278 miles under the North Sea to be processed at St Fergus, then on by buried pipeline for 138 miles to Mossmorran where it will be separated from other hydrocarbons. It will spend some time in a refrigerated storage tank. Then it will be piped to Braefoot Bay and loaded onto a tanker. By then it will be 10 o'clock the following Saturday morning.

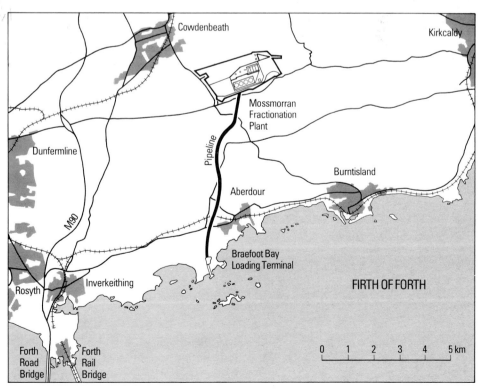

Figure 2.22
Mossmorran and Braefoot Bay in relation to the surrounding town and transport systems.

Figure 2.23
Aerial view showing the Fife Ethylene Plant beside the storage tanks.

17 Propane and butane are highly flammable and must be stored with great care. Look back at figures 2.19 and 2.20. What safety precautions are built into the Mossmorran storage tanks?

C2.4   How can molecules be broken into smaller pieces?

Cracking ethane

"Over the fence" from the gas separation plant at Mossmorran is another new installation called the Fife Ethylene Plant. This was opened in 1986 to take advantage of the ethane from the Brent Field. Here ethane molecules are broken down to the smaller molecules of ethene, which is known commercially as ethylene. Breaking down bigger hydrocarbon molecules into smaller ones is called *cracking*.

Figure 2.24
Part of the Fife Ethylene Plant.

The chemical formulae and the equation suggest that cracking ethane might be a simple process.

Ethene C$_2$H$_4$

Figure 2.25
Three ways of representing ethene molecules.

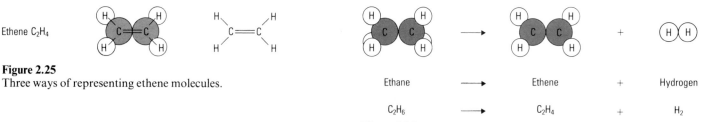

| Ethane | ⟶ | Ethene | + | Hydrogen |
| C$_2$H$_6$ | ⟶ | C$_2$H$_4$ | + | H$_2$ |

Figure 2.26
Model equations to show ethane being cracked.

Ethane is heated in a mixture with steam at a temperature high enough for it to break down to ethene and hydrogen. The reaction is then stopped by rapid cooling, and the resulting mixture of gases is passed through a series of separating columns. "Uncracked" ethene is recycled and the various by-products are used as fuel for the process. The only product leaving the plant is ethene.

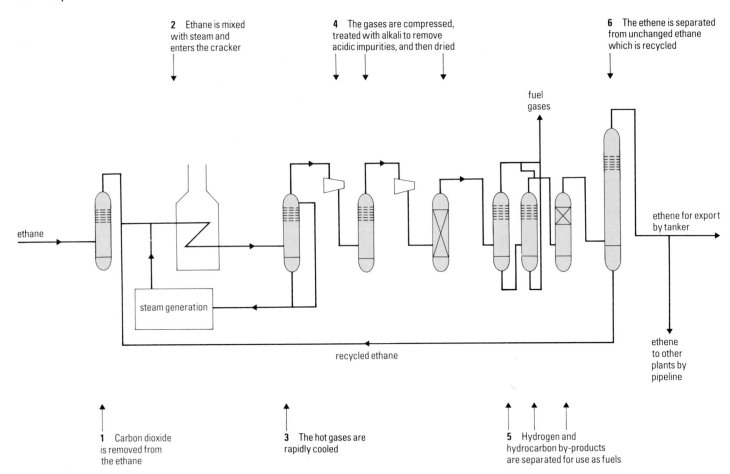

2   Ethane is mixed with steam and enters the cracker

4   The gases are compressed, treated with alkali to remove acidic impurities, and then dried

6   The ethene is separated from unchanged ethane which is recycled

fuel gases

ethane

steam generation

recycled ethane

ethene for export by tanker

ethene to other plants by pipeline

1   Carbon dioxide is removed from the ethane

3   The hot gases are rapidly cooled

5   Hydrogen and hydrocarbon by-products are separated for use as fuels

Figure 2.27
A flow diagram for the ethane cracker.

The basic chemistry may be simple but there are plenty of engineering problems.

Elaborate control systems are essential to ensure that the whole plant operates safely and efficiently.

Figure 2.28
Control room of the Fife Ethylene Plant, Mossmorran.

The ethane arriving from the separation plant contains some carbon dioxide which has to be removed before cracking.

Cracking takes place above 800 °C. The hot gases leaving the cracker are cooled rapidly in heat exchangers; the energy transferred is used to raise steam.

The cooled gases are compressed to 38 times atmospheric pressure and treated with sodium hydroxide to remove acidic impurities, and then distilled in a series of towers to remove by-products. These separation stages operate at low temperatures and require refrigeration systems which can take the temperature down to − 150 °C.

You can see from figure 2.29 that a very tall fractionating tower is needed to separate the product ethene from unchanged ethane. The tower must be tall to bring about an efficient separation of the two compounds.

Figure 2.29
The ethene fractionating tower being erected.

Some of the ethene from the cracker is taken by pipeline to the Braefoot Bay terminal for export to Europe by tanker. The rest is piped to join the grid of ethene pipelines which will take it to refineries at Stanlow in Cheshire and Carrington near Manchester.

The plant is designed to operate on a large scale, producing 500 000 tonnes of ethene per year. The plant requires more electrical power than a town of 90 000 people, and it burns more fuel than is needed by a town of that size for home heating.

18 Look up the boiling-points of ethane and ethene in table 4 of the Data section. Use the data to explain why it is difficult to separate them. Why do you think that the separation is improved if a tall distillation column is used?

19 The output of the cracker depends on the rate at which ethane arrives at Mossmorran. This in turn is linked to the quantity of methane being taken by British Gas from St Fergus. Would you expect the production of ethene to be greater in winter or in summer?

Teamwork

The cracker is run by just 350 people who are all based in a single building which includes the control centre, offices, workshop, laboratory, computer centre and medical centre. The building is designed so that all the staff share the same facilities. This makes sure that they all meet naturally in the course of their work and get to know each other.

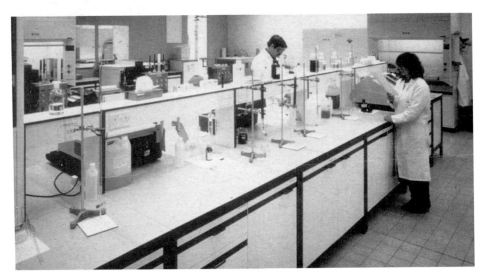

Figure 2.30
In the laboratory.

Figure 2.31
Technician checking in the plant area.

Figure 2.32
Technician checking with the control room.

The technicians who run the plant work as teams in shifts so that the cracker runs day and night. The technicians are trained to have all the skills needed to work in the control room, and to carry out maintenance of both the electrical and the mechanical equipment. They also collect samples for analysis in the laboratory to check that each stage of the process is working correctly and will give a product of the required quality.

The plant manager and the team supervisors try to make sure that every person working on the site has a varied and interesting job. Everyone on the site has a responsibility for safety. Some of the worst accidents in the chemical industry have taken place when a small incident has led to loss of control leading to a much more serious accident. At Mossmorran, emergency shut-down systems are part of the design. The control room and offices are separated from the chemical plant by a blast-proof wall to make sure that the staff can continue to run the process in the unlikely event of a serious emergency. (See figure 2.33.)

Industry and the local community

There are three main reasons why Exxon and Shell chose Mossmorran as the site for their factories.

- There was enough land. They could build the gas separation and cracking plants and still have space for further developments in the future.
- There was a safe harbour nearby at Braefoot Bay, and there were good communications by road, rail and air to other parts of the country.
- There were people living in the region with the skills needed to build the new plant. This included people who had worked in the coal industry, the naval dockyards and British Rail workshops.

Figure 2.33
The blast-proof wall on the plant side of the control room.

A big chemical works can be an opportunity for some people, especially those who hope for employment (see figure 2.34). However, it can be a threat to others who fear that it will be ugly, noisy, dirty, smelly and dangerous.

Figure 2.34
Group of junior operations technicians.

The new plant at Mossmorran was built in a region where there used to be eighteen coal mines, so the local people had some experience of the hazards of the mining industry which they could transfer to the petrochemicals industry. The new petrochemical plant was designed to make as little impact on the environment as possible.

There were some problems during the start-up of the new plant. Most spectacular was the 30-metre flare which, at night, was visible from as far away as Edinburgh. Flaring wastes gas and so is kept to a minimum. Normally small amounts of gas are burned at ground-level in a flare which is not visible from outside. The 100-metre flare tower is only needed when large amounts of gas have to be burned. Flaring is the method used to get rid of waste gas during the start-up of the plant and in emergencies.

When the gas compressors were first turned on at Braefoot Bay they were much noisier than expected. There were soon complaints from people living in the area. The compressors had to be shut down and modified by the manufacturers. Meanwhile, large sums of money were spent surrounding the pumps with acoustic hoods and silencers.

These should be short-term problems. When running normally the plant is quiet and clean. The safety and environmental adviser has day to day responsibility for making sure that nothing is done which will cause serious pollution. For example, rain water falling on the site might wash oil, dirt and chemicals off the site. This is prevented by collecting all surface water in a large pond so that it can be cleaned and checked before it is released into the Forth.

Figure 2.35
Cowdenbeath.

20 If you had lived in Cowdenbeath before the new processing plants and shipping terminal were built at Mossmorran and Braefoot Bay, would you have been in favour of the new development or against it? What reasons would you have given in support of your views?

21 What types of chemical plant might be built at Mossmorran in the future to take advantage of the chemicals available from the Shell and Exxon plants?

C2.5   How can ethene be used to build new molecules?

You will find it much easier to understand the reactions described in this section if you can make models of the molecules and then "carry out the reactions" with the models. Worksheet C2C will help you do this. Your teacher will tell you how much of the detail in this section you are expected to remember.

BOX 1   Problem
How can you crack hydrocarbons
in the laboratory?

The problem is to design an apparatus which you could use in the laboratory to crack a liquid hydrocarbon. The aim is to make some ethene so that you can investigate its properties. You will find it helpful to look at table 7 in the Data section.

If you try this experiment in the laboratory, you will probably be given a liquid oil which is a mixture of hydrocarbons. But for planning purposes assume that you are trying to crack decane. Look up the properties of decane in table 4 of the Data section.

Now decide on an arrangement of apparatus which will make it possible to pass decane vapour over a hot catalyst. A suitable catalyst is broken pieces of pottery.

Assume that the reaction works and that you make some ethene. Is ethene a solid, a liquid or a gas? How will you collect it in a test-tube? What will happen to any decane vapour which passes over the catalyst without being cracked?

Draw and label a picture of an apparatus which you think is workable.

Figure 2.36
Ball and spring model of an ethene molecule.

Ethene is special because its molecules are small and reactive. It can be thought of as one of the most important building bricks in the chemical industry.

The double bond between the carbon atoms in ethene means that there is spare bonding in the molecule. This means that ethene molecules are able to link with many other atoms or molecules. Figure 2.37 is a chart which shows just a few of the many chemicals which can be made from ethene.

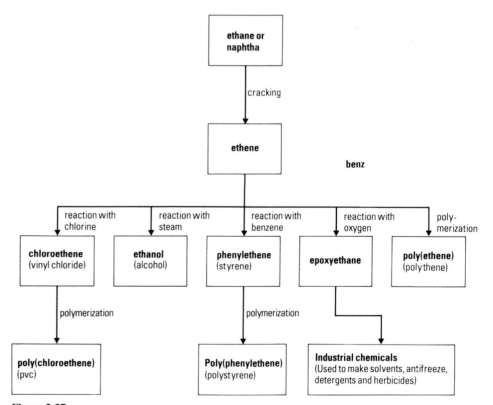

Figure 2.37
Some of the many products which can be made from ethene.

Figure 2.38

Ethene is an example of an *unsaturated compound*. The words *saturated* and *unsaturated* have a special meaning when they are used to describe carbon compounds. They are now quite commonly used, as you can see from figure 2.38. The margarine in the tub is described as being high in *polyunsaturates*. (You can read about the connection between the types of fat in our diet and heart disease in Chapter **B**6 of your Biology book.)

If you look at the graphical formulae of the alkane molecules in figure 2.10 you will see that every carbon atom has used all of its four bonds to link to other atoms. There are no spare bonds. The alkanes are *saturated*. But ethene has a double bond. The carbon atoms in its molecules have some spare bonding available. Ethene, and all other molecules with double bonds, are *unsaturated*. The fats in margarine are labelled polyunsaturated. "Poly-" means "many". The molecules of polyunsaturated fats have several double bonds in them, not just one as in ethene.

Reactions of ethene

Ethene can use its double bond to add on bits to its molecules. Three examples are given here:

● Ethene adding on bromine.
Ethene reacts rapidly with bromine and the molecules join together, as shown in figure 2.39.

Figure 2.39
Ball-and-spring models of ethene and bromine combining to make dibromoethane.

Bromine is orange, but the product of the reaction is colourless. If a little of a solution of bromine is shaken with a tube of ethene gas, the liquid becomes colourless. This reaction can be used to test carbon compounds to see if they contain double bonds.

● Ethene adding on steam.
This reaction is used industrially to manufacture ethanol (alcohol). See figure 2.40.

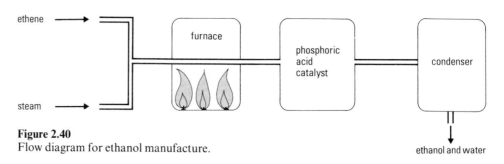

Figure 2.40
Flow diagram for ethanol manufacture.

Ethene and water are passed through a furnace which heats the mixture to 300 °C. The hot gases then go into a reactor which contains a special catalyst.

Under pressure, some of the ethene and water combine to make ethanol, as shown in figure 2.41.

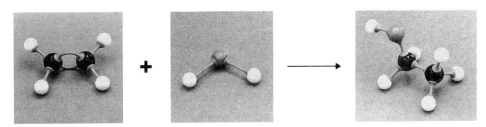

Figure 2.41
Ball-and-spring models of ethene and steam combining to make ethanol.

The catalyst used is phosphoric acid. The catalyst helps to speed up the reaction. (The use of catalysts is explained in more detail in section **C**3.4.)
Ethanol has two carbon atoms in its molecules so it is related to **ethan**e. It is one of a family of alcoh**ols**. So you can see how it gets its name.

22 Suggest a method for finding out by experiment whether a fat is saturated or unsaturated.

23 Rewrite the equations shown in figures 2.39 and 2.41 using graphical formulae instead of models. Show that both of these equations are balanced.

24 Propene is an unsaturated compound with the molecular formula C_3H_6.
a Draw the graphical formula of propene.
b Write an equation to show what you would expect to happen when propene reacts with bromine.

25a In the industrial manufacture of ethanol from ethene there are three ways in which the conditions are arranged to help to speed up the reaction. What are the three ways?
b The process illustrated in figure 2.40 produces a mixture of ethanol and water. Look up the properties of these two compounds in tables 4 and 5 of the Data section, then suggest a method of separating the ethanol from the water.

● Ethene adding to itself.
Ethene molecules will join together in long chains if they are heated under pressure with special catalysts. Figure 2.42 shows just six ethene molecules using their double bonds to join up in a chain. When this reaction is used in

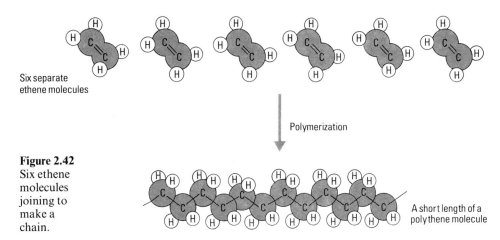

Six separate
ethene molecules

Polymerization

Figure 2.42
Six ethene
molecules
joining to
make a
chain.

A short length of a
polythene molecule

Figure 2.43
The structure of polythene.

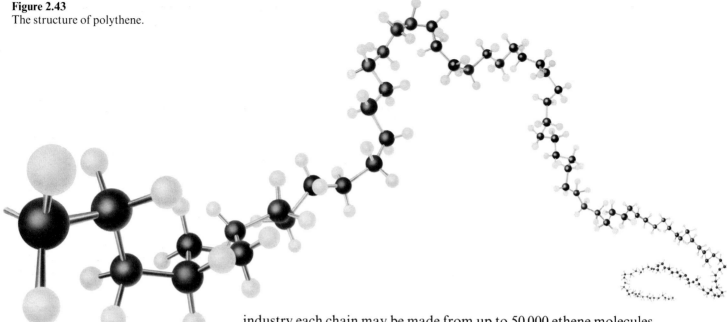

industry each chain may be made from up to 50 000 ethene molecules.
(See figure 2.43.)

The product of this reaction is polythene, which is short for poly(ethene).
"Poly-" because many ethene molecules have been joined together. The
reaction is an example of the process called *polymerization*, and polythene is
an example of a *polymer*. (Figure 2.44.)

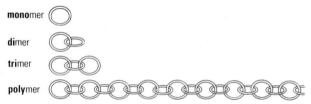

Figure 2.44
Monomer, dimer, trimer and polymer.

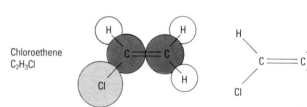

Chloroethene
C_2H_3Cl

Figure 2.45
A chloroethene molecule.

26 Approximately how many carbon atoms and how many hydrogen atoms are there in
the longest polythene molecules?

27 The graphical formula of chloroethene is shown in figure 2.45. Chloroethene
molecules can polymerize to make poly(chloroethene). Draw a short length of a
poly(chloroethene) molecule. (Note that the old fashioned name for chloroethene is
vinyl chloride and so the polymer is often called polyvinyl chloride, or pvc.)

C2.6   Why did the chickens stop laying eggs?

Do you cook? If so, have you ever looked through a cookery book and
discovered a recipe with a beautiful picture which makes you think that you
are going to prepare a wonderful meal with no trouble at all? It can all seem
so easy in the book, but when you set to work in the kitchen it doesn't always
turn out as you hope.

The same thing can happen in plants for manufacturing chemicals. Look at the equation for making polythene in section 2.5. It looks simple. Lots of little ethene molecules join up into a long chain to make polythene. The problems start when you try to make the reaction happen in a controlled way.

Figure 2.46
The high pressure polyethylene plant at Meerhout.

Figure 2.46 gives you some details of a plant for making low density polythene by the high pressure process. Powerful pumps are needed to compress the gas to about 2000 times atmospheric pressure. Compressing the gas heats it up, but it is cooled before it enters a tube. The ethene is mixed with a catalyst and polymerizes inside the tube. This is about 1 km long and has steel walls about 4 cm thick. The tube is surrounded by a water jacket so that the temperature inside can be controlled. The chemicals travel through the long tube at high speed so that the reaction is complete in a matter of seconds.

No one knows exactly what goes on inside the tube. The pressure is so high that it is impossible to put a glass window in the tube to look inside. What is known is that by varying the conditions it is possible to make over a hundred different types of polythene. Some of them are valuable, others are useless.

How can so many different types of polythene be made by the same chemical reaction? Look again at the details on the chemistry in section C2.5. You will see that in that section there is no information about how many ethene molecules join together. In fact, the chains vary in length. Altering the reactor conditions can change the average length of the molecules. Also during the high pressure process polymerization can start in the middle of a chain forming branches. Chain length and chain branching both alter the properties of the plastic.

The reactor is run from the control room which allows the team of operators to keep watch on the conditions throughout the plant. Normally the plant runs smoothly and continuously, producing the grade of polythene required. Occasionally things can go wrong. Maybe an operator accidentally turns off a valve, or a valve stops working. Usually this will mean that poor quality plastic is made.

Figure 2.47
Polymer chains of different lengths and with different amounts of branching. In these diagrams the long hydrocarbon molecules are shown as lines.

The job of the engineers at the plant is to design facilities which can handle unexpected changes in conditions and prevent accidents. When an incident does occur it is important to discover the cause and make changes which will prevent a similar accident happening again.

The plant is only likely to run smoothly if the engineers and operators are working together as a team. One of the tasks for the management is to ensure that all those involved can work together effectively.

Figure 2.48
The control room at Meerhout.

Figure 2.49
The extrusion of polythene film.

Sometimes an accident can lead to the discovery of a new set of conditions to make a special grade of polythene with valuable new properties. This is rare; usually a batch of low quality plastic creates problems for the people in marketing who have to try to find customers for off-grade polythene. One possibility is to add carbon black as a filler and use it to make dustbin liners.

You may be wondering why the reactor tube is surrounded by high concrete walls. These walls are part of the safety precautions in case the reaction gets out of control. The ethene in the tube may start to decompose into carbon and methane. This happens very fast and is highly *exothermic*. An exothermic reaction is one which gives out energy. The energy is usually transferred to the apparatus and the air surrounding the reaction which get hotter.

If the ethene starts to decompose, it is very easy to lose control of the reaction leading to an explosion. To prevent this, the safety valves open to release the compressed gas into the air and stop the reaction.

Think back to your experience in a kitchen again. Have you ever seen an impatient cook remove the pressure control on a pressure cooker before it has cooled down? If so, you know that a dangerous jet of steam rushes out. But the steam inside a pressure cooker is only at twice atmospheric pressure. You can imagine the power of the gas jet and the noise when ethene at 2000 times atmospheric pressure is released. The ethene is highly flammable. It carries hot carbon with it which ignites the mixture of ethene and air, producing an aerial explosion. The shock of the noise is enough to scare chickens in a local battery farm and stop them laying eggs for a while.

The chemical company then has to pay compensation to the farmer, so obviously it wants to try to stop these explosions if possible. The problem is that no one can predict when they will happen, and so conditions in the reactor tube are very carefully checked all the time.

Figure 2.50
Some useful products made from plastics.

28 Which piece of laboratory apparatus consists of a tube cooled by cold water?

29 What can the operators of the plant control to vary the properties of the polythene?

30 Compare the conditions near the beginning of the reactor tube where the ethene starts polymerizing with the conditions near the end. No one knows for sure, but what differences would you expect?

31 Imagine yourself employed to sell polythene for a chemical company. Apart from dustbin liners, what other uses can you think of for getting rid of poor quality plastic?

32 Write an equation for the decomposition of ethene to carbon and methane.

33 Explain why hot carbon in the jet of gas which is released when the safety valves open can set off an explosion.

34 If you were one of the engineers in charge of a high pressure polythene plant, how would you try to modify the safety valves, surrounding walls and layout of the plant to limit the damage when the safety valves open? Can you think of a way to try to prevent the ethene catching fire and exploding?

Summary

This chapter has introduced you to a number of new terms which are listed below. Write a brief explanation of each term and give an example to illustrate your meaning.

Types of compound: hydrocarbon, alkane, saturated compound, unsaturated compound, monomer, polymer.

Types of process: fractional distillation, cracking, polymerization, exothermic reaction.

Types of formula: molecular formula, graphical formula.

Chapter **C**3

Chemicals from plants

In this chapter you will study some biochemistry *which is the chemistry of living things. It took scientists a long time to work out the structures of chemicals such as carbohydrates and proteins. This chapter does not try to explain how they made these discoveries. Instead, it tells you enough about the compounds for you to be able to understand the processes of photosynthesis, respiration and digestion when you study them in Biology (see Chapters **B**3, **B**4, **B**5 and **B**6).*

*This chapter gives you more opportunities to get used to the idea of explaining things in terms of atoms and molecules. You get further practice in the use of models, symbols and formulae. In Chapter **C**2, most of the compounds were made from just carbon and hydrogen atoms. In this chapter three more elements appear – oxygen, nitrogen, and sulphur.*

C3.1   What can we get from plants?

The pictures in figures 3.1 to 3.9 show just a few of the many useful products we get from plants.

---

1 Which drugs are found in tobacco leaves?

2 Which plants are grown to make sugar?

3 Which plant foods contain starch?

4 Why do we depend on plants for the meat in our diet as well as for the vegetables?

5 Give some examples of herbs and spices obtained from plants.

Figure 3.1 (left)
Vegetable oils. In a supermarket it is possible to buy corn oil, olive oil and sunflower oil. Linseed oil is used in paints. Palm, olive and coconut oils are used to make soap. Vegetable oils are also used to make margarine. This picture shows a sunflower – the oil comes from the seeds.

Figure 3.2 (right)
Paper. The wood from coniferous forests is used to make paper. After the foliage and bark have been removed from the trunk, the wood is ground up and treated with chemicals to produce the pulp needed to make paper.

Figure 3.3
Perfumes. Perfumes are made using oils from plants. Plant essences are gathered from all over the world: lemon grass and sandalwood from India, rosemary from Spain, lavender from England and France, and rosewood from South America.

Figure 3.4
Waxes. Waxes are used in cosmetics and polishes. Carnauba wax comes from the leaves of a palm tree which grows in Brazil. It is used in lipsticks.

Figure 3.5
Fibres. Fibres come from the seeds of plants (cotton and kapok), from the leaves (sisal), from fruits (coir) and from the stem (flax, jute and hemp). This picture shows some cotton bolls.

Figure 3.6
Rubber. The latex used to make natural rubber is tapped from the bark of a tree which grows in the tropics.

Figure 3.7
Drugs. The narcotic, morphine, comes from the opium poppy shown in this picture. Mexican yams contain a chemical from which the contraceptive pill can be made. The heart stimulant called digitalis is extracted from the leaves of foxgloves. The mild stimulant in tea and coffee is caffeine. The discovery of a pain-killing drug in the bark of willow trees is described in Chapter **C**12.

Figure 3.8
Dyes. Plants were the main source of dyes for thousands of years. The most important natural dyes were the blue dye from the indigo plant and the red dye, alizarin, from the root of the madder plant. Indigo (in India) and madder (in Europe) were major agricultural crops until the discovery of synthetic dyes at the end of the nineteenth century. (You will study dyes in Chapter **C**11.) This picture shows a vat being beaten by hand on a nineteenth-century indigo plantation.

Figure 3.9
Food. We depend on plants for our foods. Sugar and starch are examples of nearly pure plant chemicals. We also rely on plants for the herbs and spices we use to flavour our food. (See Chapters **B**4 and **B**5.)

C3.2   What are carbohydrates?

*Before you read this section, you should have done some experiments with carbohydrates in the laboratory so that you have some idea of what carbohydrates look like and how they behave. Worksheet **C3B** suggests some possible investigations.*

*Carbohydrate molecules are much more complicated than water molecules, or the molecules of simple hydrocarbons such as methane and ethene. To make sense of the molecules described in this section you will need to use models. Worksheet **C3C** is designed to help you. You are not expected to remember the details of the formulae and structures of carbohydrates.*

In this section you will find references to your Biology book. Your work in Biology will show you why it is important to understand the chemistry of carbohydrates.

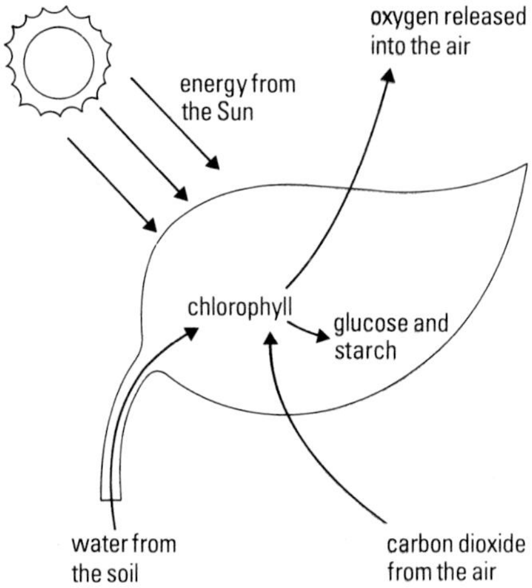

The starting point for plant chemistry is photosynthesis. You investigate this process in your Biology course (see Chapter **B**3), and figure 3.10 here sums up the main ideas.

Figure 3.10
The process of photosynthesis.

> **6** Which gas do plant leaves take in from the air during photosynthesis?
>
> **7** What else do plants need for photosynthesis?
>
> **8** What does the word *synthesis* mean?
>
> **9** Why is the process which takes place in plant leaves called *photo*synthesis?

Glucose and starch belong to a family of chemicals called carbohydrates. The leaves of plants make glucose and starch from carbon dioxide and water, so there are just three elements in carbohydrates: carbon, hydrogen and oxygen. This explains their name:

carbo-   for carbon
-hydrate   from the Greek word for water.

Carbohydrates are compounds made from carbon and the elements of water.

Some carbohydrates taste sweet and are soluble in water. These carbohydrates are called sugars. Examples include: glucose, sucrose (which is the sugar we use for cooking), and lactose (the sugar in milk).

Plants take small and simple molecules and turn them into big and complicated ones. This is clearly illustrated by the process of photosynthesis. Figure 3.11 shows what is involved in making glucose from carbon dioxide and water. This isn't a simple one-step process. Figure 3.11 shows the molecules present at the beginning and the end of the process. It doesn't tell you anything about the way in which photosynthesis happens.

> **10** Count the numbers of carbon, hydrogen and oxygen atoms being used to make glucose in figure 3.11. Now count the numbers of these atoms in the glucose and oxygen produced. How do the numbers compare?
>
> **11** If the formula of glucose is written $C_xH_yO_z$, what are the values of x, y and z?

In section **C2.5**, you may have been surprised by the large number of products made from ethene in industry. If so, you will be even more impressed when you see what plants can do with glucose molecules.

BOX 1   Problem
Are all sugars equally sweet?

There are no instruments for measuring sweetness. There are no agreed measuring units. We have to rely on our tongues and hope to get useful results by carrying out a series of tests with a team of tasters.

Suppose that you have got pure samples of several sugars such as sucrose, glucose, fructose, lactose and maltose. Plan an investigation to compare their sweetness.

Here are some of the questions you will have to think about. How will you arrange to make a fair comparison between the sugars? Will you test the sugars dry or in solution? How many tasters will you need? Does the time of day matter? Does the order they taste the sugars matter? How will you record the results? Can you devise a measurement scale? What safety precautions will you have to take?

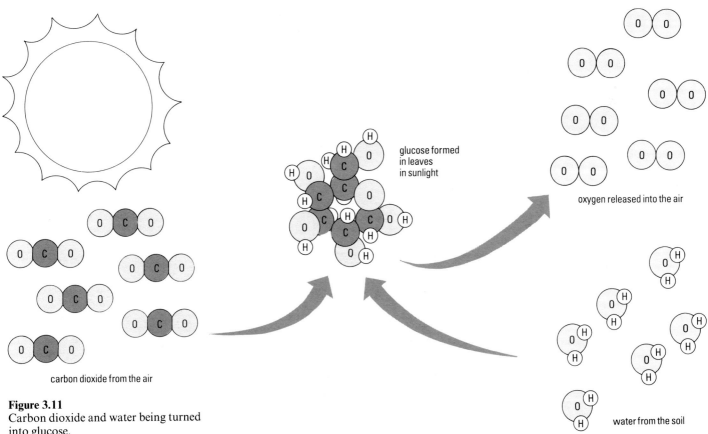

Figure 3.11
Carbon dioxide and water being turned into glucose.

Plants make starch by joining glucose molecules up in long chains as shown in figure 3.12 on the next page. Huge numbers of glucose molecules are involved. There may be up to a **million** glucose units in one starch molecule. Figure 3.12 shows part of a straight chain. The largest starch molecules have a complex network of branches.

12 What is the name of the process used to make long chains by joining up small molecules? (See section **C2.5**.)

13 What is the name of the monomer used to make starch?

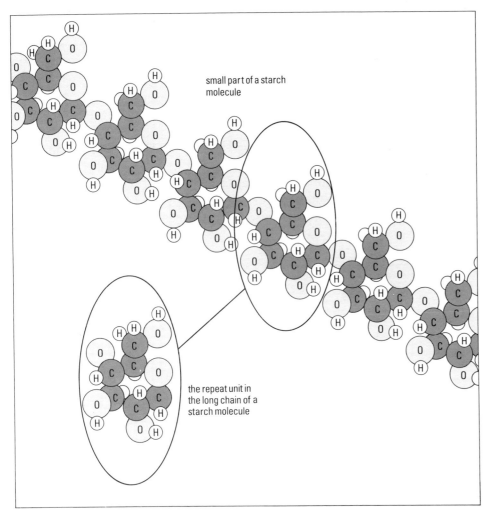

Figure 3.12
Part of a starch molecule, with the repeat unit in the chain.

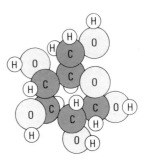

Model of a glucose molecule

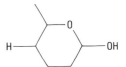

Shorthand structure of a glucose molecule

Outline diagram of a glucose molecule

Figure 3.13
Three representations of a glucose molecule.

14 The repeat unit in the starch chain shown in figure 3.12 is not exactly the same as a glucose molecule. Count up the numbers of each type of atom in the repeat unit and compare them with the numbers of atoms in the formula of glucose. What is the difference? Which common compound might be connected with this difference?

15 Measure the approximate length of the repeat unit in figure 3.12. How many metres long would the diagram have to be to show a starch molecule with just 1000 glucose units in the chain?

Figure 3.12 looks very complicated, and it is not easy to see what is happening when glucose molecules join up. Figure 3.13 shows three ways of picturing a glucose molecule. Figure 3.14 uses the simplest of the diagrams in figure 3.13 to describe how glucose molecules join together to form starch.

Now you can see that the glucose molecules join by splitting off molecules of water. This is a different kind of polymerization reaction from that used to make poly(ethene). You will meet this type of polymerization reaction again in Chapter **C**8.

The starch made by plants is a useful food store. Starch is insoluble in water and so it acts as a "bank" of "tied up glucose" which can be released when the plant needs it. Starch from plants is an important part of many foods. In some

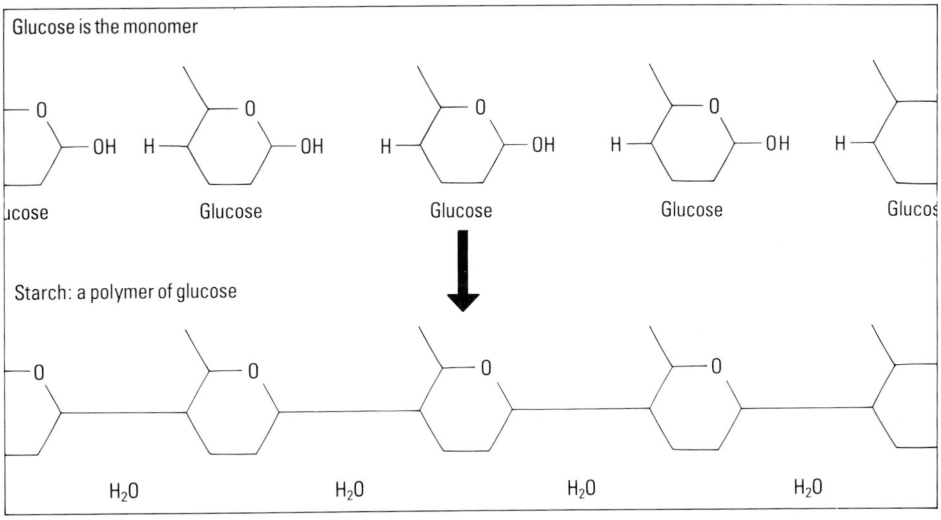

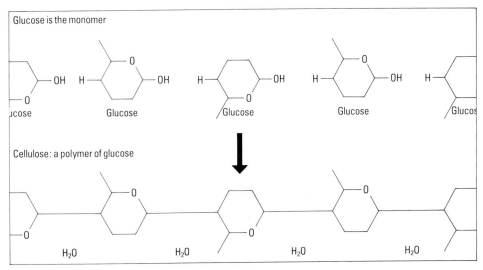

Figure 3.14
The polymerization of glucose to make starch.

Figure 3.15
The polymerization of glucose to make cellulose.

parts of the world, starch makes up 70 to 80 per cent of the "fuel" in the diet. In the United Kingdom it supplies about 25 per cent of our food energy needs.

In Biology Chapter **B**6 you will read that animals can store carbohydrates in the form of glycogen in the liver. Glycogen is another polymer of glucose. It is very similar to starch, but the chains have lots of branches.

Plants can join up glucose molecules in many other ways. One of them produces cellulose. This is very important because it is used by plants to make the walls of their cells and to give them strength. Figure 3.15 shows glucose molecules being joined to make cellulose molecules. Cellulose molecules are even bigger than starch molecules. The cellulose fibres in figure 3.19 make up part of the cell wall of an alga. They are magnified 30 000 times.

16 Why do we use the same word "cell" for part of a plant and for part of a prison?

17 Look closely at figures 3.14 and 3.15. What is the difference between the glucose chains in starch and in cellulose?

The cell walls in a potato are made of cellulose. The energy store in the cells is starch. Figure 3.20 shows some potato cells magnified about 150 times. You can see grains of starch trapped within cells whose walls are made of cellulose. When you eat a potato you can digest the starch and turn it back into glucose

BOX 2   Experiment
Molecules big and small

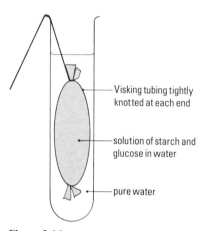

Figure 3.16
The Visking tubing experiment with starch and glucose.

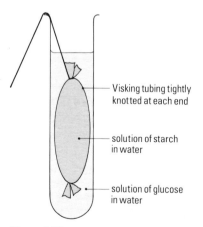

Figure 3.18
The Visking tubing experiment with different starting conditions.

The diagram in figure 3.16 describes an experiment to investigate the movement of carbohydrate molecules through a membrane.

Visking tubing is an artifical membrane made of cellulose. You can picture it as a barrier with minute holes in it. It is like filter paper but with very much smaller holes.

In this experiment the liquids inside and outside the Visking tubing are tested at the start. Then they are tested again about an hour later. The tests and the results are shown in figure 3.17.

Solution tested	Test with iodine solution		Test with Benedict's solution	
	At the start	**After 1 hour**	**At the start**	**After 1 hour**
Outside the Visking tubing	No change	No change	No change	Orange–red suspension forms on warming
Inside the Visking tubing	Blue–black colour	Blue–black colour	Orange–red suspension forms on warming	Orange–red suspension forms on warming

Figure 3.17
Results for the Visking tubing experiment.

18 What does the word "membrane" mean to you? What other examples of membranes can you think of?

19 Which compound gives a blue–black colour with iodine solution?

20 Which compound in this experiment produces an orange–red suspension when warmed with Benedict's solution?

21 According to the results of figure 3.17 which of the two carbohydrates can pass through the Visking tubing membrane?

22 Predict what would happen if you repeated the experiment with the arrangement shown in figure 3.18. Write down your predictions in the form of a table similar to figure 3.17.

23 Imagine that you could see the molecules in this experiment and the small holes in the Visking tubing. Draw a diagram to show what you think they might look like. Use your picture to explain the results of the experiment.

24 Do you think that water molecules can pass through Visking tubing? Plan an experiment to test your prediction.

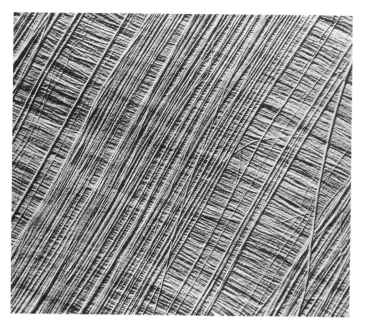

Figure 3.19
Photograph of the surface of a plant cell wall taken with an electronmicroscope ($\times 30\,000$).

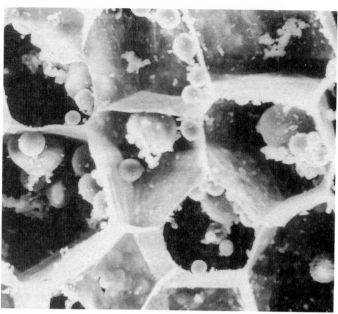

Figure 3.20
Photograph of raw potato cells ($\times 150$).

as a source of "fuel" for your body, but you cannot digest the cellulose which is part of the dietary fibre in your diet.

Why do we have to digest our food? Why can cows digest the cellulose in grass when we cannot? Why is dietary fibre important to our health? You will find the answers to these and other questions about what you eat in Chapters **B**4, **B**5 and **B**6 of your Biology book.

In box 2 you can learn that membranes will let some molecules through and not others. This is very important to the working of your body. Molecules of gas move in and out of the blood in your lungs, but the rest of the blood stays inside its capillaries (see Biology Chapter **B**7). In your intestines you absorb the molecules you need from digested food, but leave the rest behind. (See Biology Chapter **B**5.) In your kidneys your body gets rid of waste molecules while keeping the ones it finds useful (see Biology Chapter **B**12). Kidney machines use a material like Visking tubing to do the same job.

C3.3   What are proteins?

This section gives you a brief introduction to proteins. It tells you enough for you to be able to understand the chemical ideas in the Biology chapters about nutrition and digestion. You are not required to remember the details of the names and formulae.

Plants need more than just air, water and light if they are to grow. Farmers and gardeners know that they must add fertilizers, manure or compost to the soil if they are to get good crops. You can read more about what plants need to grow in Chapter **C**16 and Biology Chapter **B**15.

One of the elements which plants obtain from the soil is nitrogen. They need the nitrogen to make proteins.

Protein molecules are long chains of *amino acids*. Thousands of different proteins are made from about twenty different amino acids.

In starch, all the units in the polymer chain are the same – they are glucose molecules. Making a starch molecule can be compared to making a necklace using just one colour and size of bead. Making a protein molecule is like making a necklace with lots of different colours and sizes of bead.

The structures of three amino acids are given in figure 3.21. This also includes the simpler way of picturing the molecules which we shall use in this course. Figure 3.22 uses this simpler picture to represent a short length of a protein molecule.

Figure 3.21
The structure of three amino acids – and a simplified method of representing them.

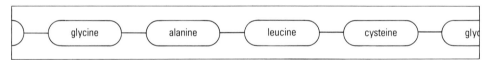

Figure 3.22
Part of a protein molecule.

---

25 Which five elements are present in the amino acid molecules shown in figure 3.21?

26 How many bonds are formed by **a** each oxygen atom and **b** each nitrogen atom in the amino acid molecules in figure 3.21? Compare your answers with the table in figure 5.14.

27 In what ways are the structure of the two amino acids in figure 3.21 the same? How do they differ?

The chemical reactions in plant and animal cells are controlled by catalysts called *enzymes*. Enzymes are proteins.

C3.4　How does all this clever chemistry happen in living things?

---

*You are introduced to enzymes in Biology Chapter **B**5, and the worksheets for that chapter suggest investigations into the properties of enzymes. This short section is included to show you how enzymes, which are proteins, are involved in one of the simpler chemical reactions involving carbohydrates. You are not expected to remember the detailed facts about inversion.*

Figure 3.23
Golden syrup contains partially inverted sugars, as you can see from the label.

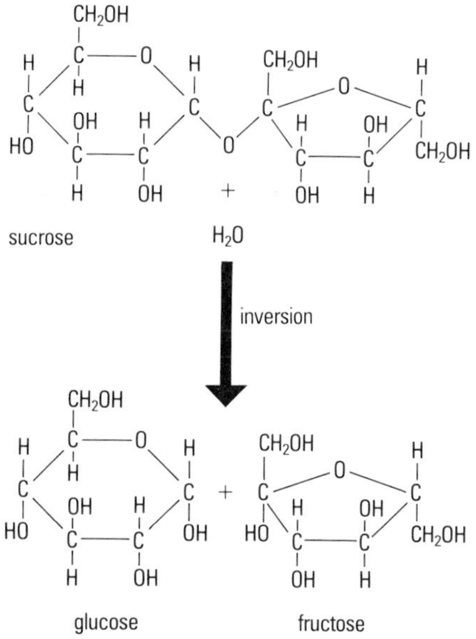

Figure 3.24
Sucrose reacting with water to give glucose and fructose.

In chemistry we use catalysts to speed up reactions. In the petrochemical industry, for example, catalysts are involved in the cracking of hydrocarbons and the polymerization of ethene. Catalysts make reactions go at a lower temperature than might otherwise be necessary.

There are many catalysts in living things too, and these catalysts are called *enzymes*. Enzymes control all the chemical reactions in the cells of plants and animals. There are enzymes for the reactions which take small molecules and join them to make bigger ones. There are also enzymes which start with big molecules and break them up into smaller ones. All the reactions in cells which enable living things to digest food, move, grow, and reproduce are controlled by enzymes. You investigate enzymes and study how they are involved in digestion in Biology Worksheets **B**5A and **B**5B.

Each reaction has its own enzyme. The enzymes which join glucose molecules together to make starch are different from the enzymes which join glucose to make cellulose.

The way in which enzymes make chemical reactions go faster at a lower temperature is illustrated by the difference between bees making honey and us making jam. Both processes involve a reaction called *inversion*. Food labels often mention inverted sugars (see figure 3.23) and you may have wondered what this means.

Bees get their sugar from the nectar of flowers. We get our sugar from cane or beet. In both cases the sugar is sucrose. Sucrose consists of two simpler sugars, glucose and fructose, linked together as in figure 3.24. The diagram also shows sucrose being split into glucose and fructose by reaction with water. This is the process of inversion.

28 What do you think is meant by the term "partially inverted sugar"?

29 With the help of figure 3.24 work out the chemical formulae of the three sugars: sucrose, glucose and fructose. Write the formulae in the form $C_xH_yO_z$.

30 In what way are the molecules of glucose and fructose the same? How do they differ?

31 In the language of carbohydrate chemistry, glucose is called a *monosaccharide*, sucrose is called a *disaccharide*, and starch is called a *polysaccharide*. Why do you think that these terms are used? (A dictionary may help you.)

32 Write a balanced symbol equation for the process of inversion.

Jam is about 60 per cent sugar; the rest is water and fruit. The high concentration of sugar is a preservative which stops it going mouldy. The mixture of fruit, water and sugar is boiled during jam making, and boiling continues until the temperature reaches about 105 °C. At this high temperature some of the sucrose reacts with water and turns into glucose and fructose. Acid from the fruit acts as a catalyst for the reaction. So the sugar in jam is a mixture of sucrose and inverted sugar. This is important, because the mixture is much more soluble in water than sucrose is on its own. This means that the sugar doesn't crystallize when the jam is put into jars and cooled to room temperature.

Bees invert sugar when they make honey (see figure 3.25). They can do this at their body temperature with the help of the enzymes they produce.

Figure 3.25
Bees collect sucrose from nectar. They produce enzymes which turn sucrose to glucose and fructose, forming honey.

33 Estimate the temperature at which bees turn the sugars in nectar into honey. How does this compare with the temperature reached in jam making?

34 This quotation comes from a cookery book about jam making:
What went wrong: the jam crystallized.
Why: there was a lack of acid in the fruit.
Cure: if you make jam from the same kind of fruit again,
add lemon juice, citric acid or tartaric acid.
Explain why a lack of acid in the fruit might cause the jam to crystallize and also explain the suggested cure.

Plants and animals can do lots of clever chemistry in their cells because of enzymes. We depend on the enzymes in bacteria and fungi to maintain the natural cycles in the environment (for example the nitrogen cycle described in Biology Chapter **B**15). We have used yeasts and their enzymes for thousands of years in baking and brewing. Recently it has been realized that there are many more ways by which we can take advantage of enzymes. For this reason, biotechnology has become very important. Among other things, research into biotechnology is aimed at producing better and cheaper drugs, alternative sources of energy, and new types of food.

Summary

1 This chapter has included a number of important types of compound which are listed below. Write a brief explanation of each one.
carbohydrate; sugar; amino acid; protein; catalyst; enzyme.

2 To complete this part of the summary you will also need to study sections **B**3.2 and **B**3.5 in your Biology book.

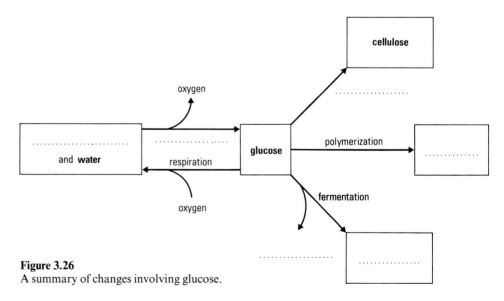

Figure 3.26
A summary of changes involving glucose.

Copy and fill in the flow diagram in figure 3.26. Write the name of the substances formed in the boxes and the names of processes beside the arrows.

Chapter **C4**

Chemicals and rocks

As you study this chapter you will have opportunities in the laboratory to plan experiments, make observations and learn new practical skills. You will find out more about the use of electrolysis and heating as ways of making new chemicals from rocks. One section starts with the heading "How much?" which is a very important question in chemistry. One way of answering the question is given in this chapter but other methods are given in Chapters C5 and C12.

This is a long chapter, but you may not be expected to study it all in detail. Your teacher will guide you and tell you which sections to concentrate on. The first section includes ideas which you may already have met in geography.

C4.1   What is the difference between a rock and a mineral?

How are rocks formed?

We usually think of a rock as being a massive thing like the cliffs which mountaineers climb. But in science we include stones, pebbles and boulders as rocks. A stone is just a small piece of rock.

Imagine that you are standing on the edge of a cliff on the east coast of England, and the waves are crashing against the rocks below you. Almost certainly, in a few hundreds of years from now no one will be able to stand where you are standing. The ground will have collapsed and been carried away by the sea. In places, whole villages that once stood on a cliff have been destroyed by the waves. An example is Dunwich on the Suffolk coast. It was once a busy port with several churches and now there are only a few houses and one church. All the others went over the cliff.

If a cliff crumbles away, some of the rock may be carried by the sea to a beach further along the coast. Some may sink to the bottom of the sea to form a layer on the sea bed which we call a sediment. Much more sediment forms from the grains of rock brought down by rivers. Gradually the layers build up on the sea bed, pushing down on the ones beneath. If this continues for millions of years, the layers of sediment become very thick. The lower layers are pressed so hard that they turn into rock. Rocks made in layers like this are called *sedimentary rocks*. Sandstone, chalk, and limestone are all examples of sedimentary rocks.

It is hard to imagine the great expanse of years which makes up the history of the Earth. You can probably picture a lifetime of about seventy years. People just like us have lived on this planet for about 700 lifetimes, around

Figure 4.1
The cliffs on the coast of Suffolk are being severely eroded! This photograph shows Pakefield which is south of Lowestoft.

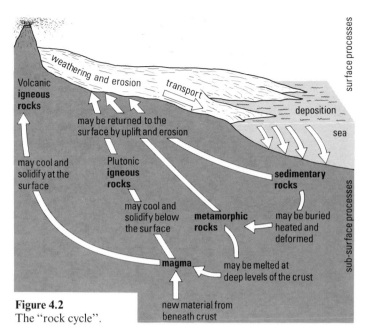

Figure 4.2
The "rock cycle".

50 000 years. But the Earth itself is about 4 600 000 000 years old. The limestone, coal and gritstone in the Pennines were formed over a period of 65 000 000 years.

Modern geology began when James Hutton (1726–97) first saw that it was possible to explain the formation of rocks and the shaping of the landscape by thinking in terms of millions of years. Given a long enough time, it is possible for wind and rain, rivers and glaciers, earthquakes and volcanic eruptions to bring about huge changes to the Earth.

Sandstone is made of grains of sand pressed and cemented together. Sandstones may be coloured red, brown or yellow by other minerals such as iron oxide. Chalk is a form of calcium carbonate. It was made from the shells of tiny animals that lived in the sea, died, and sank to form sediments, which is why you can sometimes find fossil shells in chalk.

Many of the cliffs in Britain are made of sandstone, chalk or limestone. To get from the sea bed, these rocks were pushed up by movements of the Earth's

Figure 4.3
Sedimentary sandstone rock.

Figure 4.4
This sedimentary rock has been pushed up at an angle and then most of it has been eroded away.

Figure 4.5
Sample of basalt.

Figure 4.6
Granite from Dartmoor.

crust. Sideways movements rumpled the layers of sedimentary rock into folds. Very high folds made mountains such as the Alps.

Not all rocks are sedimentary. *Igneous rocks* were formed when very hot, molten material cooled and crystallized. This hot molten material is called magma. It appears as lava when it flows to the surface during a volcanic eruption. It cools quickly as it runs down the side of the mountain, forming basalt. Magma may also cool and crystallize deep underground forming granite.

---

1 Look at the samples of basalt and granite in figures 4.5 and 4.6. Both of these rocks are made up of crystals.
a In which of the rocks are the crystals larger?
b Both rocks formed when magma cooled and crystallized. Which rock cooled faster?
c What is the connection between the size of crystals and the rate at which the crystals form?
d How many different types of crystal can you see in figure 4.6?

There is a third kind of rock – *metamorphic*, or changed, rock. Huge pressures and high temperatures can turn chalk into marble, sandstone into quarzite, or granite into gneiss. Rocks can also be altered by the action of steam, or very hot water, under pressure.

What is a mineral?

Figure 4.6 helps to explain the difference between a rock and a mineral. You can see in the photograph that the granite rock is made of several types of crystal. There are glassy grains of quartz, black crystals of mica, and large crystals of feldspar which may be pink or white. These substances are minerals. Rocks are made up of minerals.

Limestone is a rock made of just one mineral, a form of calcium carbonate which is called calcite when it is crystalline. (The importance of limestone is described in more detail in section **C4.4**.)

Mineral	Chemical name	Chemical formula
Barytes	barium sulphate	$BaSO_4$
Bauxite	aluminium oxide	Al_2O_3
Cassiterite	tin(IV) oxide	SnO_2
Calcite	calcium carbonate	$CaCO_3$
Fluorite	calcium fluoride	CaF_2
Galena	lead(II) sulphide	PbS
Gypsum	calcium sulphate	$CaSO_4$
Haematite	iron(III) oxide	Fe_2O_3
Halite	sodium chloride	$NaCl$
Magnesite	magnesium carbonate	$MgCO_3$
Quartz	silicon dioxide	SiO_2
Saltpetre	potassium nitrate	KNO_3

Figure 4.7
Some common minerals.

The table in figure 4.7 shows a variety of minerals, some of which are important as metal ores.

Figure 4.8
Haematite.

Figure 4.9
Galena.

Figure 4.10
Fluorite.

Figure 4.11
Gypsum.

2 With the help of figure 4.7, give the name of a mineral which is:

a an oxide

b a sulphide

c a carbonate

d a sulphate.

3 Name the elements which are combined in:

a quartz

b galena

c bauxite

d gypsum

e limestone.

4 Malvern water comes from underground springs. You can buy bottles of it in supermarkets. Why do you think that it is called ''mineral water''? (If you look at the label on a bottle of mineral water you may be able to see how it gets its name.)

BOX 1   Naming inorganic compounds

If you look down the list of compounds in table 5 of the Data section you will see that most of the names end either **-ide** or **-ate**.

The ending **-ide** shows that the compound contains just the two elements mentioned in the name. Sodium chloride is a compound of sodium and chlorine. Sulphur dioxide is a compound of sulphur and oxygen.

The ending **-ate** shows that the compound contains oxygen as well as the two elements mentioned in the name. Magnesium sulphate is a compound of magnesium, sulphur **and** oxygen.

C4.2   How was alum made from rocks?

This section describes a remarkable process for making alum which was carried out for hundreds of years before anyone understood the chemistry. You will probably have a chance to try to make some alum crystals from rock in the laboratory using a more modern process. You may be asked to plan your own method, or you may be guided by Worksheet C4A.

Alum was probably the first pure chemical to be manufactured in Britain. Alum was used for tanning of leather and for making good quality paper; but by far and away its most important use was for dyeing wool. During the seventeenth and eighteenth centuries the production of woollen cloth was the main industry in England. Natural dyes were used to colour the cloth, and alum was needed as a mordant. A mordant is a chemical used to help dyes to cling fast to cloth so that the colour does not fade in the light or during washing.

The main centre of the alum industry was on the North East coast of England near Whitby. The mixture of ingredients used in the process seems

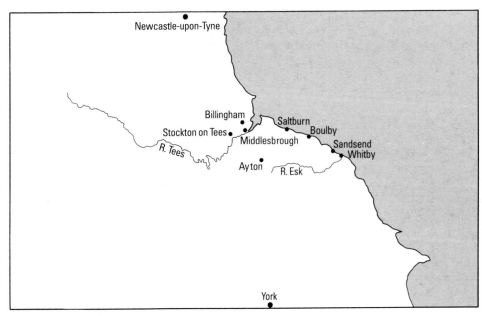

Figure 4.12
The location of Whitby and the main alum works in the North East of England.

Timeline (Figure 4.13)

1500

Copernicus publishes his theory that the Earth circles round the Sun

1600

First alum company started in the North East of England
William Harvey discovers the circulation of the blood
Robert Boyle publishes his Law of gases

Newton publishes his Laws of motion and theory of gravitation
Steam engine invented
1700
Iron smelted with coke for the first time

Discovery of oxygen by Scheele and Priestley
Lavoisier publishes his theory of burning
1800   Edward Jenner discovers vaccination
Dalton's atomic theory published
Davy discovers sodium and potassium
Faraday's first electric motor
Charles Darwin and the voyage of the Beagle
Peter Spence patents a new process for making alum
William Perkin discovers his new mauve dye
Mendeléev publishes his Periodic Table
Mendel starts the science of genetics
Last alum works closes in the North East of England
1900   Discovery of radioactivity by Becquerel

Figure 4.13
The main events related to the alum industry.

remarkable: it included rock, seaweed, coal, wood, and human urine. The industry lasted for about 270 years from 1600 to 1870, but for most of that time there was no understanding of the chemistry involved. It was a risky business, and large sums of money were lost elsewhere in England when attempts to make alum failed at places such as Alum Bay in the Isle of Wight and Alum Chine near Bournemouth.

5 Copy the time chart in figure 4.13 in such a way that you can add other events to the lefthand side of the chart. You can choose which dates to add, but you might include some of the following: the dates of the reigns of King James I, King Charles I and Queen Victoria; the dates of the Spanish Armada, the fire of London, the American declaration of independence, the battle of Waterloo, the Gunpowder plot, the voyage of the Mayflower, the sinking of the Mary Rose, and the Crimean War.

6 If a ''lifetime'' is about 70 years, how many lifetimes ago did the alum industry start near Whitby? For how many lifetimes did the industry continue?

Alum made in England had to compete with alum imported from Italy and other parts of Europe. Both James I and Charles I issued proclamations which were meant to ban foreign imports and protect the home market. This shows the economic importance of the alum industry at that time.

The alum was made from a hard but crumbly rock which the workers could dig from the cliffs near the coast. All the labouring was done with pick, shovel and wheelbarrow, but over the years they shifted huge tonnages of rock. The derelict quarries can still be seen near Whitby.

Figure 4.14
Deepgrove Quarry, Sandsend. In this old picture you can see the pits in which burnt shale was stirred with water to extract the aluminium sulphate.

We now know that this rock could be used to make alum because it contained aluminium silicate, iron pyrites (iron sulphide), and some organic material which helped to keep the fires going.

The workers laid fires of brushwood. Over the top they built huge piles of the broken rock. The heaps could be as much as 30 metres across and

Figure 4.15
A trough built in about 1750 which was used to transport liquors from the pits in Deepgrove Quarry to the boiling house (about 1 km away). Where necessary it is tunnelled into the cliffs. The gradient is about 1 in 200 (0.5%).

25 metres high. The coal to fuel the boiling-house fires was shipped from Sunderland and Newcastle. In the whole process something like 6 tonnes of coal were needed for each tonne of alum produced, so the price of coal was a major part of the production cost.

The roasting of the rock took many months. Slow burning was essential. We now understand that during the burning, some of the sulphur in the iron pyrites was turned into sulphuric acid. The acid then reacted with aluminium silicate in the shale to make aluminium sulphate.

Figure 4.16
Men at work with pick, shovel, and wheelbarrow.

7 The workers found many fossils as they dug the rock. They called them snakestones. (See figure 4.17.) According to legend, there was once a plague of snakes in the region and they were all turned to stone by the prayers of St Hilda. What do we now call these snakestones? How do we explain their presence in rocks?

8 Name the elements combined in:
a aluminium silicate
b iron pyrites.

Figure 4.17
Ammonite from Sandsend.

9 Estimate the length, breadth and height of the room you are sitting in, in metres. Approximately how many rooms of this size would be needed to contain one of the heaps of rock roasted to make alum?

10 The formula of sulphuric acid is H_2SO_4. Suggest where the hydrogen and the oxygen came from when iron pyrites was converted to sulphuric acid.

After roasting, the burnt rock was tipped into pits of water. The water was stirred with long wooden poles and then allowed to settle for several days. The aluminium sulphate dissolved while the waste material sank. The solution of aluminium sulphate was run off along wooden troughs to large boiling-pans made of lead in a central boiling-house. These pans were about 3 metres long, 2 metres wide, and 1 metre deep.

Boiling the solution for twenty-four hours made it more and more concentrated. Every so often, more of the dilute solution was added so that the pans did not boil dry.

After cooling, an extract of burnt seaweed or urine was added to the concentrated solution. The mixture was allowed to stand and then run into casks. After a day or two, crystals of alum formed. The casks were broken apart and the crystals broken up and bagged for transport to London by ship.

The ash from the burnt seaweed contained potash (potassium carbonate, K_2CO_3). A single alum works needed up to 30 000 tonnes of wet seaweed per year. To meet this demand, seaweed was harvested from as far away as Essex and Orkney. This was a major industry as potash was also needed to make soap and glass. At that time there were no easy methods for manufacturing alkali.

The only other suitable alkali was stale urine which contains ammonia. Local people stored urine in wooden pails, and it was collected by cart in large barrels. Transport by land was relatively expensive, and so much urine was needed that it was imported by sea from London. Casks were set up here and there in the streets as collection points. The ships which delivered alum to London returned laden with urine.

11 What volume of solution could be held in one of the boiling-pans? Estimate the volume of water in a typical modern bath and compare it with the volume of solution in one of the boiling-pans.

12 Which two elements are combined in ammonia? (Table 5 in the Data section will help you.)

13 Draw an outline map of the British Isles. Mark in the site of the alum industry and then mark in the places from where raw materials were imported. Draw and label arrows to show how the raw materials were brought to the alum works.

14 We now know that the alum produced was a mixture of ammonium alum, $(NH_4)Al(SO_4)_2 \cdot 12H_2O$, and potash alum, $KAl(SO_4)_2 \cdot 12H_2O$.
a Name the five elements in ammonium alum.
b Name the five elements in potash alum.
c Where did the aluminium in these alums come from?
d Where did the sulphur in the alums come from?
e Where did the nitrogen in the ammonium alum come from?
f Where did the potassium in the potash alum come from?

15 At which stages of the manufacture of alum were these processes involved:
a dissolving   **b** evaporating   **c** crystallizing?

16 Modern experiments show that 350 tonnes of wet seaweed have a mass of 100 tonnes when dry. When burned, this produces 20 tonnes of ash from which 4.4 tonnes of potash can be extracted. How much potash could an alum works obtain from 30 000 tonnes of wet seaweed?

17 Stale urine contains 2 per cent by mass of ammonia. The daily output of a person is about 1.5 kg of urine.
a Estimate the mass of ammonia, in tonnes, which could be obtained per person per year (1 tonne = 1000 kg.)
b 100 tonnes of ammonium alum can be made from 3.75 tonnes of ammonia. Estimate the number of people needed to supply an alum works producing 100 tonnes of ammonium alum per year.

Why did the old alum industry die?

In 1845, Peter Spence took out a patent for a new and more efficient process for making alum which involved heating roasted shale with sulphuric acid.

BOX 2   Problem
Can you make alum from shale?

P
PLAN

Roast the sample of shale by heating it strongly in air

↓

Mix the roasted shale with dilute sulphuric acid and boil

↓

Filter to remove shale; add dilute potassium hydroxide to the solution until the pH rises to 4

↓

Evaporate the solution until about half the water has boiled away

↓

Set aside the solution to cool and crystallize

Figure 4.18
Peter Spence's process for making alum.

The flow diagram in figure 4.18 outlines Peter Spence's process for making alum. Figure 4.19 shows some of the apparatus available to you if you try to carry out this process in the laboratory.

Figure 4.19
Laboratory apparatus.

Describe, with labelled diagrams, how you would use the apparatus shown in figure 4.19 to carry out the process outlined in figure 4.18. You will find it helpful to look at table 7 on page 328 to find out how to draw the diagrams. Assume that you are working with 10 to 20 g of powdered shale.

His works at Goole started production in 1855, and in time became the largest producer of alum in the world. This process depended on the fact that new methods for manufacturing sulphuric acid had been developed.

The traditional alum industry was further damaged when William Perkin discovered the first synthetic dye in 1856. Soon many new dyes were made which did not need alum as a mordant. (These developments in the dye industry are described in Chapter **C11**.)

The old alum works near Whitby had all closed by 1881. The sales of by-products added to the earnings and helped to keep the process going in the later years. One by-product was Epsom salts (magnesium sulphate) used in laxatives.

C4.3   Why is salt so important?

*You should already know something about electrolysis. You have perhaps used electrolysis for electroplating. In this section you will study electrolysis in more detail. Worksheet **C4C** describes a method for studying the electrolysis of solutions. You are first asked to make careful observations and you then have to interpret them. In this section you will also try to use the results of electrolysis experiments to work out rules for predicting what happens at the electrodes.*

*This section does not include any of the theory which can be used to explain electrolysis. The theory is explained in Chapters **C5** and **C18**.*

What is salt used for?

The growth of the modern chemical industry depended on the discovery of new ways of making acids and alkalis. The story of the alum industry in section C4.2 shows that the only alkalis available before the industrial revolution came from urine or the ashes of burnt plants. Nowadays the manufacture of alkalis is based on salt and limestone.

There are huge, underground deposits of salt in Cheshire. A little of the salt is mined underground by cutting, drilling and blasting (see figure 4.20). The largest mine is at Winsford. An output of up to 2 million tonnes of salt from this mine is used each year mainly to supply local authorities. They spread the crushed salt on roads in winter to melt snow and ice. Ground rock salt is also

Figure 4.20
Underground mining at Winsford.

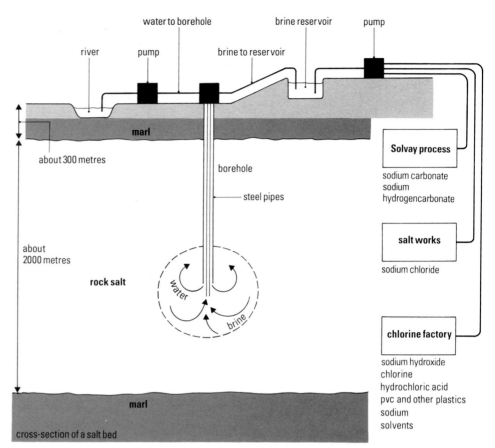

Figure 4.21
Solution mining of salt.

an ingredient of the fertilizers used to grow sugar beet. In 1982 it was estimated that there were still 35 million tonnes of rock salt which could be safely and economically mined at Winsford.

Most of the salt used in the chemical industry is not dug out; it is extracted by pumping water down into the rock. The salt dissolves and is carried to the surface as a solution called brine. This process is illustrated in figure 4.21.

Brine is the main raw material for the manufacture of two important industrial alkalis: sodium hydroxide and sodium carbonate. The demand for these alkalis is huge. Worldwide we use about 36 million tonnes of sodium hydroxide and 26 million tonnes of sodium carbonate each year. Some of the main uses of alkalis are listed in figure 4.25.

18 The Winsford salt mine opened on a small scale in 1844. At current rates of production, when can the mine be expected to close?

19 The main impurity in rock salt is clay. Suggest what happens to the clay impurities when salt is extracted in the way shown in figure 4.21.

20 Describe how you would make a sample of pure salt from rock salt in the laboratory. Draw and label diagrams of the apparatus you would use. (You may find table 7 in the Data section helpful.)

21 The salt we put on our food comes from brine obtained as in figure 4.21. What processes must take place in the factory which makes table salt from brine? Draw a flow diagram to show the main stages.

Brine is a solution of sodium chloride, NaCl, in water, H₂O. So there are just four elements in brine which can be rearranged to make sodium hydroxide, NaOH, chlorine, Cl₂, and hydrogen, H₂. The method used to rearrange the elements is *electrolysis* (see figures 4.22 and 4.23).

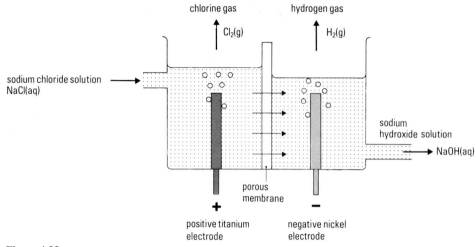

Figure 4.22
The main inputs and outputs of an electrolysis cell used to make sodium hydroxide from brine.

Figure 4.23
A membrane cell.

Chlorine is given off as a gas at the positive electrode (the *anode*). Hydrogen gas forms at the negative electrode (the *cathode*). This leaves a solution of sodium hydroxide to flow out of the cell.

The cell used for the electrolysis of brine has to be carefully designed. Chlorine and sodium hydroxide react together to make bleach. The membrane in the cell allows the solution to pass through but stops the chlorine mixing with the sodium hydroxide.

Brine and limestone, $CaCO_3$, are used to make sodium carbonate by an ingenious method called the Solvay process. It seems as if it should be easy to make sodium carbonate from these raw materials. Look at this word equation; the state symbol (aq) is short for *aqueous*, which means "dissolved in water":

sodium chloride(aq) + calcium carbonate(s) ⟶ sodium carbonate(aq) + calcium chloride(aq)

There does not seem to be anything wrong with this until you think about the white cliffs of Dover. The sea is a solution of sodium chloride, and the cliffs are made of calcium carbonate. So Dover and its cliffs would have been washed away long ago if the reaction shown in the word equation could happen.

This illustrates a very important idea: it only makes sense to write a chemical equation for a reaction which really happens. Fortunately for all those who live near Dover, the reaction of sodium chloride solution with calcium carbonate does not happen.

What makes the Solvay process so cunning is that it is based on a round-about way of turning sodium chloride and calcium carbonate into sodium carbonate and calcium chloride. As has just been explained, this cannot be done directly, but it can be done in stages. Figure 4.24 gives an impression of the scale of the process.

The alkali industry makes a vital contribution to our standard of living. Figure 4.25 illustrates just a few of the uses of each of the chemicals described in this section.

Figure 4.24
One of ICI's Solvay plants.

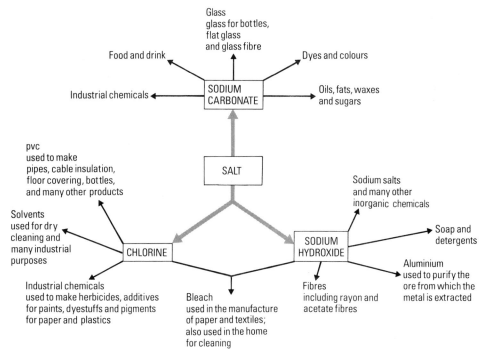

Figure 4.25
The many uses of chemicals from salt.

22 All the following sodium compounds have important uses either in industry or in the home. With the help of reference books (including dictionaries, encyclopedias and textbooks), find out one use for each of the following compounds:

a sodium borate (borax)

b sodium carbonate

c sodium hydrogencarbonate (bicarbonate of soda)

d sodium chlorate

e sodium hypochlorite

f sodium thiosulphate (hypo)

g monosodium glutamate

h sodium stearate

i sodium sulphate

j sodium hydrogensulphate.

23 What important idea about equations is mentioned on the previous page?

24 Redraw figure 4.25 and make it more interesting by adding pictures to illustrate the uses of the chemicals. You can draw these pictures yourself or cut them from magazines and newspapers.

When was electrolysis discovered?

Electrolysis is not only important as a method for making chlorine and alkalis; it is also used to extract aluminium from its ore and to electroplate metals. *Electro-lysis* means "electricity-splitting", and this name is chosen because, during the process, compounds are split up into elements by an electric current.

Electrolysis was discovered in 1800, but it was Humphry Davy who first realized the great importance of this new type of chemical change and used the process to discover new elements (figure 4.26).

Figure 4.26
The work of Humphry Davy.

Humphry Davy   1778-1829

1   Davy was born in Penzance in 1778. He went to the local school but he did not work hard and was often in trouble. Nearby there were copper and tin mines and it was the mining industry which helped to interest him in what things were made of and how they could be put to use.

Soon after he left school, at the age of fifteen, his father died and he had to support his mother and her four other children. He was apprenticed to a local surgeon. At the age of nineteen he began his experimental study of chemistry working in his spare time in an attic.

2   In 1801 Davy moved to the Royal Institution in London and it was there that he established himself as one of the foremost scientists of his day. The Royal Institution had been founded in 1799 to apply science to such everyday needs as the preparation of 'cheap and nutritious food for feeding the poor' and 'improving the construction of cottages, cottage fireplaces and kitchen utensils'. The Institution also aimed to create interest in science and its applications.

3   Davy was appointed as a lecturer's assistant but soon became a professor. From his first lectures in 1801 Davy was a great success. Hundreds of people packed into the lecture room to hear him talk about chemistry.

4   Davy was full of high spirits and when the day's work was over, he was to be found at the supper tables of his many fashionable friends. As well as being a scientist, he was interested in many aspects of life. He helped to found the London Zoo. He was passionately fond of fishing. He knew well many of the literary men of the time including Wordsworth, Coleridge and Scott, and he wrote many poems himself.

5   This apparatus was used by Davy to investigate carbonates in the soil. His laboratory was in an underground room below the Royal Institution. He was neat and methodical in the lecture theatre but in the laboratory he worked very fast in a state of apparent chaos as he often carried out several unconnected experiments at the same time.

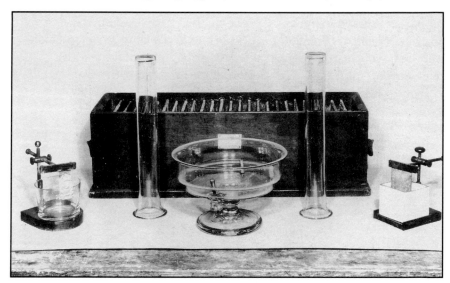

6   Davy carried out his first experiments to investigate electrolysis before he moved to London. It was in 1807 that he crowned his electrical work by using electrolysis to obtain potassium from potassium hydroxide. This picture shows some of Davy's original apparatus. The box at the back is a battery made from zinc and copper plates in a wooden box coated with resin.

7   To help his work Davy persuaded the Royal Institution to build a bigger and more powerful battery. With this he gave a spectacular demonstration of an electric arc and demonstrated the possibility of electric lighting.

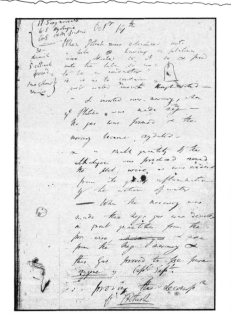

8   In his notes Davy called the discovery of potassium a 'capital experiment'. Within three days he obtained sodium by a similar method and a year later, after a serious illness, he went on to identify barium, strontium, calcium and magnesium as elements. Six new elements in two years was good going!

9   Davy's other important researches showed that chlorine and iodine are both elements. He also invented the miners' safety lamp and suggested the use of zinc to prevent the corrosion of copper on the hulls of ships.

25   Look at figure 4.26 which describes the elements which Davy discovered. Then answer these questions.
a   Where do sodium and potassium appear in the Periodic Table?
b   Where do barium, strontium, calcium and magnesium appear in the Periodic Table?
c   Are these six metals high or low in the activity series?
d   Why was it not possible to obtain these elements before Davy's time?

26   Look at figure 4.26, which shows the apparatus Davy used to study carbonates in the soil. A measured quantity of soil was put in the lefthand flask, and acid was dripped onto it through the funnel. The gas formed blew up the balloon in the middle container. This container was full of water. As the balloon got bigger it displaced water into the measuring glass on the right.
a   Which gas is formed when an acid reacts with carbonates?
b   Explain how Davy could use his apparatus to measure the volume of gas formed from a sample of soil.
c   How could you repeat Davy's investigations using modern apparatus? Draw a diagram with the help of table 7 in the Data section.

BOX 3   Experiment
The electrolysis of molten
compounds

INTERPRET

The apparatus shown in figure 4.27 is being used to investigate the electrolysis of lead(II) bromide. Lead(II) bromide is a white powder which does not conduct electricity. The bulb does not light until the lead(II) bromide melts. This is true of all compounds of a metal with a non-metal; they do not conduct electricity when they are solid.

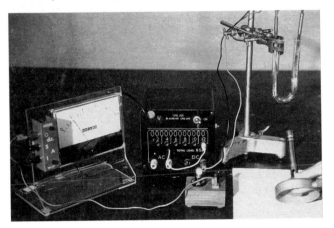

Figure 4.27
The apparatus for the electrolysis of molten lead(II) bromide.

When lead(II) bromide conducts electricity, the most obvious change is that the liquid round the positive electrode starts to bubble and an orange–brown vapour fills the U-tube. After electrolysis has continued for about half an hour, it is possible to show that lead has been formed by pouring the liquid out of the tube into a mortar. The liquid soon solidifies.

If you gently crush the remaining lead(II) bromide with the pestle, you will discover a shiny lump of lead among the white powder.

The table shows the results obtained by electrolysing several molten substances consisting of a metal combined with a non-metal.

Compound	At the negative electrode	At the positive electrode
Sodium chloride	Sodium	Chlorine
Lead bromide	Lead	Bromine
Potassium iodide	Potassium	Iodine
Copper(II) chloride	Copper	Chlorine
Aluminium oxide	Aluminium	Oxygen

27 Draw and label a line diagram of the apparatus shown in figure 4.27. (See table 7 in the Data section).

28 Which type of element is formed at the negative electrode during electrolysis?

29 Which type of element is formed at the positive electrode?

30 What is the rule which can be used to predict what will happen at the electrodes when these molten compounds are electrolysed?

31 Are there any exceptions to the rule you have stated in answer to question **30**?

32 Predict the products at the electrodes when these molten compounds are electrolysed: sodium iodide; magnesium chloride; calcium bromide.

BOX 4 Experiment
The electrolysis of compounds in solution

INTERPRET

Figure 4.28
The apparatus which can be used to electrolyse solutions.

Figure 4.28 shows an apparatus being used to investigate the electrolysis of compounds in solution. The gases formed at the electrodes can be collected and identified. The table shows the results observed using the apparatus in figure 4.28. All the compounds tested are salts consisting of a metal combined with one or two non-metals.

Compound dissolved in water	At the negative electrode	At the positive electrode
Copper(II) chloride	copper	chlorine
Potassium chloride	hydrogen	chlorine
Sodium sulphate	hydrogen	oxygen
Zinc sulphate	zinc and hydrogen	oxygen
Potassium iodide	hydrogen	iodine
Sodium bromide	hydrogen	bromine
Lead(II) nitrate	lead	oxygen
Silver nitrate	silver	oxygen
Copper(II) sulphate	copper	oxygen

33 Draw and label a line diagram of the apparatus shown in figure 4.28.

34 Which tests are used to identify the gases hydrogen, oxygen and chlorine?

35 How do the results for electrolysing compounds in solution differ from the results for electrolysing molten compounds?

36 Which four elements are present in a solution of sodium sulphate in water? Where does the hydrogen come from when it is formed at the negative electrode during the electrolysis of sodium sulphate solution?

37 Which of the metals in the compounds in the table are low in the activity series? What happens at the negative electrode if the metal in the compound is low in the activity series?

38 Which of the metals in the compounds in the table are high in the activity series? What happens at the negative electrode if the metal in the compound is high in the activity series?

39 Which of the non-metals in these compounds are halogens? What happens at the positive electrode if the compound contains a halogen? What happens if the non-metal part of the compound is not a halogen?

40 Predict the products at the negative and positive electrodes when solutions of these salts are electrolysed: sodium iodide; copper(II) nitrate; potassium sulphate.

41 Try to write down a set of rules which can be used to predict what will happen at the electrodes when solutions of salts are electrolysed.

Davy discovered sodium by electrolysing molten sodium chloride. However the electrolysis of sodium chloride in water produces hydrogen instead of sodium. Why is there a difference? Can you predict what will happen during electrolysis? Study the results of the experiments in boxes 3 and 4, and try to make sense of the results by answering the questions.

C4.4   Why is limestone important?

Electrolysis is one way of breaking up chemical compounds. Heating is another. In this section you can read about the use of heat to make new chemicals from limestone. You will find out why limestone is such an important raw material for the chemical industry. You will also have a chance to think about the social and environmental problems involved in quarrying minerals. Worksheet C4E suggests one way of discussing these issues. The effect of industry on the environment is also mentioned in Biology Chapter B17.

What is limestone used for?

Limestone is calcium carbonate, $CaCO_3$, and it is quarried on a huge scale for use in the chemical industry, in agriculture, and in the construction industry. Much of the limestone is converted to quicklime, CaO, by heating in furnaces fired by coke, oil or gas. When it is mixed with water quicklime reacts to form calcium hydroxide $Ca(OH)_2$, which in industry is called hydrated lime. Adding more water produces a white suspension of hydrated lime called milk of lime. These changes are summarized in figure 4.29. Figure 4.29 also

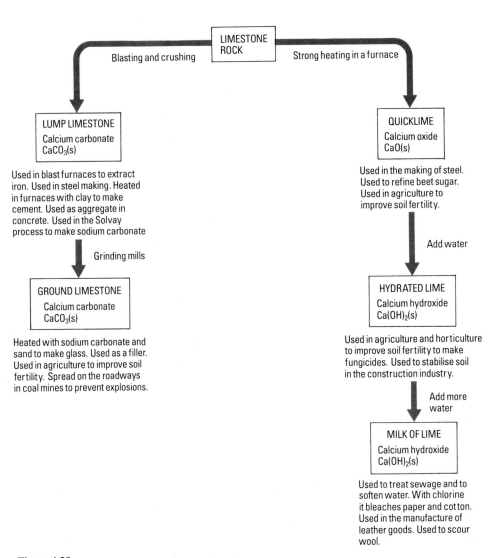

Figure 4.29
The conversion of limestone to quicklime, hydrated lime and milk of lime, and some uses of these products.

Figure 4.30
Cement and limestone form the basis of concrete, and lime products are used for brick mortars and plasters.

Figure 4.31
The new blast furnace at Redcar. Blast furnaces use limestone to make iron. Quicklime is used to convert pig iron into steel.

mentions just a few of the many uses of limestone and lime products, some of which are illustrated in figures 4.30 to 4.31.

---

42 Write word equations for the following changes. (Use the chemical names for the substances involved and include state symbols.)
a limestone to quicklime, and
b quicklime to hydrated lime.

43 Use the examples of milk of lime and brine to explain the differences between a solution and a suspension.

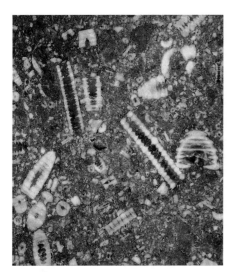

Figure 4.32
Polished limestone specimen showing fossils.

Must we ruin the landscape?

Many of the rocks which are of economic importance are found in parts of England which are protected as National Parks. There are important deposits of limestone in the Peak District and the Yorkshire dales. Fluorspar is extracted in the Peak District, potash in the North York Moors, and china clay in Dartmoor.

There is a long history of mining and quarrying in the National Parks, and on a small scale this need not spoil the landscape. But now in some areas we have huge quarries which make a major impact on the surrounding countryside forming scars which will take years to heal. Figure 4.33 shows one of the largest limestone quarries near Buxton in Derbyshire.

The limestone in the Peak District was formed about 300 million years ago. Most of Britain was then under a shallow, sub-tropical sea. In this sea there lived corals and many other creatures. Over a period of about 20 million years the shells and skeletons of these creatures built up layer upon layer on the sea floor, forming sediments which turned into limestone rock up to 1000 metres thick.

Figure 4.33
Limestone quarry near Buxton. You can see quarry faces near the top of the picture, the crushing plant near the middle, and the kiln for converting limestone to quicklime at the bottom on the right. You can also see rail links.

The seas over Britain were then calm and clear so that the limestone formed was unusually pure. It was not contaminated by silt washed into the sea by rivers from the surrounding land. Only later in the period did huge deltas from the north cover the limestone with sediments which became millstone grit. Finally the area which is now the North and Midlands of England became a vast tropical swamp in which grew giant reeds, mosses, ferns and trees. For millions and millions of years these plants grew, reproduced and died. The dead plants formed layers of decaying vegetation many metres thick which later turned into coal, as described in Chapter C13. Later still, gigantic earth movements folded and lifted the rock. Erosion then exposed the limestone in areas such as Derbyshire, Yorkshire and North Wales.

For many industrial purposes it is very important that the limestone used should be pure. Unfortunately the purest limestone happens to be found in areas of great natural beauty. There is a conflict between the industries which extract and use the limestone and those who wish to preserve the countryside. Buxton limestone has been supplied to the chemical industry in Cheshire and South Lancashire since early in the nineteenth century.

There are three conditions which must be met before planning permission can be given to extract minerals in a National Park.

- The use of the mineral must be essential in the national interest.
- There must be no reasonable alternative source of supply.
- There must be satisfactory plans for the restoration, or after-use, of the land.

44 Imagine that you live in a village close to a large limestone quarry such as the one shown in figure 4.33. How do you think that the quarry might benefit you, your family and your friends? What effect do you think that the quarry operations might have on living conditions in the village?

45 What reasons would you give if you were the owner of a large limestone quarry near Buxton and wanted to show that your industry is in the national interest?

46 Look again at figure 4.33. What plans do you think could be made for the restoration or after-use of the quarry once it has been worked out?

C4.5   Where do metals come from?

You probably know that iron is extracted from its ores in a blast furnace using coke. In this section you can read about the extraction of aluminium, which is an industrial application of electrolysis. You will be able to find out more about the useful properties of metals in Chapter C7.

What is an ore?

Metals have been used for thousands of years, but only a few of them are found as free elements. They are more likely to be found combined with other elements in mineral compounds. Some of the minerals from which we get metals are listed in figure 4.7 and illustrated in figures 4.8 and 4.11. Rocks which contain useful minerals are called ores. After mining the rock, the first step is to process it to separate the useful mineral from the rubbish. The mining industry then has the problem of disposing of its rubbish, and this can produce ugly spoil heaps.

Figure 4.34
Bauxite extraction.

The commonest metal in the Earth's crust is aluminium. However, the metal used in the greatest quantity is iron, which is the main element in all steels. Other common metals include copper, lead, zinc, and magnesium as well as nickel, cobalt and chromium. Overall there are about forty metals which are produced and used regularly in industry.

Aluminium is an excellent conductor of electricity and so it is used to make wires, transmission cables and electrical equipment. Aluminium is strong but has a low density; this makes it very suitable for making parts of ships, aeroplanes and cars. Aluminium is easily shaped and can be rolled into thin foil, drawn into fine wire, or extruded to give complex shapes. It is non-magnetic, a good conductor of heat, and can be used as a reflector of light.

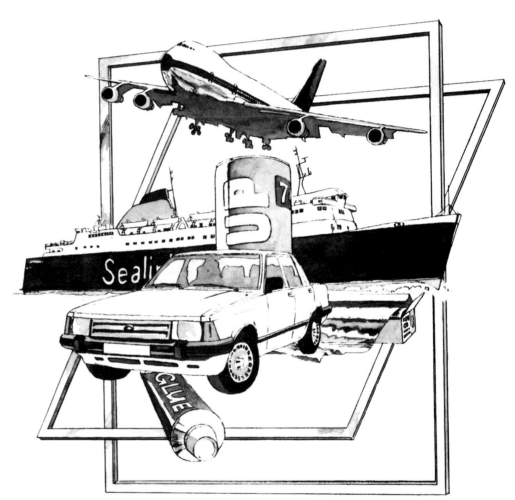

Figure 4.35
Aluminium in use.

The extraction of aluminium is one of the most important industrial uses of electrolysis. The main aluminium ore is called *bauxite* (see figure 4.34). Aluminium is obtained from bauxite in two stages. In the first stage, the bauxite is purified to produce aluminium oxide, Al_2O_3, in the form of a white powder. This stage uses sodium hydroxide to dissolve the aluminium oxide from the ore and separate it from the impurities.

In the second stage, the metal is extracted by electrolysis of aluminium oxide. If you look up the melting-point of aluminium oxide in table 5 of the Data section, you will understand why it took a long time to find a way to electrolyse this compound. Compounds of metals with non-metals will only

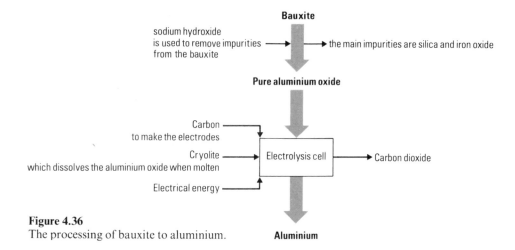

Figure 4.36
The processing of bauxite to aluminium.

Figure 4.37
Alumina, processed from bauxite, awaiting shipment to an aluminium smelter.

conduct when molten or when in solution. Aluminium oxide has a very high melting-point and it does not dissolve in water. This means that it is normally impossible to electrolyse it. Industrial extraction of the metal became possible when it was discovered in the 1880s that aluminium oxide will dissolve in a molten mixture of cryolite, Na_3AlF_6, and aluminium fluoride, AlF_3. Electrolysis takes place in rectangular steel tanks lined with carbon. The carbon lining of the tank is the negative electrode. The positive electrodes are blocks of carbon dipping into the molten mixture as shown in figure 4.38.

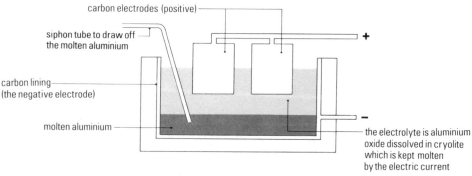

Figure 4.38
How aluminium is extracted by electrolysis.

Figure 4.39
The potroom in an aluminium smelter. You can see the metal being siphoned from a pot.

The carbon blocks which form the positive electrodes burn away and have to be replaced regularly. The whole process uses huge amounts of electricity. In a modern aluminium smelter the production of 1 kg of aluminium requires about 15 kWh of energy. For this reason smelters are usually built where large amounts of electricity are available relatively cheaply. Hydro-electricity is often the best source of low cost power.

Figure 4.40
An aluminium smelter in the north-west United States. Note the electricity power lines (bottom right).

BOX 5   Oxidation and reduction

A substance is oxidized when it combines with oxygen. Iron is oxidized when it rusts and turns into iron oxide. Carbon is oxidized when it burns and turns into carbon dioxide.

Reduction is the opposite of oxidation. A substance is reduced when oxygen is removed from it. Aluminium oxide is reduced when aluminium metal is extracted from it.

Figure 4.41
Aluminium metal being extruded.

47 What is sodium hydroxide used for in the extraction of aluminium?

48 How long after Davy obtained potassium and sodium by electrolysis was aluminium extracted by this method? Compare the melting-points of potassium hydroxide, sodium hydroxide and aluminium oxide, and so explain why it was so difficult to get aluminium from its oxide.

49 What is the product at the negative electrode when aluminium oxide is electrolysed?

50 What is the product at the positive electrode when aluminium oxide is electrolysed? Use your answer to explain why the carbon electrodes burn away. Write a word equation, with state symbols, for the reaction.

51 For how long could you run a 100-W bulb using the energy needed to extract 1 kg of aluminium? (See Chapter **P**17 in the Physics book.)

52 With the help of figure 4.35 and your everyday experience, make a list of ten uses of aluminium. For each of the uses you list, suggest which of the properties of aluminium make it a suitable metal for the purpose.

53 Why does the use of aluminium in the construction of cars help to save fuel?

C4.6 How much?

Analysis is an important part of chemistry. Chemical analysts in the mining industry may be asked to find out if there is a useful metal in a rock sample. Then, if there is, the analysts may be asked to measure how much metal is in the rock. This is important because the answer will show whether or not it is likely to be worth while to mine the ore.

You have probably already investigated malachite to find out which elements it contains. If so, you will know that it is a copper ore. In this section you will be asked to plan and carry out an analysis to find out how much copper there is in malachite.

Figure 4.42
Using an absorption spectrometer to analyse samples of copper in Papua New Guinea.

Malachite is one of the minerals in copper ores. Figure 4.43 reminds you of some of the substances which can be made from malachite.

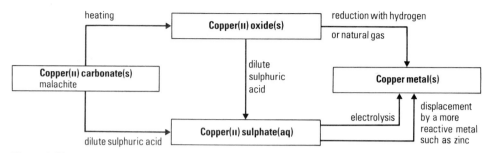

Figure 4.43
The reactions of malachite.

BOX 6   Problem
How can malachite be analysed?

You may try to solve this problem of chemical analysis in the laboratory. You will be supplied with pure, powdered malachite. You have to try to find out how much copper it contains. You will start with 2 to 3 g of the mineral and somehow you must get all the copper out of it so that you can measure it.

Make your plans with the help of figure 4.43. You can see that there is more than one possible route from malachite to copper. You may find it useful to attempt questions 54 to 58 while thinking about this problem.

54 Consider the three or four possible routes from malachite to copper. Which is going to be the easiest and safest for you to try? Will you choose a two-step or a three-step route?

55 When you have decided what to do, draw a series of labelled diagrams to show how you will carry out the steps in the laboratory.

56 How can you be sure that you will not lose any of the copper while doing the experiment?

57 How can you make sure, at the end, that the copper is pure and dry?

58 What measurements will you have to make to work out the quantity of copper in malachite? How will you state your answer: as a percentage, as a fraction, or as a ratio?

In the alum industry described in section **C**4.2, the managers of the works had little idea of how much alum they were going to get. Sometimes they made as much as 3 tonnes of alum from 100 tonnes of rock, but at other times they got as little as 1 tonne. With full chemical knowledge they might have been able to produce as much as 14 tonnes of alum from 100 tonnes of rock, but they did not know how the process worked and they had no method for analysing the rock to see what it was made of.

One of the reasons for using chemical formulae and equations is that they help to answer the question "How much?". The formula of alum is complicated, so the example given here is based on the extraction of aluminium from bauxite.

BOX 7   How to calculate the mass of an element in a compound

Example
How much aluminium is there in aluminium oxide?

Answer
If you look in table 5 of the Data section you will see that the formula of aluminium oxide is Al_2O_3. This means that in aluminium oxide there are two aluminium atoms to every three oxygen atoms.

The number ratio of aluminium atoms to oxygen atoms is 2 : 3.

Table 3 in the Data section shows that the:
atomic mass of aluminium = 27 u
atomic mass of oxygen   = 16 u
The formula mass of aluminium oxide is found by adding up the atomic masses.

The formula mass of aluminium oxide $= (2 \times 27\,u) + (3 \times 16\,u)$
$= 54\,u + 48\,u$
$= 102\,u$

The table shows how to use this information to calculate the mass of metal in a certain amount of aluminium oxide. The answer is often given as a percentage.

Elements present	Aluminium	Oxygen
Number ratio by atoms	2	3
Mass of each element in the compound	Two atoms of Al $= 2 \times 27\,u$ $= 54\,u$	Three atoms of O $= 3 \times 16\,u$ $= 48\,u$
	Total formula mass $= 102\,u$	
Mass of each element as a fraction of the total	$Al = \dfrac{54\,u}{102\,u}$	$O = \dfrac{48\,u}{102\,u}$
Mass of each element as a percentage of the total	$Al = \dfrac{54}{102} \times 100\,\%$ $= 53\,\%$	$O = \dfrac{48}{102} \times 100\,\%$ $= 47\,\%$

Thus there is 53 % by mass of aluminium in aluminium oxide.

---

59 What is the percentage of:
a sodium in halite, $NaCl$
b iron in haematite, Fe_2O_3
c lead in cerussite, $PbCO_3$
d tin in cassiterite, SnO_2
e copper in pyrites, $CuFeS_2$?

Summary

1 Copy the table in figure 4.44 and use the information in this chapter to complete it. Include at least five examples of rocks or minerals. Possibilities include: rock salt, limestone, bauxite and malachite.

Rock or mineral	Chemical(s) in the rock or mineral	Products made from this raw material	Uses of the products

Figure 4.44

2 Copy figure 4.45 and complete it by adding three examples in each of the spaces provided.

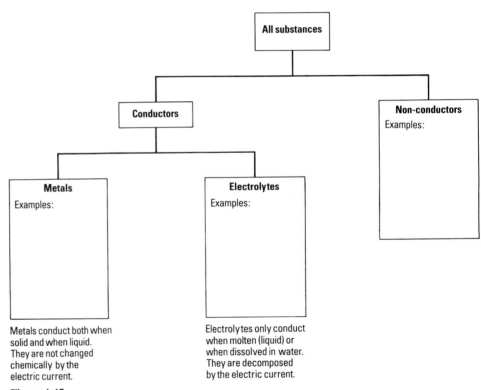

Figure 4.45

3 Heating and **electrolysis** are ways of changing things to make new chemicals. Make lists of useful substances which are made by these two methods. You will find some examples in this chapter. You will find other examples in your previous work in science. Reference books will also help.

Topic **C**2

Materials in use

Chapter **C5** **Materials and structures** 86
Chapter **C6** **Glasses and ceramics** 113
Chapter **C7** **Metals and alloys** 129
Chapter **C8** **Polymers** 145

You are surrounded by things made of glass, pottery, metals and plastics. In this topic you will be studying these materials. There are two main questions for you to think about:

- *What is the connection between the uses of these materials and their properties?*
- *How can we explain the properties of the materials using theories of structure and bonding?*

*This topic follows on from Chapters **P**1 and **P**2 of your Physics book. In Chapter **P**1 you can read about the materials used to build bridges. Chapter **P**2 shows how the differences between solids, liquids and gases can be explained. Now you are asked to take these ideas further and think more about the materials in the world around you.*

Recycling a useful material – aluminium.

Chapter C5 | Materials and structures

Why are things like they are? Here you will meet some chemical theories which have been invented to answer this question. In the laboratory you will try to use the theories to explain the properties of the substances you test with the help of Worksheets C5A and B.

This is the first chapter in a topic about materials and it introduces all of the main ideas which you are expected to understand. You will have more time to become familiar with the ideas as you use them in Chapters C6, C7 and C8 which are about glasses and ceramics, metals, and polymers.

C5.1 What are materials?

Figure 5.1
Sculptor working with modelling clay, which becomes a ceramic after firing.

Figure 5.2
Pound coins from the Royal Mint.

Figure 5.3
Cut glass vessel being decorated with a diamond wheel.

Figure 5.4
Helium-filled airship.

Figure 5.5
Modern tennis racquet made from an epoxy/aramid composite.

Figure 5.6
Concorde windscreen – designed to withstand the impact of a 2-kg bird at 830 km per hour, the windscreen consists of toughened glass, laminated with a thin layer of a transparent polymer. The front screens include a gold-film heating element for deicing and demisting.

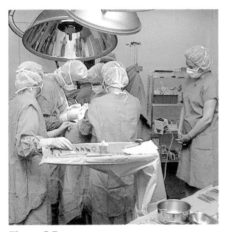

Figure 5.7
Surgical team at work in an operating theatre.

Material is a word with many meanings. Look at the clothes you are wearing. What are they made of? If you read the labels sewn into them you will probably find that several different types of fabric have been used including: cotton, wool, polyester and nylon. These fabrics are the materials from which your clothes are made. This is a very common use of the word, material, but it is not the only one.

Strong materials are needed to make bridges. How many bridges are there near where you live? You can probably think of bridges you know which are made of wood, steel, concrete and stone, but can you think of one made of glass or plastic? In Chapter **P**1 of your Physics book you can read about the materials used to make bridges. Figure **P**1.1 shows how iron and steel, concrete and stone have been used to build two bridges over the River Forth.

Chapter **P**1 explains the ways in which we can investigate the strength and stiffness of materials by trying to bend, stretch and squash them. Words such as strong, weak, stiff, and flexible are used to describe how materials behave when forces are applied.

You drink so often that you may never have bothered to think about the design of cups, mugs, beakers and glasses. How many different things will you drink out of today? What are they made of? Figure 5.8 gives some examples of the range of materials which designers can choose from when they decide to make something for you to drink from.

Figure 5.8
Drinking vessels – how many different materials can you spot?

Materials science is the study of the properties of all the substances we make things from, whether we sew them together as clothes, use them to build bridges, or shape them into drinking vessels.

There are specialist branches of this science for each of the main classes of materials. Some scientists research into the behaviour of ceramics and glass. The name ceramic comes from a Greek word meaning pottery, or burnt stuff. Potters have fashioned and fired clay to make containers, bricks and ornaments since prehistoric times. Most of the common ceramic materials are still made from clay. Metallurgists are experts in the behaviour of metals and alloys. Polymer scientists investigate plastics and fibres.

1 Look at figure 5.8 on the previous page and decide which mug, glass or beaker you think might be chosen for drinking:

a water in a school canteen

b cola from an automatic machine

c wine at a dinner

d tea at camp

e hot cocoa.

Try to explain your choices and to give reasons for them based on what you know about the properties of the materials involved.

2 The enamel mug in figure 5.8 is an example of something made from a *composite material*. What does this term mean? Why are composite materials often preferred? Which composite materials are used to make:

a the skeleton of your body

b bridges

c canoes

d furniture

e tennis racquets.

3 Think again about the materials shown in figure 5.8. Consider each type in turn: metals, ceramics, glass, and polymers. What are the advantages and disadvantages of each of these materials when they are used to make the containers we drink out of?

4 Here are some property words which can be used to describe materials. The words are given in alphabetical order:

bendable	hard	rotproof
biodegradable	impermeable	shiny
brittle	inelastic	soft
conducting	insulating	strong
dull	opaque	transparent
elastic	porous	weak

a What do the following words mean: biodegradable, opaque, porous?

b The words in the list can be grouped in pairs of opposites. Pick out, and write down, each of the nine pairs.

c Choose and name a material which is well described by each word in the list (*e.g.* biodegradable – paper).

Figure 5.9
Three different types of model house.

C5.2   Why use models?

Molecular models are used in Chapters C2 and C3. In this section you are asked to think about the problems of using models to give a picture of things you cannot see. Compare these chemical models with the models discussed in Physics Chapters P2 and P10.

Children make and play with models. Architects draw plans and make models when designing new buildings. The computer program for an adventure game is a model of an imaginary world. The models of science help us to have a picture of what theories mean.

The problem with models is that they are not exactly like the real thing, and so you have to understand the model to know its limitations. Look at figure 5.9 which shows three different model houses. You can immediately recognize the models because you know what houses are like, but there are plenty of ways in which each of the models is not like a house you might live in.

---

5 For **each** of the models in figure 5.9, pick out three ways in which it is like a real house. Then pick out three ways in which it is not like a real house.

There are examples of molecular models in Chapters **C**2 and **C**3. The problem with chemical models is that you will never see an atom or molecule. This makes it difficult to know which parts of the model have real meaning and which parts are just accidents to do with the way in which the models are made. We have theories about atoms and molecules and use chemical models to give us a picture of our theories. So you could say that in chemistry we are using models of models!

Figure 5.10 shows three different types of chemical model of ethanol. The first is an example of a *ball-and-spring* model which shows clearly the number of chemical bonds formed by each atom. The second is a *space-filling* model in which the distances between the atoms are shown to scale. This model perhaps gives a better impression of the shape of an ethanol molecule. The third is a very theoretical model which shows the way in which we now think that the atoms in the molecule are held together by chemical bonding. In this book most of the models used are of the ball-and-spring type.

It is quite common to use several models to show what we know about something. As an example consider the three maps shown in figures 5.11, 5.12 and 5.13 which give information about the countries of Europe. Notice the use of colour codes in the maps; these are similar to the codes used for the different types of atom in chemical models. Each map can only tell you a little about Europe and you have to know how to read the maps before you can make sense of them. Not one of the maps gives you a picture of what the countries of Europe really look like. The same is true of chemical models.

Figure 5.10
Three models of an ethanol molecule.

Figure 5.11
Political map.

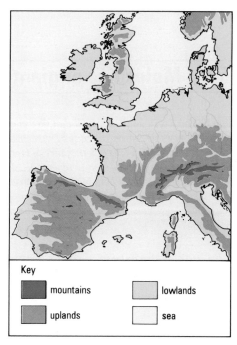
Figure 5.12
Physical map.

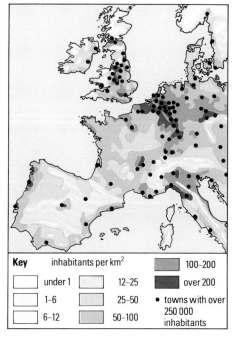
Figure 5.13
Population map.

Element	Colour code	Number of bonds formed	Symbol
Carbon	black	four	C
Hydrogen	white	one	H
Oxygen	red	two	O
Nitrogen	blue	three	N
Chlorine	green	one	Cl
Sulphur	yellow	two	S

Figure 5.14
The colour codes for atoms and the number of bonds formed.

Figure 5.15
Molecular models for some of the compounds listed in the Data section.

6 Figure 5.14 shows the colour code for atoms in chemical models, and the number of bonds normally formed by each type of atom. Figure 5.15 illustrates a number of molecules. Draw the graphical formulae (see figure 2.10) of these molecules using lines for the bonds. Work out the name and formula of each compound with the help of tables 4 and 5 in the Data section.

The theories of chemistry try to explain the properties of materials. The explanations are often given in terms of *structure* and *bonding*. The structure of a substance is a description of the way in which the atoms are arranged. Theories of bonding are attempts to account for the forces which hold the atoms together.

C5.3   Molecules or giant structures?

Atoms are very very small; so small that they cannot be seen with the most powerful light microscope. For this reason you may be wondering how it is possible to find out the structure of materials. Section C5.6 tells you the story of how Lawrence Bragg and his father discovered a method for finding out the structure of crystals. This section shows how our knowledge of structure can explain the differences between some materials.

This section assumes that you have already some understanding of the difference between solids, liquids, and gases. You can remind yourself about the main ideas by reading Chapter P2 in your Physics book.

If you look at the lists of melting-points and boiling-points of the elements in table 3 of the Data section you will see that there are some big differences. Compare the values for chlorine and bromine with the values for iron and carbon.

Now if you look at structures listed in table 2 of the Data section you will see that some of the elements are *molecular* and some have *giant structures*. How can a knowledge of structure explain the differences in properties?

BOX 1   Solid, liquid, or gas?

Example
What is the state of gallium **a** at room temperature and **b** at the temperature of boiling water?

Answer
a Room temperature is usually near to 20 °C.
Table 3 in the Data section shows that gallium melts at 30 °C.
A temperature of 20 °C is not high enough to melt gallium.
Gallium is a solid at room temperature.

b Water boils at 100 °C.
This temperature is high enough to melt gallium.
Table 3 in the Data section shows that gallium boils at 2237 °C.
A temperature of 100 °C is not hot enough to turn gallium into a gas.
Gallium is a liquid at the temperature of boiling water.

---

7 Choose five elements with giant structures and five elements which are molecular from table 2 of the Data section. Look up their melting-points and compare them. How do the melting-points of elements with giant structures compare with the melting-points of elements which are molecular?

8 Use the tables of melting- and boiling-points in the Data section to decide whether the following substances are solids, liquids or gases at room temperature:
fluorine, vanadium, gallium, krypton, decane, eicosane, methanoic acid, propanone, ammonia, hydrogen bromide, silicon dioxide, nitric acid, sulphur trioxide.

Molecules

In most non-metal elements the atoms are joined together in small groups called molecules. The bonds holding the atoms together **within** the molecules are strong. So the molecules do not easily break up. However, the forces **between** the molecules are weak. This means that molecular substances are often liquids or gases at room temperature. Molecular solids are usually easy to melt or evaporate.

Bromine is an example of an element with a molecular structure. It is a liquid at room temperature and it evaporates easily. Figure 5.16 is a picture of a gas jar containing both bromine liquid and bromine gas. The diagram also shows how we think of the arrangements of the molecules in the jar.

As you know, there are many compounds with molecular structures. All the hydrocarbons described in Chapter **C2** are molecular and so are the

9a What is the molecular formula of bromine?
b Describe how you picture the movement of the molecules in bromine liquid.
c How do you picture the movement of the molecules in bromine gas?
d Figure 5.16 shows nothing between the molecules in liquid bromine. Do you think that this is a true picture?

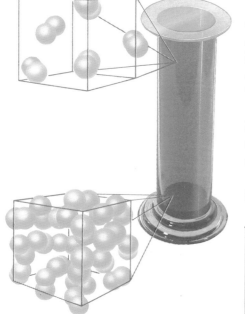

Figure 5.16
Bromine and its structure.

sugars and amino acids in Chapter **C3**. Small molecules can be joined in chains to make bigger molecules. Chapter **C8** in this topic tells you more about the long chain molecules we call polymers.

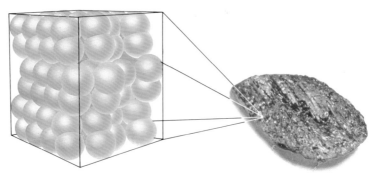

Figure 5.17
The structure of iodine.

Figure 5.18
The effect of gentle heat on an iodine crystal in a glass flask.

10 Figure 5.17 shows the structure of iodine which is molecular.
a Where are bromine and iodine to be found in the Periodic Table?
b In what way is the structure of iodine similar to the structure of bromine? In what ways is it different?
c Do you think that the molecules are still or moving in an iodine crystal at room temperature?
d Why is it easy to crush an iodine crystal?
e Explain what happens to an iodine crystal when it is warmed gently. (See figure 5.18.)

Giant structures

In a giant structure there is a continuous network of atoms which are strongly held together. This means that it is usually very difficult to melt and evaporate a substance with a giant structure. Metals have giant structures and so do a few non-metals. The structures of metals are described in Chapter **C7**.

Diamond is one of the two main forms of the element carbon: the other form is graphite (figures 5.19 and 5.25). Diamond is one of the hardest substances known (figure 5.22), while graphite is so soft that it is used as a lubricant and in pencils (figure 5.23). Graphite gets its name from a Greek word meaning "I write". For a time it was mistaken for lead and called "black lead". This is why we still talk about the "lead" in a pencil.

Figure 5.20 gives a dramatic picture of the structure of diamond. This is truly a *giant structure* in which each carbon atom is strongly bonded to four other atoms to build up a continuous three-dimensional network in which there are no weak links. A tiny diamond crystal no bigger than a grain of sand contains many millions of atoms – perhaps 10^{20} of them.

Small fragments of the diamond giant structure are drawn in figure 5.21. You can see the connection between the two figures if you compare the basic unit lettered ABCDE in figure 5.20 with figure 5.21 part *d*.

Look again at figure 5.21. The atoms are packed tetrahedrally as shown in part *a*. What this means is that there is an atom at the centre with four other atoms around it at the corners of a tetrahedron, as in part *b*. The arrangement is shown from three views in parts *c*, *d* and *e*. When built up, as in part *f*, it gives the giant structure shown in figure 5.20.

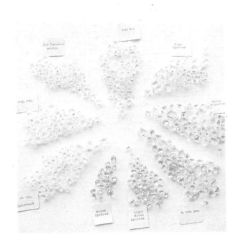

Figure 5.19
Diamonds.

Figure 5.20
Diamond – a giant structure!

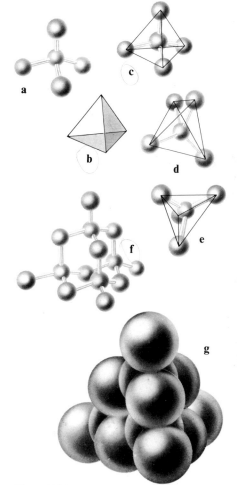

Figure 5.21
Details of the diamond structure.

Figure 5.22
Diamonds in use in a cutting wheel.

If you trace diagram *f* and lay it over the top of *g* you will have a better impression of the structure than you get from either diagram on its own. You can see from *g* that the atoms really merge with one another. All the atoms are held strongly together and cannot slide over one another. This explains why diamond is so extremely hard.

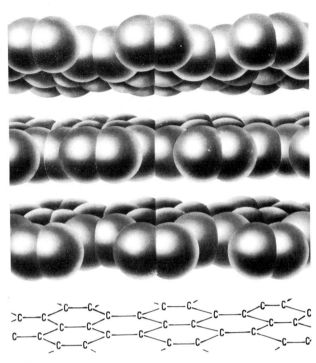

Figure 5.23
Graphite is a soft mineral.

Figure 5.24
Graphite structure.

Figure 5.25
Graphite.

Graphite also has a giant structure, as you can see from the side view in figure 5.24. You have to imagine that the layers are huge extended sheets of atoms, piled on top of each other like cards in a pack. The bonds holding the atoms together in the layers are strong, just as strong as in diamond; but the forces between the layers are weak so that they can slide over each other. This is why graphite feels greasy when you touch it. Flakes of carbon easily rub off onto your fingers, or onto the paper when you write with a soft pencil.

Figure 5.26 includes three ways of modelling the arrangement of atoms in a layer of the graphite structure. In parts *b* and *c* the bonds between the atoms are shown as lines. In part *a*, the bonding is suggested by the way in which the atoms overlap.

There are compounds with giant structures too. Sometimes the atoms are in three-dimensional networks as in diamond; sometimes they are in vast, two-dimensional sheets as in graphite. You may be able to tell the difference just by handling the materials. Among the more important compounds with giant structures are the minerals in igneous rocks including the quartz and mica crystals in granite. (See figure 4.6.)

Quartz is an example of a three-dimensional network of atoms. Quartz is a crystalline form of silica, a compound of silicon and oxygen. Sand is impure quartz.

If you look at figure 5.27, you can see that the oxygen atoms are forming two bonds to silicon just as they form two bonds in a molecule such as ethanol (see figure 5.10). In quartz, the bonding continues on and on alternately from silicon to oxygen, so that there are no weak links. In ethanol there are only strong bonds **within** the molecules because hydrogen atoms can only form one bond. The forces **between** the molecules are weak. Quartz is very hard, as you will know if you have ever suffered from a grain of sand in your eye, or if you have used sandpaper to smooth wood. Ethanol is a runny liquid which easily evaporates.

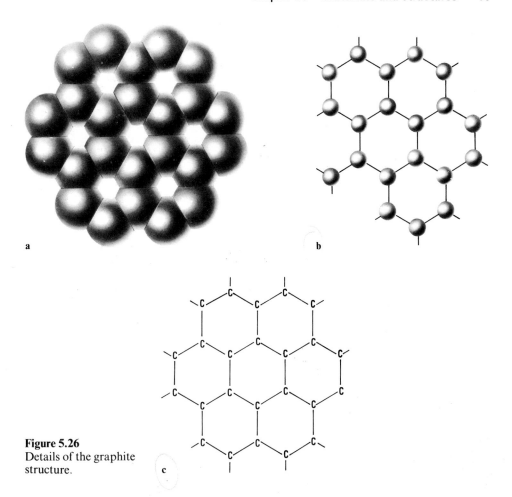

Figure 5.26
Details of the graphite
structure.

Figure 5.27
Quartz crystal
with a drawing
of part of its
structure.

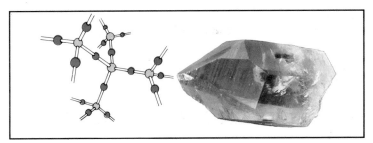

Mica is an example of a mineral in which the atoms are in sheets. Mica is a compound of silicon and oxygen with potassium and aluminium. You can easily spot the mica in a lump of granite if you turn it in the light because the plate-like crystals show up brightly as they reflect the light. Mica can easily be split into flakes (figure 5.28).

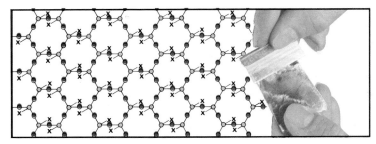

Figure 5.28
Mica, with a drawing of part of its structure.

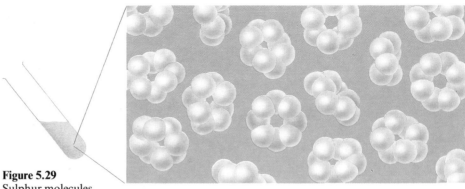

Figure 5.29
Sulphur molecules.

11 How many bonds does a carbon atom form with other atoms in the diamond structure, or in ethanol? How many bonds does a silicon atom form with other atoms in quartz? To which group in the Periodic Table do both carbon and silicon belong? (See table 1 in the Data section.)

12 Look up the formulae and the melting- and boiling-points of carbon dioxide and silicon dioxide (silica) in table 5 of the Data section. Compare them and explain the differences from what you know about the structures of these two compounds.

13 Figure 5.29 shows the arrangement of atoms in sulphur when it has just melted.
a Does sulphur have a giant or a molecular structure?
b Compare the melting-points of graphite and sulphur (table 3 in the Data section) and explain the difference.
c Write the chemical formula for the sulphur particles in figure 5.29.

C5.4   Why is sodium chloride so different from sodium and chlorine?

This section introduces ionic theory in chemistry. Ions were first thought of to explain electrolysis. This is described in Chapter C4. The idea of ions is important in both Physics and Biology. Chapter P3 explains that radiation can turn atoms and molecules into electrically charged particles called ions. Chapter P16 describes the electrostatic forces between charged objects. Biology Chapter B15 describes the ions which plants take up from the soil.
  This section does not explain how atoms turn into ions. You can find out about this in Chapter C18.

Sodium is a dangerous and reactive metal. Chlorine is a very poisonous gas. Yet these two elements combine to make sodium chloride which you can safely sprinkle on your food as salt.

How can we explain these changes? Presumably the particles in sodium chloride are somehow different from the atoms in sodium or the molecules in chlorine.

We now think that during this reaction the atoms become electrically charged and turn into particles which we call *ions*. Sodium forms ions which are positively charged, symbol Na^+. Chlorine forms ions which are negatively charged, symbol Cl^-. So we can write the word and symbol equations for the reaction as follows:

Figure 5.30
Sodium burning in chlorine.

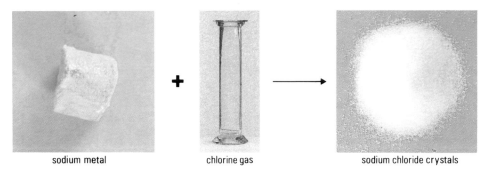

Figure 5.31
A chemical equation in pictures.

$$sodium(s) + chlorine(g) \longrightarrow sodium\ chloride(s)$$
$$\mathbf{2}Na(s)\ +\ Cl_2(g)\ \longrightarrow\ \mathbf{2}Na^+Cl^-(s)$$

The figures in bold type are included to balance the equation.

Ionic theory was first thought of by Michael Faraday, who made a detailed study of electrolysis. He used the idea of ions to explain the changes at the electrodes during electrolysis. (See page 98.)

One test of a theory, or model, is to see what it can explain. Answer questions **14** to **19**, which relate to the following facts.

- The crystals of sodium chloride are cubes (see figure 5.32).
- Sodium chloride has a high melting-point (check this in table 5 of the Data section).
- Sodium chloride does not conduct electricity when solid but does conduct when liquid. It also conducts when dissolved in water (see section **C4.3**).
- When liquid sodium chloride conducts electricity, sodium is formed at the cathode and chlorine is formed at the anode.

Figure 5.32
A large sodium chloride crystal.

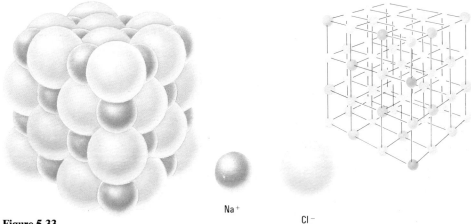

Na⁺

Cl⁻

Figure 5.33
Space-filling and ball-and-stick models of sodium chloride.

14 Use figure 5.33 to explain why sodium chloride crystals are cubes.

15 What force would you expect between sodium ions and chloride ions in a crystal of sodium chloride? (See Chapter **P16**.)

16 Use figure 5.33 and your answer to question **15** to explain why sodium chloride has a high melting-point.

Michael Faraday   1791-1867

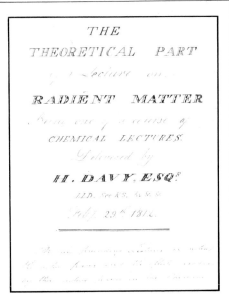

THE
THEORETICAL   PART
of a Lecture on
RADIENT   MATTER
from one of a course of
CHEMICAL   LECTURES.
Delivered by
H. DAVY, ESQ^R
LLD. Sec.R.S. &c. &c.
Feby. 29th 1812.

1 Michael Faraday came from a poor family. At the age of thirteen he became an errand boy to a bookseller. He did well and was accepted as an apprentice bookbinder. In 1812 a customer of the shop gave him tickets to hear Humphry Davy lecture at the Royal Institution. Faraday took careful notes and bound them. He sent a copy to Davy and asked for a job.

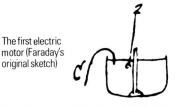

The first electric motor (Faraday's original sketch)

3 When Faraday was thirty he made his first big discovery. He invented the first electric motor. You can find out more about Faraday's studies of electricity and magnetism in Chapter P18.

2 One day in 1813 one of the laboratory's assistants at the Royal Institution was sacked for fighting. Humphry Davy recommended Michael Faraday for the post. Davy reported: 'His habits seem good, his disposition active and cheerful and his manner intelligent.'

4 At the age of 39 Faraday was appointed Director of the Laboratory at the Royal Institution. He started the Christmas Lectures for children which continue today. In this picture you can see Faraday talking to an audience which includes Prince Albert who is sitting in front of the bench on the left.

5 In 1833 Faraday started his studies of the chemical effects of electricity. He invented methods of measuring and comparing electric currents. He discovered two laws which are still the basis of modern theories of electrolysis.

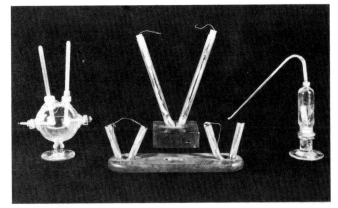

6 Faraday had to look for new words to describe his discoveries about electrolysis. He consulted a Greek scholar and together they decided on the terms: anode, cathode, electrode and ion.

Figure 5.34
Michael Faraday's studies of electrolysis.

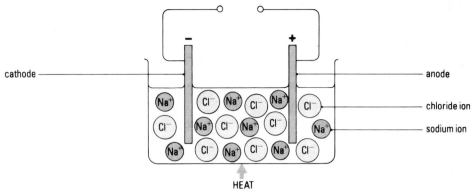

Figure 5.35
Ions in a molten electrolyte between charged electrodes.

---

17 How do you picture the arrangement and movement of ions:
a in solid sodium chloride, and
b in liquid sodium chloride?

18 Use your answer to question **17** to explain why sodium chloride does not conduct electricity when it is solid but does when it is liquid.

19 What would you expect to happen to:
a the sodium ions, and
b the chloride ions
when charged electrodes are dipped into liquid sodium chloride? (See figure 5.35.)

If you have answered questions **14** to **19** you will see that the ionic theory and crystal model can account for a number of the properties of sodium chloride. The theory can also be used to explain and predict the formulae of ionic compounds.

BOX 2   Charges on ions and the formulae of compounds

Every ionic compound contains positive ions combined with negative ions. The total positive charge must equal the total negative charge so that overall the compound is uncharged.

Example
What is the formula of magnesium bromide?

Answer
Table 6 in the Data section shows that the magnesium ion is Mg^{2+} and the bromide ion is Br^-.

A magnesium ion with a 2+ charge needs two bromide ions each with a 1− charge to make the compound electrically neutral.

Ions present:	Mg^{2+}	Br^-
		Br^-
Total charge	2+	2−

The formula of magnesium bromide is $Mg\,Br_2$.

BOX 3   Why are there sometimes brackets in formulae?

Example
What is the formula of aluminium nitrate?

Answer
Table 6 in the Data section shows that the aluminium ion is Al^{3+} and the nitrate ion is NO_3^-.

An aluminium ion with a 3+ charge needs three nitrate ions, each with a 1− charge, to make an electrically neutral compound.

Ions present:	Al^{3+}	NO_3^-
		NO_3^-
		NO_3^-
Total charge	3+	3−

We put the symbol for the nitrate ion inside a bracket with a 3 outside to show that there are three whole nitrate ions with each aluminium ion.

The formula of aluminium nitrate is $Al(NO_3)_3$.

Inside the bracket there is one nitrogen atom linked to three oxygen atoms. The "3" inside only refers to the oxygen atoms.

The "3" outside the bracket refers to everything inside – giving a total of three nitrogen atoms and nine oxygen atoms in the formula.

---

20 Use table 6 in the Data section to work out the formulae of:
a sodium iodide
b calcium bromide
c potassium sulphide
d magnesium nitride

21 Use table 6 in the Data section to work out the formulae of:
a zinc nitrate
b aluminium sulphate
c calcium phosphate

22 Look up the formulae of the compounds in table 5 of the Data section, and then use table 6 to work out the charge on:
a barium ion in barium chloride
b manganese ion in manganese(IV) oxide
c manganese ion in manganese(II) sulphate

We still have not answered the question in the heading for this section: "Why is sodium chloride so different from sodium and chlorine?" We have the beginnings of an answer if we suppose that ions have different properties from atoms and molecules. Sodium atoms are dangerously reactive and so are chlorine molecules, while the ions of these elements are safe. But why?

We need to know more about atomic structure (Physics Chapter **P3**) before we can explain why ions are different from atoms and molecules; then we can also describe in more detail what happens during electrolysis. These ideas are developed further in Chapter **C18**.

C5.5 Atoms, molecules or ions?

This section shows you how to test a substance in the laboratory to decide on its structure. Alternatively you can look up the properties in a set of data tables. To help you the section includes a key. You can find another example of a key on Biology Worksheet B1C.

Figure 5.36 is a key which can be used to work out the structure of the elements and most of the compounds you will meet in a school laboratory.

23 Use the information in figures 5.37 to 5.40 and the data in tables 3, 4 and 5 of the Data section to work out the structure of these substances: titanium, lead(II) bromide, benzene, and boron.

24 Which of the elements and compounds which are classified as molecular in tables 2, 4 and 5 of the Data section have boiling-points above 500 °C?

25 How many of the elements and compounds with giant structures in tables 2, 4 and 5 of the Data section have boiling-points below 500 °C?

26 What is the range of melting-points for the elements and compounds which are molecular?

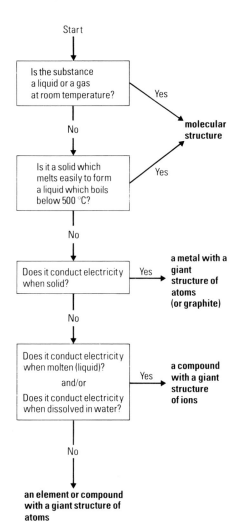

Figure 5.36
Key for determining the structure of substances.

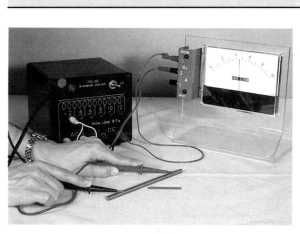

Figure 5.37
A lump of titanium being tested to show that it conducts electricity as a solid.

Figure 5.38
A bottle containing benzene.

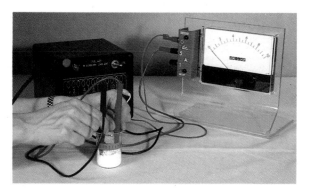

Figure 5.39
A sample of solid lead(II) bromide showing that it does not conduct electricity. (See also figure 4.27 – lead(II) bromide does conduct when molten.)

Figure 5.40
Boron does not conduct electricity when solid or when liquid.

Figure 5.41
Sir Lawrence Bragg at the time of his
Nobel award in 1915.

C5.6   How can we discover the structure of substances?

Study this section if you are interested in reading Lawrence Bragg's own story of how he and his father discovered the most important method of finding out about the structure of substances. You are not expected to understand how the method works and you do not have to remember the details.

*Note that the use of X-rays to investigate crystals is very different from the use of X-rays in medicine (see Biology Chapter **B**5).*

It is only in the last seventy years that a way has been found to find the position of the atoms in a crystal. To do this, scientists use X-rays, which are waves of the same nature as light but with a wavelength that is ten thousand times shorter (see Physics Chapter **P**15). Sir Lawrence Bragg was the first person to realize that X-rays could be used to find out the structure of substances. His father, Sir William Bragg, invented the first instrument to be used to investigate crystal structure. Together they discovered the structure of many substances and started the science of *X-ray crystallography.*

This is how Sir Lawrence Bragg described their work:

"My interest in science started when I was at school, and I think the main reason was that my chemistry teacher taught in an interesting way. I went to school in Australia. I was born in Adelaide and my family lived there till we all came to England when I was eighteen years old.

"When we were in Adelaide, my father used to talk to me about his scientific ideas. He had gone to Adelaide as a young man to be Professor of Mathematics and Physics at the University. He was so busy building up the courses in the physics laboratory, and helping to develop the new University that he never thought of doing research for nearly twenty years. But when he was in his forties he was asked to give a lecture to the Australian Association for the Advancement of Science. He chose to talk about the exciting new discoveries which were being made in radioactivity.

"Preparing his lecture he began to wonder whether the explanations of the way the rays from radium behaved were right. He decided to check some of the properties for himself. So he got the University to buy some radium, and started experiments on alpha rays [see Physics Chapter **P**3]. They were brilliantly successful. Ernest Rutherford was tremendously interested because they fitted in so well with his theory that radioactive decay resulted from the breakdown of the atoms of radioactive elements. At this time Rutherford was trying to convince doubting scientists that one element was changing into another during radioactive processes. This idea did not fit with the existing theory that atoms were unchangeable.

"In two or three years my father became world famous as a pioneer in radioactivity. In 1908 we all came home to England because my father was invited to become Professor of Physics at Leeds University. In England I went to Trinity College, Cambridge. I started with mathematics but after a year my father thought it would be better if I switched over to physics. So that was how I became a physicist.

"My father went on to study the other radiations coming from radioactive substances. He was particularly interested in gamma rays, which like X-rays, could make gases conduct electricity. He came to the conclusion

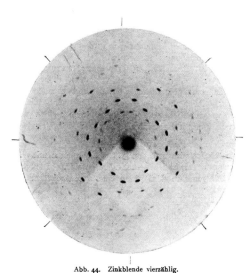

Abb. 44.   Zinkblende vierzählig.

Figure 5.42
One of von Laue's original diffraction photographs.

that gamma rays and X-rays were not waves but more like a stream of little bullets. I remember the first time he told me of his ideas, just as we were boarding the old horse tram which ran down the main street near our home.

"My father went on developing his theories and had scientific fights with those people who believed in waves. So there was great excitement when, in 1912, a German scientist called von Laue published a paper with some beautiful photographs which, he claimed, showed that X-rays were definitely waves (figure 5.42). He obtained these photographs by sending a narrow beam of X-rays through a crystal, and placing a photographic plate on the far side. When the plate was developed, it showed a number of spots in a pattern of the same symmetry as the crystal. Von Laue explained that the effect was caused by 'diffraction' of waves by the regular structure of the crystal.

"My father and I read about von Laue's explanation while we were on summer holiday on the Yorkshire coast. I had just taken my degree at Cambridge and was then twenty-two. I, of course, was a warm believer in my father's theories and we tried to find a way of explaining von Laue's photographs by something other than waves.

"However, when I returned to Cambridge for the autumn term I became convinced that von Laue was right to think that they were produced by waves. At the same time, I had an inspiration which led me to believe that von Laue was on the wrong track when he tried to explain the peculiarities of his diffraction picture. He thought that the odd results were due to a complex set of wavelengths coming from the X-ray tube. I decided that the peculiarities were due to the way the atoms were arranged in the crystal. If this were so, then X-rays could be used to find out the arrangement of the atoms.

"It is worth while describing this inspiration in some detail because it shows how scientific ideas often arise. They come because one hears about a piece of knowledge from one source, and happens to have a quite separate piece of knowledge from another source, and somehow the two just click together and there is a new idea. In my case, it was a kind of treble-chance.

"First J. J. Thomson lectured us about X-rays, and explained them as a wave-pulse caused by the electrons hitting the target in the X-ray tube and being stopped suddenly. Second C. T. R. Wilson had given us a very stimulating lecture which showed that one could think of white light either as a series of quite irregular pulses or a continuous range of wavelengths. Third at a meeting of my college's scientific society a member discussed a theory that in crystal structures the atoms were packed together like spheres with volumes proportional to the number of chemical bonds they form. This theory proved in the end to be quite wrong, but it suggested some very useful ideas. In science a wrong theory can be very valuable, and much better than no theory at all.

"Hearing this paper, I realized that the atoms in crystals were arranged in parallel sheets. Anyone thinking about crystal patterns would see this at once, but I had never thought about it before. So these three bits of knowledge were part of my background. When I was walking one day along the Backs at Cambridge – I can remember the place behind St John's College – suddenly the three bits came together with a click in my mind. I suddenly realized that von Laue's spots were the reflections of the X-rays in the sheets of atoms in the crystals.

"Excited by my idea I did some experiments to get von Laue patterns of alkali–metal chlorides. I was able to show that there was a simple law connecting the wavelength of the X-rays, the spacing of the atoms in the crystal and the angles at which the diffraction spots were obtained. I was able to use the law to work out the crystal structure of several compounds.

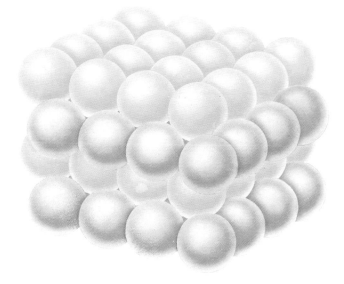

Figure 5.43
The structure of caesium chloride, showing the ions in layers.

"When I told my father about my results, he was of course very interested, and he at once started experiments to find out whether the rays which I had found to be reflected from crystal faces were in fact X-rays. When I wrote my paper I did not want to take this for granted, so I called the paper 'The diffraction of short electromagnetic waves'. I avoided mentioning X-rays, having been teased at Cambridge for upsetting my father's own theory!

"To make an accurate study of the waves, my father built what he called an *X-ray spectrometer* (figures 5.45 and 5.46). It was a beautiful instrument. My father was very good at designing scientific apparatus and he had at Leeds a genius of an instrument maker named Jenkinson.

"My father's instrument proved invaluable for use on crystal structure. I was at that time trying to interpret the X-ray photographs produced by a diamond, and was quite bogged down. With my father's spectrometer it was possible to measure the reflections of the X-rays from the crystal planes, and this led at once to a solution. The diamond structure aroused a great deal of interest and had a strong influence in convincing scientists of the value of X-ray crystallography.

"I had a grand time in the holidays. My father's interests were still mainly in X-ray spectra and he let me examine crystals with the X-ray spectrometer, and use his measurements to try to work out their structures. We worked furiously in 1913 and 1914, going back in the evenings to the deserted university to get more measurements. It was like discovering a gold field with nuggets just lying there to be picked up. One could not resist the temptation to pick up more and more, without a rest. I was very lucky. If it had not been my father who developed the X-ray spectrometer, I should never have been able to work with it."

Figure 5.44
Sir William Bragg.

Figure 5.45
Bragg X-ray spectrometer.

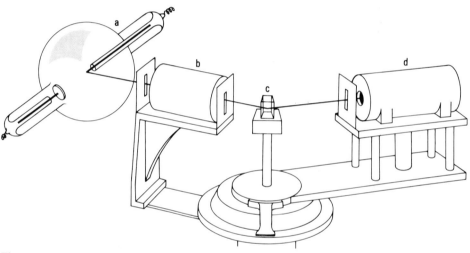

Figure 5.46
Diagram of the X-ray spectrometer. X-rays generated in the tube **a** pass through the slit system **b** on to the inclined face of a crystal **c**, and the reflection of the X-rays is measured in an ionization chamber **d**.

Figure 5.47
Dorothy Hodgkin.

The work of the Braggs has been followed up by many other scientists. One of them, Dorothy Hodgkin (figure 5.47) graduated in chemistry from Oxford in 1932, then spent two years doing research at Cambridge before returning to continue her career at Oxford. She married in 1937 and, along with her research studies, brought up a family of three.

Dorothy Hodgkin was the third woman to win a Nobel Prize in Chemistry which she was awarded in 1964 for determining the structure of important biological molecules.

"Crystals are the most obvious thing to like when you're starting chemistry and I've kept with them ever since." Dorothy Hodgkin realized, during the lectures she attended at Oxford, that there was very little evidence for the chemical formulae of complex organic compounds. At that time the geometrical arrangements of the atoms were virtually unknown. Here, it seemed, were problems that X-ray analysis might be able to solve.

She took an X-ray photograph of insulin as long ago as 1935. This was the second protein crystal ever to be X-ray photographed. She remembers it as

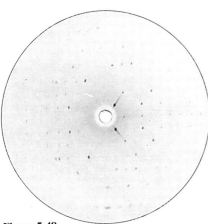

Figure 5.48
X-ray photograph of a crystal of
vitamin B_{12}.

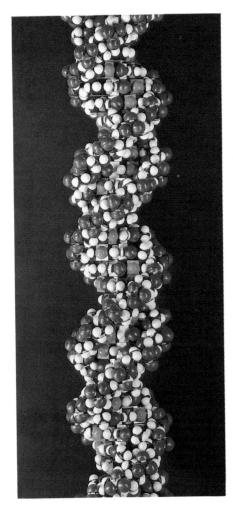

Figure 5.49
DNA double helix.

the most exciting moment in her career. "I developed the photographs late at night and walked elated round the streets of Oxford before going to bed. Then I woke early, worried that the crystals might not be insulin after all. I slipped round to the laboratory before breakfast to test that I really had protein crystals." At the time too little was known about proteins for her to be able to interpret the photograph.

Of the many complex structures that Dorothy Hodgkin has examined, two in particular have established her reputation: that of penicillin which she worked out during the Second World War, and that of vitamin B_{12} which she discovered during the early fifties (figure 5.48).

Now analyses of this kind have been made much easier by the greater use of computers. Dorothy Hodgkin dismisses much of her achievement as "pig-headedness" which led her to try to analyse difficult structures before the techniques had been developed that were to make the calculations easier.

The kind of skill needed in X-ray analysis is a feeling for the way in which atoms are arranged in a molecule. To be successful you have to be able to look at a series of X-ray photographs and see how the atoms might fit together. It needs imagination to picture the various possible arrangements, and it takes experience to choose the possibilities which might lead to an answer. Sometimes the answer comes out very quickly: but with complex molecules, such as vitamin B_{12}, it may take years – working out the arrangement of atoms piece by piece and then seeing how the pieces fit together in the whole molecule.

One of the greatest successes of X-ray analysis was the discovery of the structure of DNA (deoxyribonucleic acid). DNA is present in all living cells and it plays a vital part in the mechanism of heredity. In 1954, Francis Crick and James Watson managed to work out the structure of this molecule while at the Cavendish Laboratory in Cambridge. They could not have done this without the X-ray analysis carried out by Maurice Wilkins and Rosalind Franklin in London; nor would it have been possible without developments in computer science. They discovered the remarkable double-helical structure shown in figure 5.49. Much of the best work in this field has been done by groups of young people, gathered from all over the world, working as a team.

C5.7   How many atoms, molecules or ions?

This section starts with some more guidance about writing chemical equations and then it tells you about a method of measuring amounts of chemicals which is very important to chemists. You should find out from your teacher how much of this section you will have to study in detail.

Atoms, molecules and ions in equations

In Chapter **C1** you can find the rules for writing symbol equations. The examples used in that chapter, and in Chapter **C2**, all involve molecules. The same rules can be used to write equations for reactions involving atoms and ions.

After you have written the word equation (see step 1 in box 2 on page 17) the problem is to write down the correct symbols for the reactants and products (see step 2 on page 17). This is what you do:

- If the substance is molecular you write the formula for the molecule.
- If the substance is an element with a giant structure of atoms you write the symbol for a single atom.
- If the substance is a compound with a giant structure of atoms or ions you write the simplest formula for the compound.

The correct formulae for use in equations are shown in tables 2, 4 and 5 of the Data section.

---

27 Write balanced symbol equations, with state symbols, for these reactions:
a the conversion of limestone to quicklime (see figure 4.29)
b the conversion of quicklime to hydrated lime (see figure 4.29)
c magnesium burning to make magnesium oxide
d sodium reacting with water to make sodium hydroxide and hydrogen
e potassium reacting with chlorine
f glucose fermenting to form carbon dioxide and ethanol.

Amounts of chemicals

In section **C5.6**, Lawrence Bragg mentions a law which he discovered. This law can be used to calculate the sizes of atoms. So we now know that atoms are almost unbelievably small. Anything which you can see must contain millions and millions of atoms. This is a problem for chemists who often want to know that they are making a fair test of different substances by comparing **equal numbers** of atoms, molecules or ions.

Your school has a similar problem when it buys paper. One sheet is such a small quantity that it would be very inconvenient to order, store and distribute paper one sheet at a time. So amounts of paper are measured in reams. Your school may order a ream of plain paper, or a ream of graph paper or a ream of lined paper. In every case it gets the same number of sheets of paper. The number of sheets of paper in a ream has been chosen so that the pack is a convenient quantity to handle.

In a similar way chemists have chosen a unit of amount for substances. The unit is called the *mole* and its symbol is *mol*. Figure 5.51 shows one mole of atoms of several elements. Each of the piles contains the **same number** of atoms. One mole of atoms of carbon contains the same number of atoms as one mole of atoms of sulphur.

Like the ream, the mole has been chosen so that the amounts are convenient to handle. You can see how convenient by looking at table 3 in the

Figure 5.50
1 ream packs of several types of paper. Each pack contains the same number of sheets of paper.

Figure 5.51
1 mol amounts of some elements.

Data section. One column in the table gives you the atomic masses in atomic mass units. The next column gives you the molar masses of the elements in grams per mole. You will notice that the numbers are the same. So the number of atoms in a mole has been chosen so that the molar mass (in grams per mole) has the same number value as the atomic mass (in atomic mass units).

Your school buys paper in reams, but you may well find that few people know how many sheets there are in a ream. Similarly, chemists measure amounts in moles without needing to know how many particles there are in one mole.

X-ray measurements make it possible to calculate the number of particles in a mole. The result is called the Avogadro constant. (This is named after a famous scientist, like so many other scientific quantities and laws.) The value of the Avogadro constant is approximately 6×10^{23} particles per mole. So there are 600 000 000 000 000 000 000 000 atoms in each of the piles shown in figure 5.51.

You do not have to remember the value of the Avogadro constant. The important idea is that **there is the same number of atoms in one mole of atoms of any element**.

---

28 How many moles of carbon atoms, C, contain the same number of atoms as:
a 1 mol of nickel atoms, Ni   **c** 0.01 mol of aluminium atoms, Al
b 2 mol of uranium atoms, U   **d** 0.1038 mol of gadolinium atoms, Gd?

29 What is the mass of:
a 1 mol of potassium atoms, K   **d** 0.5 mol of copper atoms, Cu
b 2 mol of silver atoms, Ag   **e** 0.1 mol of iron atoms, Fe?
c 10 mol of nitrogen atoms, N

30a What mass of calcium contains the same number of atoms as 12 g of carbon?
b What mass of oxygen contains the same number of atoms as 160 g of bromine?
c What mass of lead contains the same number of atoms as 2 g of hydrogen?
d What mass of magnesium contains the same number of atoms as 4 g of sulphur?

It is also convenient to measure amounts in moles for molecules and ions as well as atoms. Figure 5.52 shows one mole samples of several molecular elements and compounds. Each sample contains the **same number** of molecules.

To find molar mass of a molecular compound you first find the molar mass

Figure 5.52
1 mol amounts of some molecular substances.

BOX 4   Finding the molar mass of a molecular compound

Example
What is the molar mass of ethanoic acid?

Answer
Table 4 in the Data section shows that the formula of ethanoic acid is CH_3CO_2H. So in the molecule there are 2 carbon atoms, 4 hydrogen atoms and 2 oxygen atoms.

The atomic masses are given in table 3 of the Data section.

The molecular mass of ethanoic acid $= (2 \times 12\,u) + (4 \times 1\,u) + (2 \times 16\,u)$
$$= 60\,u$$

The number value of the molar mass of a substance is the same as its molecular mass.

The molar mass of ethanoic acid $= 60\,g/mol$

by adding up the atomic masses in the formula. Then the same rule applies as with atoms. The molar mass of a molecular substance has the same number value as its molecular mass. This makes it possible to carry out calculations for reactions which involve molecules.

It is always important to show which particles you are referring to when stating amounts in moles. One mole of oxygen atoms, O, has a mass of 16 g. One mole of oxygen molecules, O_2, has a mass of 32 g. So it is always better to give the formula as well as the name.

31 What is the molar mass of these molecular substances:
a nitrogen, N_2
b chlorine, Cl_2
c sulphur dioxide, SO_2
d nitric acid, HNO_3?

32 What is the mass of:
a 1 mol of hydrogen atoms, H
b 1 mol of hydrogen molecules, H_2
c 2 mol of carbon dioxide, CO_2
d 0.5 mol of hexane, C_6H_{14}?

Figure 5.53 shows one mole samples of some ionic compounds. One mole of sodium chloride, Na^+Cl^-, consists of one mole of sodium ions with one mole of chloride ions. The molar mass of an ionic compound has the same number value as its formula mass.

Figure 5.53
1 mol amounts of some ionic compounds.

BOX 5   Finding the molar mass of an ionic compound

Example
What is the molar mass of silver nitrate?

Answer
Table 5 in the Data section shows that the formula of silver nitrate is $AgNO_3$.
   The atomic masses are given in table 3 of the Data section.

The formula mass of silver nitrate = 108 u + 14 u + (3 × 16 u)
$$= 170\,u$$

The number value of the molar mass of a compound is the same as its formula mass.

The molar mass of silver nitrate   = 170 g/mol

---

33 **What is the molar mass of these ionic compounds:**
a **sodium chloride, NaCl**
b **calcium bromide, $CaBr_2$**
c **sodium sulphide, Na_2S**
d **aluminium oxide, Al_2O_3?**

   Measuring amounts in moles is useful to chemists who want to make things, because it means that they can work out the "recipe" from the chemical equation.

BOX 6   Calculations from equations (1)

Example
How much ethanol (alcohol) can be made from 9 g of glucose by fermentation?

Answer
Step 1: Write the equation for the reaction

$$C_6H_{12}O_6(aq) \longrightarrow 2C_2H_5OH(aq) + 2CO_2(g)$$

The equation shows that one molecule of glucose forms two molecules of ethanol.

Step 2: State the amounts in moles for the substances mentioned in the question.

1 mol of glucose and 1 mol of ethanol contain the same number of molecules.
So in this reaction 1 mol of glucose produces 2 mol of ethanol.

Step 3: Write down the masses of the amounts.

The molar mass of glucose   = 180 g/mol
So the mass of 1 mol of glucose   = 1 mol × 180 g/mol
$$= 180\,g$$

The mass of 1 mol of ethanol   = 46 g/mol
So the mass of 2 mol of ethanol   = 2 mol × 46 g/mol
                                  = 92 g

So the equation shows that fermenting 180 g of glucose produces 92 g of ethanol.

Step 4: Scale the masses to the quantities stated in the question.

Let the mass of ethanol formed be m. (The symbol m includes the number value and the units.)

Then: $\dfrac{m}{92\,g} = \dfrac{9\,g}{180\,g}$ hence $m = \dfrac{9\,g}{180\,g} \times 92\,g = 4.6\,g$

(Notice how the units are included in the calculation. They cancel to give a mass in grams at the end as expected.)
So 9 g of glucose will produce 4.6 g of ethanol by fermentation.

BOX 7   Calculations from equations (2)

Example
What mass of water is needed to convert 28 g of calcium oxide to calcium hydroxide?

Answer
Step 1: Write the equation for the reaction

$$CaO(s) + H_2O(l) \longrightarrow Ca(OH)_2$$

Step 2: State the amounts in moles for the substances mentioned in the question.

In this reaction, 1 mol of calcium oxide combines with 1 mol of water.

Step 3: Write down the masses of the amounts.

The molar mass of calcium oxide   = 56 g/mol
The mass of 1 mol of calcium oxide   = 1 mol × 56 g/mol
                                    = 56 g

The molar mass of water   = 18 g/mol
So the mass of 1 mol of water   = 1 mol × 18 g/mol
                               = 18 g

So the equation shows that 18 g of water are needed to react with 56 g of calcium oxide.

Step 4: Scale the masses to the quantities stated in the question.

Let the mass of water needed be m.

Then: $\dfrac{m}{18\,g} = \dfrac{28\,g}{56\,g}$ hence $m = \dfrac{28\,g}{56\,g} \times 18\,g = 9\,g$

So 9 g of water are needed to convert 28 g calcium oxide to calcium hydroxide.

34 What mass of water, H_2O, is formed when 10 g of hydrogen, H_2, burn in oxygen, O_2? (See section **C1.4**.)

35 What mass of ethanol, C_2H_5OH, can be made from 7 g of ethene, C_2H_4? (See section **C2.5**.)

36 What mass of bromine, Br_2, is needed to react with 0.28 g of ethene, C_2H_4? (See section **C2.5**.)

37 What mass of calcium oxide, CaO, can be made from 1000 g of calcium carbonate, $CaCO_3$? (See figure 4.29.)

Summary

Make a summary of the structures of elements and compounds by copying and completing figures 5.54 and 5.55. The information in this chapter will help you, as will the results of experiments you have done in the laboratory and tables 3 and 4 of the Data section.

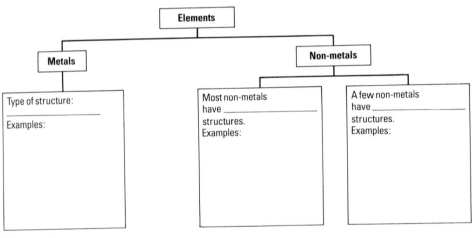

Figure 5.54

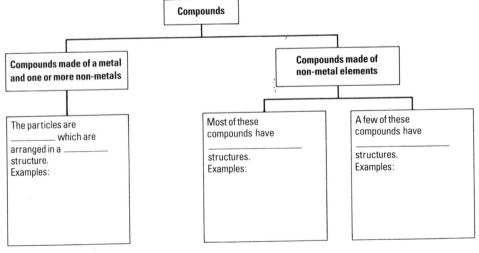

Figure 5.55

Chapter C6

Glasses and ceramics

There are two main themes to the chapter. One shows you how the uses of materials depend on their properties. The other shows how the properties can be explained in terms of structure.

In the laboratory you will concentrate on glass. Worksheet C6A tells you how to make glass. Worksheet C6B introduces you to some of the techniques of glass bending and blowing. In a short time you will gain only limited skill at handling glass, but the experience will give you the background you need to study this chapter.

If you have done some pottery you will have some knowledge of the behaviour of clay, ceramics, and glazes. Bear this in mind as you read this chapter.

Figure 6.1
Skilled glass working is an ancient craft: these objects were both made by the Romans.

C6.1 What are glasses?

Making glass is not a new technology, but now we have many kinds of glass. For instance, the zoom lens of a television camera is made of up to twenty different types of glass.

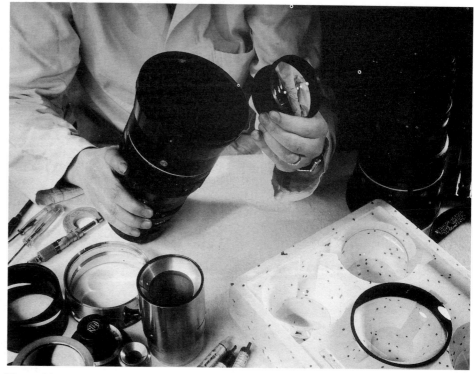

Figure 6.2
A modern camera has many different glass lenses.

Figure 6.3
Objects made from soda-lime glass.

Figure 6.4
Modern lead crystal decanters.

Figure 6.5
Objects made of borosilicate glass.

Figure 6.6
Traditional glass blowing.

We use various types of glass in our homes. Windows, bottles and ordinary drinking glasses are made of soda-lime glass. Lead glass is used for cut-glass dishes. Ovenware is made from borosilicate glass because it is heat-resistant.

Glass is made by heating a mixture of oxides and carbonates in a furnace. The liquid glass is cooled until thick enough to mould. It is then shaped and cooled further until it sets solid.

A lump of glass is rigid. It behaves like a solid but it is not crystalline. The arrangement of atoms in a solid glass is disorganized as in a liquid. But unlike in a liquid, the atoms in a glass cannot move around. They vibrate about fixed positions.

Glass can be made from pure silicon dioxide which melts above 1700 °C to give a very thick and sticky liquid. Such thick and sticky liquids are said to be *viscous*. As the liquid cools back to its melting-point, the atoms cannot move

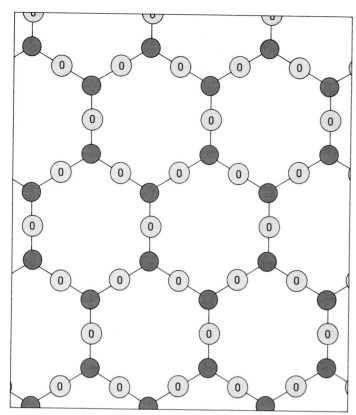

Figure 6.7
The structure of a crystalline glass-forming oxide.

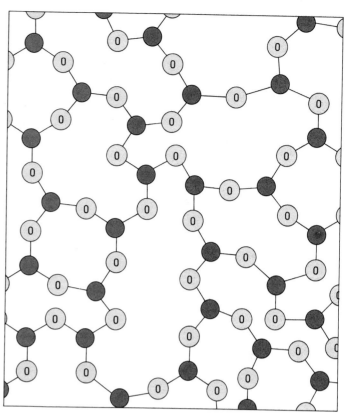

Figure 6.8
The structure of a glassy oxide.

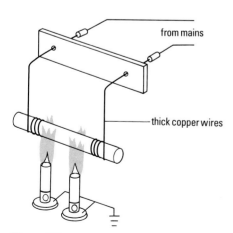

Figure 6.9
Apparatus to show that glass conducts electricity.

freely enough to return to the ordered arrangement in a crystal (figure 6.7) and are "frozen" into a disordered state (figure 6.8).

Crystalline and glassy forms of silicon dioxide consist of three-dimensional *giant structures* of atoms. Each silicon atom forms four bonds with oxygen atoms. It is difficult to draw these structures clearly, and so in figures 6.7, 6.8 and 6.10 the giant structures are given in two dimensions. In these diagrams you can see that the atoms shown as black circles are forming three bonds with oxygen atoms. Boron atoms do this in borosilicate glass.

The change from liquid to solid does not happen at a definite temperature. It takes place gradually as the melt stiffens on cooling. Brittle toffee is a glass, and the changes when molten toffee is cooled are very similar to those seen when silicate glasses solidify.

It is difficult to make and mould glass at the very high temperature of molten silicon dioxide, and so most glasses are made from a mixture of silicon dioxide with metal oxides. The metal oxides lower the melting-point.

Soda-lime glass is made from sand (silicon dioxide), limestone (calcium carbonate) and sodium carbonate (which is made from salt). The carbonates decompose to oxides during melting, so the final result is the same as if a mixture of silicon oxide, sodium oxide and calcium oxide had been heated. In the glass the ingredients combine to a giant structure of atoms and ions.

The apparatus in figure 6.9 gives evidence that there are ions in soda-lime glass. The lamp begins to glow when the glass rod is red hot showing that an electric current is flowing.

Sodium ions carry a 1+ charge while the charge on calcium ions is 2+. The properties of glass can be modified by varying the proportions of sodium and calcium ions in the structure. The example in box 1 shows how to work

BOX 1
The composition of glass

Soda-lime glass is made from a mixture of sodium carbonate, calcium carbonate and silicon dioxide. The carbonates decompose to oxides in the furnace and so the composition of glass is often quoted in terms of oxides. Glasses used to make bottles and jars have a range of compositions as follows:

silicon dioxide, SiO_2	70–74 %
sodium oxide, Na_2O	12–16 %
calcium oxide, CaO	5–11 %

Small amounts of magnesium and aluminium oxides may also be included.
    Looking at the list of percentages, it is hard to tell what the figures mean in terms of atoms and ions. It is easier to understand the figures if the quantities are converted to amounts in moles.

Example
A sample of glass was made by melting together sand, limestone and sodium carbonate to give a mixture with this composition:

silicon dioxide	420 g
sodium oxide	93 g
calcium oxide	36 g

What is the ratio of the numbers of silicon atoms and sodium ions in the glass?

Answer
Use this formula:

$$\bullet\,\text{Amount of substance (mol)} = \frac{\text{mass of sample (g)}}{\text{molar mass (g/mol)}}$$

You can check that the formula is correct if you look at the units.

For silicon dioxide, SiO_2 the mass of the sample in the glass is 420 g.
The formula mass = $28\,u + (2 \times 16\,u) = 60\,u$
The molar mass   = 60 g/mol

$$\text{Amount of silicon dioxide in the glass} = \frac{420\,g}{60\,g/mol}$$

$$= 7\,mol$$

For sodium oxide, Na_2O, the mass of the sample in the glass is 93 g.
The formula mass = $(2 \times 23\,u) + 16\,u = 62\,u$
The molar mass   = 62 g/mol

$$\text{Amount of sodium oxide in the glass} = \frac{93\,g}{62\,g/mol}$$

$$= 1.5\,mol$$

So for every 7 mol of SiO_2 there are 1.5 mol Na_2O.
But there are two sodium ions in the formula of sodium oxide, so for every 7 mol of Si atoms there are 3 mol of Na^+ ions.

Thus in this glass the ratio of silicon atoms to sodium ions is 7:3.

Figure 6.10
Diagram to represent
soda-lime glass.

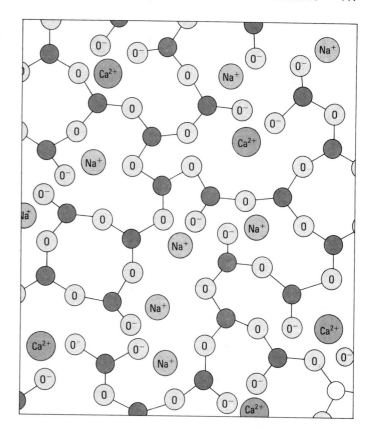

out the ratios of the numbers of atoms and ions in a glass by calculating the amounts in moles (see section **C5.7**).

1 Which elements are present in soda-lime glass?

2 How is sodium carbonate manufactured from salt? (See Chapter **C4**.)

3 What does the word ''decompose'' mean in chemistry? What other meanings of the word are there?

4 Write a chemical equation for the decomposition of calcium carbonate during the manufacture of glass. Use your equation to explain why gas bubbles form as the ingredients of glass are melted together in a furnace.

5 Look carefully at figure 6.10, and compare it with figure 6.8. Use the diagrams to suggest an explanation for the fact that soda-lime glass has a lower melting-point than pure silicon dioxide glass. Which ingredient do you think has the bigger effect on the lowering of the melting-point: the sodium carbonate or the calcium carbonate? Try to give reasons in support of your answers.

6 Study the worked example in box 1 opposite.
a Use the method shown to calculate the ratio of the numbers of silicon atoms and calcium ions in the glass described in the example.
b Work out the percentages by mass of the three oxides in the glass described in the example.
c Which is more helpful: the glass composition given as percentages by mass, or the composition worked out as amounts in moles:
   i if you are trying to make some glass,
   ii if you are trying to understand the structure of glass?

C6.2   What are ceramics?

The name *ceramic* comes from a Greek word meaning pottery, or burnt stuff. Potters have fashioned and fired clay to make bricks, containers and ornaments since prehistoric times. Clay can be dug from the ground, purified, moulded when wet, dried and then heated in a fire to harden it.

Figure 6.11
Excavating china clay.

Figure 6.12
Throwing a pot.

Clay is similar to silicate minerals such as mica (see Chapter **C5**, section **C5**.3). Most clay is a mixture of several materials, but many of the clays used to make pottery contain a mineral called kaolinite, $Al_2Si_2O_5(OH)_4$, in which the atoms are arranged in layers in a giant structure. The crystals are minute, flat flakes as shown in figure 6.13.

Clay can be moulded when wet because the kaolinite crystals slide over each other with water acting as a lubricant. When the clay is dried it becomes rigid because the crystals stick to each other.

Although dry clay is rigid, it is still very easily broken; it must be heated to a temperature of around 1000 °C to harden it. During this firing in a furnace, a complicated series of changes takes place: new minerals are formed and some of the substances present in the clay combine to form glass. The ceramic produced consists of many minute crystals of silicate minerals bonded together with glass.

You can see that there are similarities between ceramics and glasses. Everyday ceramic and glass objects are made from common materials from the earth. Both ceramics and glasses need to be fired at a high temperature in a furnace at some stage during their manufacture. Chemically they are similar too, because they both consist of giant structures of silicon and oxygen atoms.

Figure 6.13
Electronmicrograph of kaolinite crystals (× 30 000).

7 The kaolinite crystals in figure 6.13 are magnified 30 000 times. Measure the distance (in mm) across one of the crystals in the photograph, then use the magnification to calculate the actual distance. Give your answer first in millimetres, then in nanometres. (1 mm = 1 000 000 nm.)

8 The diameter of a silicon atom is approximately 0.1 nm. About how many silicon atoms could be placed in a line across the kaolinite crystal you measured in question **7**?

9 Use five or six microscope slides as a model of how kaolinite crystals behave in clay. Make a pile of the slides, when dry, and try to slide them over each other. Now wet the slides and pile them up. How easily do the slides slip over each other now? What happens if you try to pull the wet slides apart? (Compare this with the behaviour of the pleural membranes described in Chapter **B7** of your Biology book.)

10 In what ways are the crystal structures of graphite, mica and kaolinite similar?

C6.3 How do we use glasses and ceramics at home?

*Ceramics and glass are mentioned in Physics Chapters **P**1 and **P**9. Concrete is sometimes classed as a ceramic. Chapter **P**1 refers to concrete being weak in tension but strong in compression. Chapter **P**9 mentions the insulating properties of glass fibre in the sections about saving energy.*

Figure 6.14 on the next page illustrates some of the ways in which we use ceramics and glasses at home.

Some of the advantages of ceramics and glasses are as follows:

- they are hard
- they are strong in compression
- they are chemically inert
- they are easily cleaned
- they are heat-resistant
- they are electrical insulators.

Glasses have the additional advantage that:

- they are transparent.

These materials also have some disadvantages:

- they are weak in tension
- they are brittle
- they may crack if there is a sudden temperature change.

11 Use the list of advantages of ceramic materials to explain the way they are being used in figure 6.14.

12 Use your everyday experience and figure 6.14 to give examples to illustrate the disadvantages of ceramics and glasses.

13 Much chemical apparatus (including test-tubes, beakers and flasks, evaporating basins, pipe-clay triangles, mortars and pestles) is made from glasses or ceramics. Show how the advantages and disadvantages of these materials are illustrated by your experience of using them in a laboratory.

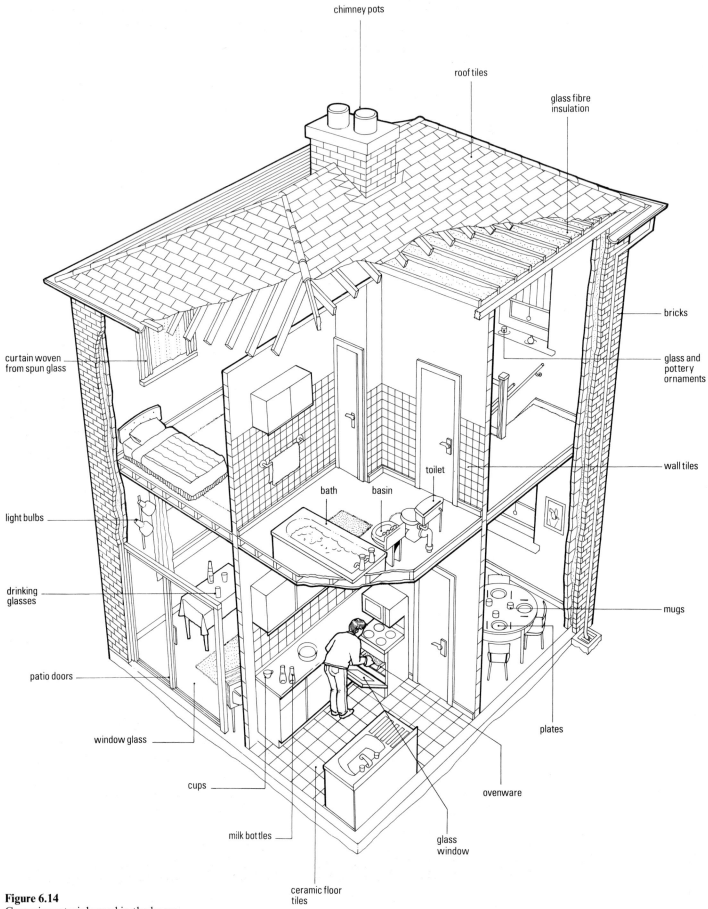

Figure 6.14
Ceramic materials used in the home.

chimney pots

roof tiles

glass fibre insulation

bricks

glass and pottery ornaments

curtain woven from spun glass

wall tiles

toilet

bath

basin

light bulbs

mugs

drinking glasses

patio doors

plates

window glass

cups

ovenware

milk bottles

glass window

ceramic floor tiles

C6.4 How are glasses and ceramics used in industry?

Glasses and ceramics are very important in industry. This section discusses their use as refractories and insulators.

Glasses and ceramics have high melting-points and are chemically inert. They consist of oxides so they do not burn when heated in air. They differ from metals and polymers which combine readily with oxygen when heated. Materials which are difficult to melt or change in any way by heating are said to be *refractory*.

Many industrial processes are carried out at high temperatures. Examples are the firing of ceramics, glass making and metal extraction, as well as cement manufacture and all the processes which use high pressure steam. The furnaces used by these industries have to be lined with refractory ceramic materials because anything else would melt or burn away.

Figure 6.15
A furnace lined with refractory bricks.

Most refractories are made from fireclay, which in Britain is found in deposits under coal seams. Fireclays can be used at a higher temperature than other clays because they contain a bigger proportion of aluminium oxide combined with silica. Bricks made from fireclay are used on a large scale to line blast furnaces, lime kilns, cement and ceramic kilns, and steam-raising boilers.

Ceramics do not react chemically at low temperatures, but at high temperatures they may be attacked by the contents of a furnace. Iron oxide, and the slags produced in steel making, are very corrosive when hot and can destroy furnace linings. So in the steel and glass making industries more specialized refractories are used to build the furnace walls to give them a longer life. The bricks may be made of almost pure silicon dioxide or magnesium oxide.

The transmission of electricity from power stations to homes and industry (see Physics Chapter **P**11) depends not only on the metal conducting cables, but also on the electrical insulators which support them. The insulators are made of glazed porcelain or glass, because these materials are cheap and strong. They are also weather-resistant and can be fashioned into intricate

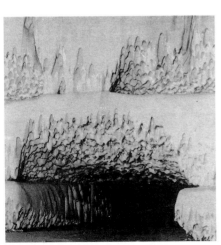

Figure 6.16
The bricks in this furnace have been attacked by the molten glass.

Figure 6.17
Glass insulators used in electricity
transmission.

shapes to make sure that, in wet weather, the whole surface of the insulator
does not get covered with a film of water which would cause the insulation to
break down.

14 With the help of this section and a dictionary, explain the meaning of these words:
inert, refractory, slag, corrosive, and insulator.

15 Explain why it is that metals and polymers will burn but ceramics and glasses will
not.

16 Why do you think that we are asked to remove the metal caps and tops from bottles
and jars before putting them into a bottle bank?

17 Why do the insulators used on power lines have to be much bigger and more
complex than insulators used in our homes? (See Chapter **P**11 in your Physics book.)

18 Use Plasticine, or clay, to make a model of an electrical insulator as shown in
figure 6.17. Study what happens to water as it runs over the surface of your model.

C6.5   How are glass and ceramic objects manufactured?

Making and moulding glasses

Glass articles are moulded from hot molten glass. The mixture of oxides and
carbonates, together with about 20 per cent of crushed scrap glass, is heated
in a furnace lined with thick refractory walls. In the first stage of heating the
ingredients react together to make glass. Bubbles of gas form during the
reaction (see question **4**) and so, in the second stage, the glass is heated until it
is thin and watery. This allows the bubbles to escape. Finally the glass is
cooled until it is viscous enough to be moulded.

Hot molten glass is formed into bottles, jars and bulbs by blowing and
moulding. Some liquid glass is drawn into a mould and then blown into shape
by air pressure.

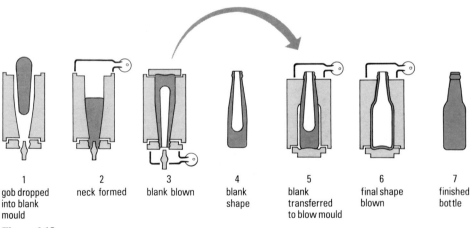

| 1 | 2 | 3 | 4 | 5 | 6 | 7 |
| gob dropped into blank mould | neck formed | blank blown | blank shape | blank transferred to blow mould | final shape blown | finished bottle |

Figure 6.18
How glass bottles are made.

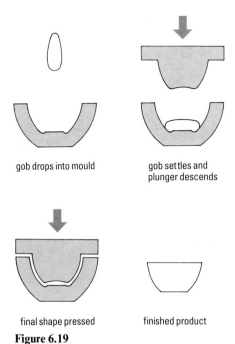

gob drops into mould

gob settles and plunger descends

final shape pressed

finished product

Figure 6.19
How glass objects are made by pressing.

Table and oven glassware is made by pressing the hot glass into a mould. The moulds have to be carefully designed so that they do not react with the hot glass. The moulds also have to be shaped very accurately so that the surface of the moulded glass is smooth.

Figure 6.20
Glass articles made by pressing.

Flat glass is now used not just for windows, doors and mirrors, but also for the curtain walls covering the outside of steel frame buildings.

When making flat glass it is difficult to form the hot glass into sheets which are uniform in thickness, smooth and shiny. A new method for making sheet glass was invented by Pilkington Brothers and introduced in 1959.

The glass leaves the furnace at 1500 °C and floats along the surface of a bath of molten tin. The glass is kept at a high enough temperature on the molten metal for long enough for the surfaces to become flat and parallel. The sheet of glass is then allowed to cool as it continues across the bath of tin until, at about 600 °C, it is hard enough to be lifted onto rollers without damaging its underside.

Figure 6.21
The Float Process.

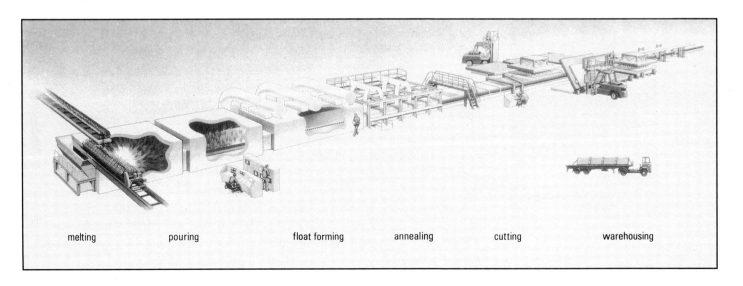

melting   pouring   float forming   annealing   cutting   warehousing

Figure 6.22
Continuous glass sheet travelling through the annealing lehr in the Float Process.

Next the continuous sheet of glass moves through an oven where it cools slowly under controlled conditions. This is called *annealing*. Glass contracts on cooling. It has low thermal conductivity so if it is cooled quickly the outer surfaces cool and shrink more rapidly than the inside. This produces strains in the glass which mean that the glass is likely to crack and break in use. Annealing evens out the strains.

Finally the glass sheet is cut and stacked ready for distribution to customers.

19 Why must the sheet glass used to make windows and mirrors be uniform in thickness, smooth and shiny?

20 Why do you suppose that Pilkingtons called their new method for making flat glass "The Float Process"?

21 Explain how it is that the Float Process makes sure that both sides of the glass are flat, and that the glass is even in thickness.

22 Suggest a reason for making sure that there is no air above the bath of molten tin.

23 Suggest reasons for using tin, in preference to other metals, in the Float Process. (Your knowledge of the activity series, and the data in table 3 of the Data section, may help you.)

Fashioning and firing clay

Wet clay can be moulded by gentle pressure. Materials which have this property are said to be *plastic*. The family of materials called plastics (described in Chapter C8) get their name from the fact that they are easily shaped.

Traditionally domestic pottery was shaped by hand, or "thrown" on a potter's wheel, but mass-produced pottery is now moulded by machines. These use a scraper to shape the lump of clay by pressing it against a rotating plaster mould.

Figure 6.23
Plate making.

Figure 6.24
Pouring slip into a mould.

Figure 6.25
Removing a cast basin from a mould.

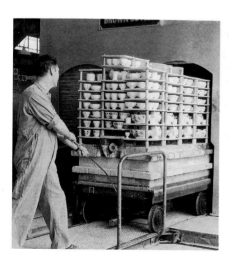

Figure 6.27
Removing china from the kiln.

Hollow objects such as teapots, ornaments and basins can be made by a method called "slip casting". Slip is very runny clay which can easily be poured into moulds. The moulds are made of porous plaster which absorbs water from the clay so that a solid crust forms on the inside walls of the mould. When the crust is sufficiently thick, the remaining slip is poured out, leaving a hollow object to be removed from the mould for drying and firing.

If clay is forced through a nozzle, or die, it forms a continuous column. This process is called *extrusion*. A more familiar example of extrusion is the method of decorating cakes by squeezing the icing through a nozzle.

Extrusion is very useful for making pipes and some bricks, because the column can then be cut into the required lengths. Other bricks are made by pressing clay into steel moulds. Tiles are also made by pressing an almost dry mixture of clay, sand and limestone in steel moulds.

Figure 6.26
Extruded clay slabs being cut into bricks. You can see them uncut in the foreground, and cut and being carried away by conveyor belt in the background.

After shaping the clay must be dried so that no water remains, otherwise the water would turn to steam and shatter the clay during the high-temperature firing. Most commercial pottery is now fired in continuous tunnel-shaped furnaces (kilns) and the clay articles are carried slowly through the kilns on trolleys running on rails. The burners which heat the kiln are placed near the middle of the tunnel. The clay is gradually raised to its firing temperature as it moves towards the centre, and then steadily cools as it travels on to the exit.

Apart from bricks and roof tiles, most clay articles are covered with a glaze. This glassy layer gives the ceramic a smooth surface which is waterproof and easy to clean.

Glazes are composed of the oxides of silicon and various metals just like other glasses. The finely powdered mixture of oxides is suspended in water and then the pottery to be glazed is either dipped into the mixture or sprayed with it. After drying, the powdery oxides are converted to glass by firing in a furnace.

Figure 6.28
Dipping pots in glaze.

Figure 6.29
Decorating a glazed plate with a transfer.

To make high quality tableware the clay is fired first, and then the pottery is glazed and fired again. But cheaper tableware and sanitary ware may be coated with glaze after drying and then the clay and glaze can be fired together.

C6.6   Recycling glass

*Quarrying the materials needed to make glass affects the landscape. Section **C4.4** mentions some of the issues raised when limestone is quarried in a national park. In Chapter **B**17 of your Biology book you can read about the effect of farming on the environment. Physics Chapter **P**12 considers some of the issues involved in energy conservation. In this section you are invited to think about the problems of waste disposal and recycling.*

In recent years we have become more and more aware of the fact that the Earth's resources of energy and minerals are limited, and that we have to do something to reduce the amount of waste we throw away.

In 1980 about six and a half thousand million bottles and jars were sold in Britain, mainly as food and drink containers, and most of them were thrown away after only being used once.

Yet glass is a packaging material which can be reused by being cleaned, sterilized and refilled. Glass can also be recycled by being crushed and added to the furnace to make new glass containers over and over again. At the moment only about 20 per cent of the raw materials used to make glass is scrap and most of this is waste produced during manufacture and moulding. There is no reason why this proportion should not increase to 50 per cent if more schemes for recycling used glass could be implemented.

Glass is made from raw materials which are relatively cheap. There is plenty of sand, limestone and salt in Britain, but they sometimes have to be extracted in areas where quarrying spoils the landscape. Another problem is that much energy is involved in the manufacturing processes. About 80 per

	Re-usable bottle	**Non-return bottle**
Energy needed to manufacture the bottle	7.2 MJ	4.7 MJ
Energy needed to wash the bottle, fill it and deliver it	2.5 MJ	2.2 MJ

Figure 6.30
The energy cost of returnable and non-returnable containers.

24 Keep a record for a week of all the glass containers which enter and leave your home. What products come into your home in glass bottles or jars? How many glass containers are thrown away, how many are reused and how many are recycled? Can you estimate the proportion of glass in the rubbish thrown out by your household? (The average figure is 10 per cent by mass.)

25 Reusable bottles have to be stronger than non-returnable bottles, so they cost more to make and transport. However, each time they are reused they only have to be washed, filled and returned to the shops. Compare the energy cost of the one trip of a non-returnable bottle, with the average energy used per trip of a reusable bottle after one, two and five trips. The information you need is in figure 6.30.

cent of the energy used to make glass containers is used in glass factories, mainly for melting the glass. The rest is needed for the production and delivery of raw materials. The main reason for trying to reuse or recycle glass containers is to save money by saving energy.

Using recycled waste glass can save up to a quarter of the energy needed to produce glass.

- Energy is saved by reducing the energy needed to manufacture and deliver raw materials to the factory.
- Energy is saved because less is needed to melt waste glass in the furnace.
- Energy is saved because the amount of domestic rubbish to be disposed of by local councils is reduced.

Figure 6.31
A bottle bank.

Energy costs per tonne

Energy needed to collect waste glass	= 312 MJ
Energy needed to process waste glass	= 97 MJ
Energy needed to deliver processed waste to the glass manufacturer	= 49 MJ

Energy savings per tonne

Energy saved in the manufacture and delivery of raw materials	= 4273 MJ
Energy saved in the glass furnace	= 1650 MJ
Energy saved by not having to dispose of waste glass	= 87 MJ

Figure 6.32
The energy costs and savings involved in recycling waste glass.

26 The energy saved by recycling waste glass has to be set against the energy used in collecting and processing the waste. The figures are in figure 6.32.
a What is the overall energy saving per tonne of waste glass recycled?
b If one litre of fuel oil produces 41 MJ of energy when burned, what volume of fuel oil is saved per tonne of waste glass recycled?

27 Plans for reusing or recycling containers have to take note of the way in which people behave and what they are willing to do. Do you think that it is likely to be easier to persuade people to use bottle banks regularly rather than to return refillable bottles? How might people be encouraged to reuse or recycle glass?

Summary

1 Make a comparison of household articles made of pottery with objects made from glass. How are they similar? How do they differ? What are their advantages and disadvantages? Consider the following: the raw materials used to make them, the methods of manufacture, their chemical composition, and their properties.

2 Copy and complete the diagram in figure 6.33 to show how soda-lime glass is manufactured.

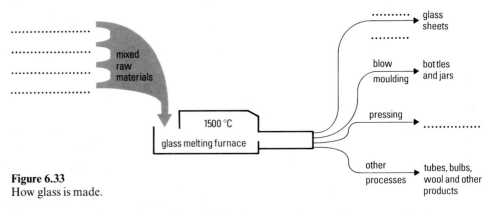

Figure 6.33
How glass is made.

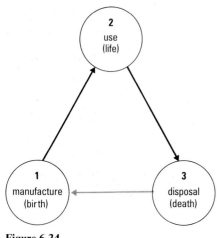

Figure 6.34
The "life cycle" of a glass bottle.

3 Imagine the "life" of a glass bottle as having three stages, as shown in figure 6.34. Make a list of some of the economic and environmental problems which can arise at each stage.

Chapter **C**7

Metals and alloys

*Metallurgy is a science which requires a knowledge of physics, chemistry and engineering as well as an ability to use mathematics. Metallurgy has to do with extracting metals from ores, shaping metals, and controlling the properties of metals and alloys. The extraction of aluminium is described in Chapter **C**4.*

When studying this chapter you will examine the connection between the properties of metals and alloys and their uses. You will think about the structure of metals and try to understand how a knowledge of structure can explain the properties of metals. You will carry out an experiment to investigate alloys, and you will be asked to plan an investigation into the methods used to prevent rusting.

*The properties and uses of metals are also described in several of the chapters in your Physics book including Chapters **P**1 and **P**16.*

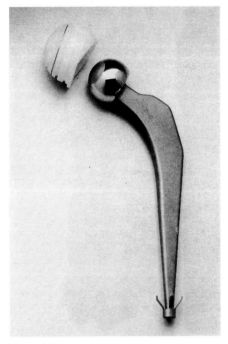

Figure 7.1
Artificial hip joint made of stainless steel with a high density polyethylene cup.

C7.1   Why are metals so useful?

Hundreds of things that we use every day are made of metal – bicycles and cars, knives and forks, sinks and saucepans, paper-clips and tin cans – the list is almost endless. Synthetic plastics have been known for less than a hundred years, but metals have been part of human history for thousands of years.

In the last hundred years or so, some of the most exciting developments in technology have depended on the discovery of new uses for metals and alloys. Examples are illustrated in figures 7.1 to 7.6.

Figure 7.2
Strips of a zinc-titanium alloy are used in the roof of this energy-efficient building.

Figure 7.3
A printed circuit uses lead in solder.

Figure 7.4
Galvanized steel is used in building a commercial greenhouse.

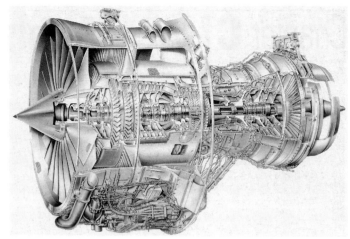

Figure 7.5
Cut-away section view of RB211 engine which powers the Boeing 757.

Figure 7.6
A pack of rhodium – platinum catalyst gauzes being installed in a nitric acid production plant.

1 Worksheet **C7A** lists a number of the useful properties of metals. Fill in the table. Decide for yourself whether the properties listed are useful. Also give at least one example of the way in which each property is put to use.

2 Produce an illustrated guide to show the importance of metals, using pictures from magazines and catalogues, or your own drawings.

C7.2   How can we explain the properties of metals?

This section introduces you to a new type of giant structure. As you read, notice that our knowledge about metals is based on observation and experiment. The theories of structure and bonding are attempts to explain what we know about metals.

Are metals crystalline?

Section **C**6.1 shows that silica can either be crystalline or glassy depending on

Figure 7.7
Zinc crystals on galvanized iron.

how fast it cools. What about metals? If you look at figures 7.1 to 7.6 it is hard to believe that metals are crystalline. Where are the straight edges and flat faces of alum or copper(II) sulphate crystals?

Metals **are** crystalline, but the crystals are hard to see because they are normally very small, usually too small to be seen without the help of a microscope. There are some exceptions. Large zinc crystals can be seen on galvanized iron, as shown in figure 7.7.

You can see that the crystals of zinc do not have straight edges because they have grown into each other, but their internal structure is just as regular as other crystals.

There is another reason for not seeing the crystalline structure of a metal. When you look at a metal you may not see the metal itself, but a layer of metal oxide, or other compound, which has formed on the surface by reaction with the air. Rust on iron and the green coating on weathered copper hide the metals underneath. Other metals, such as aluminium, are protected by an invisible layer of oxide.

So to see metal crystals we first have to remove the oxide layer and then examine the surface with the help of a microscope. The oxide layer can be removed using acid. The metal is first polished until it has a scratch-free surface. It is then dipped in a solution which attacks it chemically and highlights the crystal structure. This process is called *etching*.

Figure 7.8
Polished door knocker.

Figure 7.9
The same door knocker – etched.

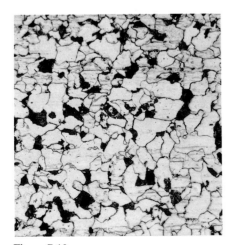

Figure 7.10
Grains in etched steel.

Figure 7.8 shows a brass door knocker which has been polished. Figure 7.9 shows the same door knocker after it has been dipped in dilute nitric acid. In this example, the metal crystals are large enough to be seen with the unaided eye. Similar, but much smaller crystals are revealed on the surface of steel when it is examined with a microscope after polishing and etching.

The small metal crystals which make up metals are called *grains*. The grains stick together very well and are not easily broken apart. Sometimes, however,

Figure 7.11
Grains of a copper-zinc alloy (× 10).

Figure 7.13
Rough diamond crystal.

Figure 7.14
Native copper crystals.

it is possible to separate the grains, as in figure 7.11. This shows some of the crystal grains from a copper–zinc alloy magnified ten times.

How are the atoms arranged in metal crystals?

Figure 7.12
The Great Pyramid of Cheops at Giza.

Figure 7.12 shows that the Great Pyramid of Cheops has a similar shape to the top half of the diamond crystal in figure 7.13. We can see that the pyramid has a regular shape. When we get closer, we can also see that this is because the blocks of stone were carefully placed in an ordered pattern by the builders. So we imagine that the atoms in diamond are arranged in a similarly ordered way.

Figure 7.14 is a photograph of copper crystals in a specimen of the metal which was found free in nature. Figure 7.15 is a drawing which shows more clearly the shape of the crystals. Figure 7.16 is a drawing of a model to show the arrangement of atoms in a copper crystal. This arrangement was discovered by using X-ray analysis to study copper crystals, as described in Chapter **C5**. The model can account for the shape of the crystals.

Looking at the model in figure 7.16, you can see that we imagine that metal atoms

● are tiny spheres
● are arranged in a regular pattern
● are packed close together in a crystal.

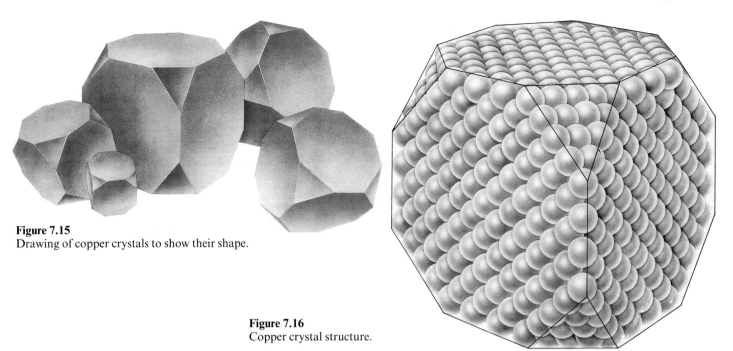

Figure 7.15
Drawing of copper crystals to show their shape.

Figure 7.16
Copper crystal structure.

---

3 Are metal elements denser than solid non-metal elements? Find out with the help of table 2 in the Data section. Examine the densities of the elements with these symbols: Al, B, C, Ca, Cr, Co, Cu, Au, I, Fe, Pb, Mg, Ni, P, K, Se, Si, Ag, Na, S, Sn, Ti, W and Zn. Display the data in the form of a bar chart, keeping the metals separate from the non-metals. What conclusions can you come to?

4 Look again at the data in your answers to question **3**. Where in the Periodic Table are the metals with high densities? Where in the Table are the metals with the low densities?

Why are metals strong?

Look at figure 7.17. This is a photograph of a famous experiment to show

Figure 7.17
This bus is standing on six tea cups.

Figure 7.18
The cables which support the Humber bridge are bundles of many wires.

Figure 7.19
Building a reinforced concrete bridge.

that a bus can be supported on six china tea cups. The cups are strong enough to hold up a bus but if you drop them they will shatter. Glasses and concrete are similar: they are strong so long as you are compressing them, but they are brittle if you try to bend them.

Metals are different. If you drop a metal mug it may dent but it will not break. You can bend metal wire and it will not crack – at least not the first time. Only if you keep on bending it back and forth will it break in the end.

Metals are very important because they can support a big pull. We say that they are strong in *tension*. The tensile strength of metals is illustrated by the use of metal wires to make the cables of a suspension bridge. The tensile strength of steel is combined with the compressive strength of concrete to make beams for bridges. (See Physics Chapter **P**1.)

Metals can also be bent without breaking. This is put to use when metals are shaped by forging and pressing.

Figure 7.20
Forging a motor shaft part for a steam turbine in an electricity generator.

The theories of structure and bonding for metals have to explain why they are so different from glasses and ceramics. Looking at figure 7.16 you must realize that this is a small fragment of a giant structure. As in diamond, there is a continuous network of strong bonding throughout the crystal. If you look at table 1 in the Data section you will see that all metal elements have giant structures.

Unfortunately, the model shown in figure 7.16 cannot show what it is that holds real atoms together. The model is made by glueing together plastic spheres, and is only designed to illustrate the arrangement of the atoms. We think that, in metal crystals, the atoms are bonded to each other by strong forces called *metallic bonds*. We do not explain anything by giving the bonds in metal crystals a special name, but it shows that we realize that the bonding in metals is different from the bonding in other materials. Metallic bonds must be strong bonds to account for the strength of metals.

Figure 7.21
Inspecting copper wire after manufacture.

Figure 7.22
A view from the control pulpit of the Slabbing Mill at Llanwern Works, Newport, South Wales.

5 Look at the model of the structure of copper in figure 7.16. Here are some ideas which someone looking at the model might think to be true of copper.

- Copper crystals are naturally cubic because the atoms are packed cubes.
- There is air in between the atoms in a crystal of copper.
- Copper is dense because the atoms are closely packed.
- The atoms in a copper crystal are not moving at room temperature.
- Copper has a high melting-point because all the atoms are strongly bonded in a giant structure.
- Copper melts when heated strongly because the atoms melt.

Which of the ideas in this list are true and which are false? Give your reasons for deciding on your answers.

Why can metals be bent and stretched?

Figure 7.20 shows a huge piece of steel which has been forged in a powerful press. Copper can be drawn out to make fine wire (figure 7.21). Hot rolling is used to reduce slabs of metal to thin plates (figure 7.22). Sheets of steel, or aluminium, are forcibly shaped to make cans for food and drink (figure 7.23).

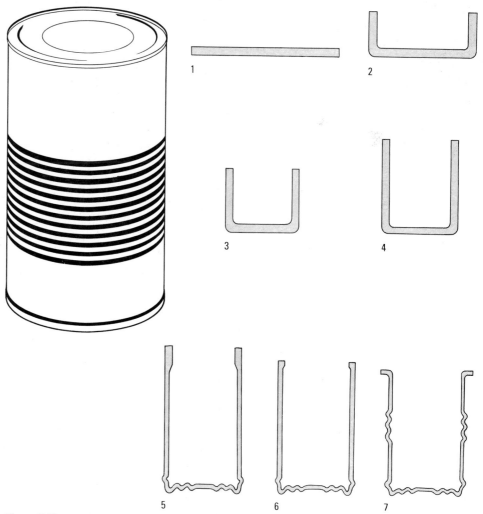

Figure 7.23
How two-piece metal cans are made by drawing and wall ironing.

Structure and properties

Figure 7.24

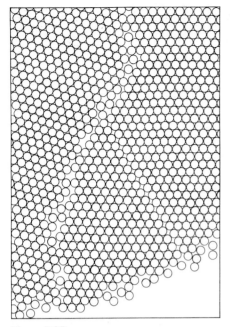

Figure 7.25
Bubble raft.

Figure 7.26
A bubble raft in a Petri dish. Notice that some bubbles are larger than the rest.

We need a model which can explain the fact that metals can be bent and stretched. The model has to show that the metal atoms can slide over each other without the bonds breaking. Sir Lawrence Bragg, famous for his work on X-ray analysis, came up with an answer to this problem. He blew layers of small soap bubbles on water and noticed that they took up a regular, close-packed arrangement. He suggested that the atoms in a metal might be represented by soap bubbles.

In figure 7.25 you are looking down on a layer of bubbles. All the bubbles are the same size, just like the atoms in a pure metal. You can see three areas with regular arrangements of bubbles. These can be taken to represent three metal crystal grains. Pushing on the edge of the layer makes it change shape, and as it does so the rows of bubbles slide over each other. The bubbles move but stay stuck together.

We can explain the properties of metals if we imagine that metallic bonds do not break as the atoms slide over each other. Diamond and silicon dioxide are hard and brittle. To explain this we assume that the bonds in the giant structures of diamond (figure 5.20) and silicon dioxide are quite different from the bonds in metals. The bonds in diamond and silicon dioxide are strong, but they have a definite direction and if you try to force them out of that direction they break.

6 The bubble raft model can be used to make predictions. In the dish in figure 7.26 most of the bubbles are the same, but a few of them are larger. This means that the bubbles in the dish are a model of an alloy in which one metal is mixed with another metal with larger atoms. Will the presence of a few larger bubbles make it easier or more difficult for the rows of bubbles to slide past one another? Do you predict that an alloy will be easier, or more difficult, to bend than a pure metal?

C7.3   What are alloys, and how do they differ from pure metals?

If you have made things with metals in CDT, you will already know that the properties of metals can be changed in several ways: by alloying, by heat treatment, and by the methods used to shape the object being made. Choosing the right material and selecting the correct method of fabrication are two of the problems which designers have to solve when making things from metals and alloys.

Analysis is an important part of chemistry. Chromatography is one of the most sensitive analytical methods. You may already have used chromatography to analyse inks, or food colours or grass. While studying this section you may have the opportunity to use chromatography to find out which metals are present in an alloy.

Alloys are more useful than pure metals. They are harder and stronger, and it is easier to control their properties.

Most alloys are mixtures of metals (figure 7.27) but steel, which is the most important alloy of all, is a mixture of a metal with a non-metal. Mild steel is an alloy of iron and carbon.

Figure 7.27
A molten alloy being cast at the Royal Mint. Note the two strips of metal leaving the plant at the left of the picture.

Every kind of steel is an alloy, but they are not all *alloy steels*. Alloy steels consist of iron and carbon together with one or more of these metals: nickel, chromium, tungsten, molybdenum and vanadium. One of the best known examples is stainless steel which contains 12 to 18 per cent by mass of chromium.

The importance and variety of alloys are illustrated in the series of pictures in figures 7.28 to 7.34.

Figure 7.28
Mild steel consists of iron with 0.12–0.25 % carbon. It is cheap and easily manufactured in sheets which can be pressed into shape.

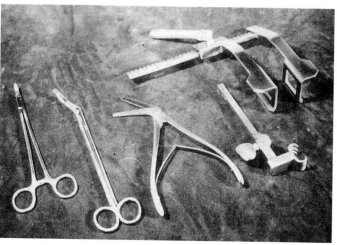

Figure 7.29
Brain surgery instruments. Stainless steel is an alloy steel. It consists of iron with 12–18% chromium and 0.2–1% carbon. It resists corrosion.

Figure 7.30
A nickel aluminium bronze alloy was used to make this ship's propeller.

Figure 7.31
An alloy of aluminium and lithium is used to make the skin of this aircraft. It has high strength and low density, and is resistant to fatigue.

Figure 7.32
Bronze sculpture of Mary and Jesus at Rochester Cathedral. Bronze is made from copper with up to 12 % tin. It is a strong, hardwearing alloy with good corrosion resistance. It is used to make gear wheels, bearings, ships' propellers, coins and statues.

Figure 7.33
Solder is an alloy of tin and lead. It melts at a low temperature and can be used to join metals. Here it is being used in the making of a support frame for stained glass.

Figure 7.34
"Silver" coins are made from cupro-nickel. These alloys have good corrosion resistance and are also used to make chemical plants.

7 Consider these five alloys:

Alloy	Metals in the alloy	Special properties
Wood's metal	bismuth, lead, tin and cadmium	it melts at 70 °C
An alloy steel	iron, chromium and vanadium	very strong and can resist strain much better than ordinary steel
Gun metal	copper, zinc and tin	easy to cast and has good corrosion resistance
High speed steels	iron with up to 20% tungsten	maintains a sharp edge at high temperatures
Brass	copper with less than 20% zinc	easily worked, has a gold colour and does not corrode

Which alloy would you use to make:
a the tip of a masonry drill
b the fittings for the outside of a ship
c cheap jewellery
d the ''plug'' in an automatic fire sprinkler
e a heavy duty spanner or wrench?
For each one, try to give reasons to explain your answer.

8 With the help of figure 7.35 and section 7.2, suggest an explanation for the fact that alloys are generally stronger and harder than pure metals.

9 Analysis of a stainless steel shows that in a 200-g sample there are 168 g of iron, and 26 g of chromium with small quantities of carbon and other elements. What is the ratio of the number of iron atoms to the number of chromium atoms in this alloy? (Use the method shown in box 1 on page 116 and refer to table 3 in the Data section for the molar masses of the metals.)

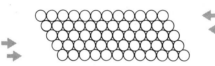

In a pure metal all the atoms are the same size.
What happens to the layers when forces are applied?

Figure 7.35
Atoms in a pure metal and in an alloy.

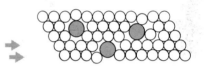

How will the presence of larger metal atoms in the structure of an alloy affect the ease with which the layers of atoms slip over each other when forces are applied?

C7.4 How can we stop corrosion?

We all face the problem of stopping corrosion because so many things that we use are made of metal. Many cars fail the MOT test because they have rusted away underneath. If you can discover the cure for rusting you will make your fortune. In this section you have an opportunity to plan an investigation to study the methods used to stop rusting.

What is corrosion?

Painting the Forth Bridge is a never-ending job. It is like doing the housework. As soon as you have worked through all the rooms it is time to start again at the beginning.

Painting steel structures to stop them rusting is part of the fight against corrosion which is a never-ending battle. In Britain, corrosion costs thousands of millions of pounds each year.

The metals we use are in a chemically unnatural state. Most metals are found naturally in the ground combined with oxygen or other elements. Much energy has to be used to extract the metals from their ores, as for example in the extraction of aluminium described in section **C4.5**. When metals corrode, they react with oxygen in the air and turn back to oxides. This cycle is illustrated for iron in figure 7.37.

Figure 7.36
Painting the Forth Bridge.

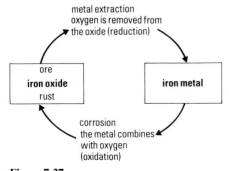

Figure 7.37
The cycle of extraction and corrosion for iron.

Figure 7.38
Thin steel plate is coated with tin for
making food cans. If the food has a high
acid content a layer of lacquer may be
added to prevent direct contact between
the food and the tinplate.

More than six thousand million food
cans are bought every year. Some of them
are collected and recycled, as you can see
from the picture at the beginning of
Topic C2.

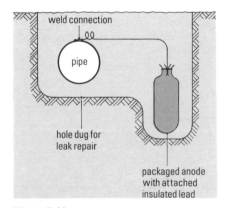

Figure 7.39
Gas pipe protected by a magnesium
sacrificial anode at a point where a leak
has been repaired. Magnesium is used to
protect steel pipes from corrosion
underground. The lumps of magnesium
corrode instead of the steel, and can be
replaced much more easily than the pipes.

10 Have a look at a bicycle, or car, which is beginning to suffer from rusting. Where is
the corrosion worst? Is the corrosion uniform or in patches? Where does the corrosion
start – in the middle of a piece of metal or at the edges and joints? What steps has the
manufacturer taken to minimize corrosion? What can the owner do to reduce corrosion?

11 In earlier years at school you may have done an experiment to show that iron nails
only rust in the presence of both air and water.
a Describe the experiment and show how it is designed to demonstrate that air alone,
and water alone, do not make iron go rusty.
b How might you modify the experiment to show that it is the oxygen in the air which
reacts with the iron during rusting?
c How might you modify the experiment to find out whether sea water makes iron rust
faster than rain water?

12 Read the following passage:
   "The metals which are most easily extracted from their ores, or which occur
   uncombined in nature, are the ones which are most resistant to corrosion.
   Historically these are also the earliest metals to have been discovered and used."
Do you think that the statements in the passage are true or false? Give evidence to
support your argument.

How can rusting be controlled?

It is impossible to stop corrosion, but it can be kept under control. Good
design, the correct choice of materials, the use of protective coatings and
careful maintenance can all help. The problem is that we are often not willing
to pay the extra initial cost of goods of high enough quality to last, and we
may not look after our property carefully. Some of the methods used to stop
corrosion are illustrated in figures 7.38 to 7.42.

Figure 7.40
Galvanizing is the process of covering steel with a layer of zinc to prevent corrosion. The zinc
corrodes to form an oxide layer that resists further corrosion. If the zinc layer is scratched it is
the zinc which corrodes not the iron. The structural steel frames for this multi-storey car park
at Gateshead are protected in this way.

Figure 7.41
Car bodies are given a phosphate treatment as part of a series of anti-corrosion processes.

Figure 7.42
Coating a steel pipe with a layer of plastic to protect the metal from air and water.

BOX 1   Problem
Which is the best way to stop rusting?

Rusting is normally a slow process and it may not become visible until some time after it begins to form. However, there is a special indicator which can quickly show when iron is corroding.

Rust consists of iron oxide formed in the presence of water. Iron oxide is a compound of a metal with a non-metal. Compounds of this sort consist of ions, as explained in Chapter **C5**. When iron rusts, iron atoms turn into iron ions.

Corrosion indicator contains a chemical which reacts with iron ions to produce a deep blue colour. The indicator is made up as a hot solution in water with gelatine. It sets when cold, so that the experiment can be observed for some hours or even days without being spoiled when the apparatus is moved.

Experiments are conveniently set up in test-tubes or Petri dishes.

You are asked to design experiments which you could carry out to investigate one or more of the following questions. Write down your plan and illustrate it with labelled diagrams.

- Are the corrosion inhibitors sold for use on cars effective?
- Does putting oil, or grease, on steel stop or slow rusting?
- Does paint stop rusting completely? Which type of paint works best?
- Is steel protected from rusting if other metals are connected to it?
- Does zinc plating, or tin plating, continue to prevent rusting even if the metal coat is scratched so that the steel underneath is exposed?
- Is iron more likely to rust if it has been shaped by hammering or by bending?
- Does stainless steel rust at all?
- If a piece of steel is bent into a narrow V-shape, will it rust faster in the crevice or in the open?

Figure 7.43
Part of a rusty car exhaust system.

13 The mild steel exhaust systems used for most cars last only for about two years. Stainless steel systems are also made. They are much more expensive but last as long as the rest of the car and never need replacing. Some car owners and many business users sell their cars every two to three years and buy new ones. Other people keep their cars for as long as possible. Who benefits from cheap exhaust systems which rust away in about two years?

14 Ceramics and glasses do not corrode, nor do plastics. Why are car exhausts not made from ceramics, glasses or plastics?

Summary

1 Some of the important words and terms used in this chapter are listed below. Write a brief explanation of each and, where possible, give an example to show what you mean.

grain, giant structure, tensile strength, metallic bonding, alloy, steel, corrosion, galvanizing.

2 Figure 7.44 is an outline of the life cycle of a metal. Copy the diagram and label it very fully to include as much chemistry as you can, choosing aluminium as the example. Refer back to Chapter **C4** for information about the extraction of the metal. Illustrate your diagram with drawings or pictures cut from magazines.

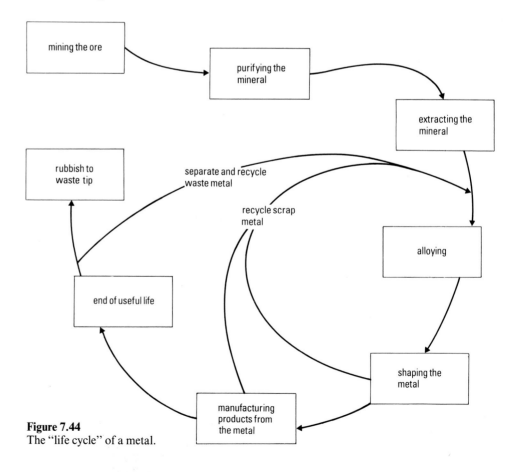

Figure 7.44
The "life cycle" of a metal.

Chapter **C**8   **Polymers**

*Chapters **C**6 and **C**7 in this topic have described materials with giant structures. In this chapter you will read about materials which are molecular and are all polymers. Section **C**2.5 explains that the molecules of polymers are very long.*

This chapter continues to emphasize the two main themes of this topic which are:

- *the connection between the properties of materials and their uses, and*
- *the theory which explains the properties in terms of structure and bonding.*

In the laboratory you will see some polymers being made and you may have first-hand experience of the properties of polymers if you use a key to identify plastics. You will learn about some of the processes used to manufacture plastic products and get some impression of what it can be like to work in the industry.

C8.1   What are polymers?

Could we survive without polymers?

There are at least two good reasons why you could not be reading this book if there were no polymers. There would be no book since its pages are made of cellulose which is a polymer. There would be no you! Take away all the polymers from your body and you would cease to be.

Look at figure 8.1 to see how much of your body is made of polymers, particularly proteins.

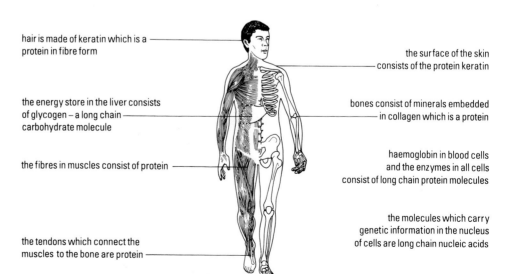

Figure 8.1
There are many polymers in the human body.

hair is made of keratin which is a protein in fibre form

the energy store in the liver consists of glycogen – a long chain carbohydrate molecule

the fibres in muscles consist of protein

the tendons which connect the muscles to the bone are protein

the surface of the skin consists of the protein keratin

bones consist of minerals embedded in collagen which is a protein

haemoglobin in blood cells and the enzymes in all cells consist of long chain protein molecules

the molecules which carry genetic information in the nucleus of cells are long chain nucleic acids

Artificial polymers are now so common that it is easy to forget that they are a recent invention compared with glass, ceramics and metals. Polythene was not in common use until the 1950s. Figure 8.2 shows how everyday objects were changed by the discovery of new polymers.

Figure 8.2a and b
Household articles before and after plastics.

Figures 8.3 to 8.7 illustrate a few of the many uses of polymers today.

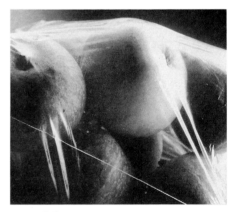

Figure 8.3
Polystyrene used for building. The walls are filled with concrete and have high insulation qualities.

Figure 8.4
Rubber tyres – elastomers in use in a car.

Figure 8.5
Food packaging – plastic film.

Figure 8.6
Plastics in use in agriculture: these low tunnels of plastic film in Southern Iraq are protecting tomatoes from frost.

Figure 8.7
Plastics and synthetic fibres for leisure: this windsurfer is made almost entirely of polymers.

The board of the windsurfer is constructed from a polyethylene "skin" filled with rigid polyurethane foam, a polypropylene tail fin, and a strong, flexible, glass-reinforced polyester mast.

Two types of polymerization

The term *polymer* is explained in figure 2.44. Section **C2.5** shows how small ethene molecules can join up in long chains because they have a double bond in their structure. Figures 2.42 and 2.43 give a picture of the way in which very long-chain molecules are made just by adding ethene molecules together. Making a new chemical by joining together simpler substances is called *synthesis*. So polythene is an example of a synthetic polymer.

Figure 2.42 shows that the molecules of ethene and polythene are made up of carbon and hydrogen atoms held together by chemical bonds like the atoms in other molecules. However, polymer molecules are so long and are made up of so many atoms that it is easier to use simpler diagrams to describe them.

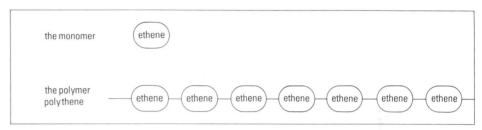

Figure 8.8
The addition polymerization of ethene.

There are no other products when polythene is formed by joining together lots of ethene molecules. This way of making long-chain molecules is called *addition polymerization*.

Like many other substances, polythene has a chemical name as well as a common name. The chemical name is poly(ethene). In this chapter we will use both chemical names and the more familiar common names. One of the advantages of the chemical names is explained in box 1. A further complication is that many manufacturers have their own trade names for polymers.

BOX 1   Chemical names

The advantage of chemical names is that once you know the rules you can work out the formula and structure of a compound from its name. You will only have to learn the rules if you go on to more advanced chemistry courses.

One rule is that simpler carbon compounds are regarded as relations of the alkanes. The alkane with two carbon atoms in its formula is called ethane. All other substances with two carbon atoms in their molecules have chemical names starting with "eth-". Here are some examples:

ethane   ethene   ethanol   ethanoic acid
                (ethylene)   (alcohol)   (acetic acid)

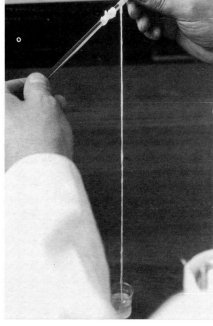

diaminohexane
(soluble in water)

hexanedioyl dichloride
(soluble in organic solvents)

when these two compounds form nylon by
condensation polymerization they join by splitting
off molecules of hydrogen chloride

Figure 8.9
The structures of diaminohexane and
hexanedioyl dichloride.

Figure 8.10
Making nylon.

Starch and cellulose are polymers made by plants. They are both known to
be polymers of glucose as explained in section **C**3.2. Figures 3.14 and 3.15
show that two different polymers of glucose can exist because there are two
ways of joining the smaller molecules to make a chain.

You can see from figures 3.14 and 3.15 that water molecules are split off as
the glucose molecules join. This type of polymerization is called *condensation
polymerization.*

Proteins are also condensation polymers. They are made from small
molecules called amino acids. In natural proteins there are twenty different
amino acids. As you can see in section **C**3.3, making proteins can be compared
to making a necklace with lots of different colours and sizes of bead.

Nylon is an example of a synthetic condensation polymer. You may already
have seen, or done, an experiment to make nylon. Two monomers are used
which have complex chemical names. One, called diaminohexane, is dissolved
in water. The other, called hexanedioyl dichloride, is dissolved in a dense
organic solvent. Figure 8.9 shows you the chemical structures of the two
chemicals, but you are not expected to remember the names or the formulae.

The two solutions are carefully poured into a beaker so that the water layer
floats on top of the organic layer. The chemicals react at the surface between
the two solutions. A continuous thread of nylon can be drawn from the
surface.

Figure 8.11 gives a simplified description of the way the molecules
polymerize by condensation during this experiment. The conversation in
figure 8.12 (overleaf) may help you to understand how the two types of
molecules combine to form a long chain.

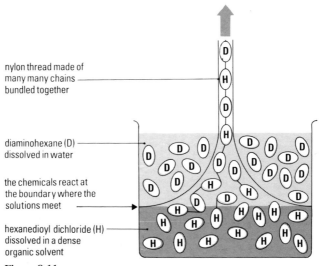

nylon thread made of
many many chains
bundled together

diaminohexane (D)
dissolved in water

the chemicals react at
the boundary where the
solutions meet

hexanedioyl dichloride (H)
dissolved in a dense
organic solvent

Figure 8.11
How the nylon experiment works.

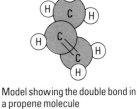

Model showing the double bond in
a propene molecule

propene

Simplified model of a
propene monomer molecule

Figure 8.13
Two ways of representing
propene.

HO ─⬭ B ⬭─ OH   H ─▭ E ▭─ H

benzenedicarboxylic acid   ethanediol

Figure 8.14
Some different ways of representing the
monomers used to make polyester.

1 Two ways of representing propene are shown in figure 8.13. Propene polymerizes in
the same way as ethene. The product is polypropylene. Draw a length of a
polypropylene molecule, either in the style of figure 2.42 or in the style of figure 8.8.
[The chemical name of polypropylene is poly(propene).]

2 Polyester is a condensation polymer made from benzenedicarboxylic acid and
ethanediol. The two monomers can be represented simply as shown in figure 8.14.
Draw a diagram to show these monomers joining together to form a polyester chain.

C8.2   How do we use polymers?

While studying this section you will read about the difference between natural and synthetic polymers. Starch and cellulose, proteins and nucleic acids are all natural polymers.

As you read this section, it will help you to remember what you may know about plastics from other subjects you are studying at school. If you are studying Home Economics you may have some knowledge of different types of fibres and the way in which they are spun to make yarn. If you are taking a course in CDT you may have worked with plastic materials and shaped them by heating and pressure.

Natural polymers

The first polymers to be used were those which occur naturally and which can be made into fabrics. Cotton, wool, and silk are all natural polymers. The cellulose in cotton is a carbohydrate. Wool and silk are animal proteins.

Figure 8.15
Silk moth larva spinning its cocoon. The "silkworm" is not really a worm. It is the larva of the silk moth. When preparing to pupate, the larva protects itself by spinning a cocoon. It forces out a liquid through the two small holes which make up its spinneret. The liquid hardens on contact with the air to form a continuous thread.

The length of the thread produced by a single silkworm may be up to 1000 metres long or more. It is a skilled job to unwind the threads from a number of cocoons and combine them to make a yarn. Many people have to be employed to make silk, and so it is an expensive fabric.

Figure 8.16
A stage in processing wool: rolling the wool fibres into flattened balls called "tops" for further processing into yarn for spinning.

Figure 8.17
Unwinding silk cocoons in 1607.

Were polymers discovered by a conversation like this?

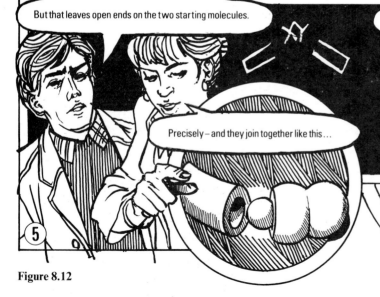

Figure 8.12

Figure 8.18
Alexander Parkes.

Modified natural polymers

The word "plastic" is both an adjective and a noun. As an adjective, the word describes anything which can be moulded. Wet clay is plastic and so is Plasticine. As a noun, the word plastic is used for the family of polymers which can be moulded when soft and then set.

The first plastics were made by changing natural polymers. A start was made by Alexander Parkes in the 1850s. He used nitric acid to turn cellulose into cellulose nitrate (sometimes called gun cotton or nitrocellulose). He then dissolved this in a mixture of ethanol and ether. When the solvent evaporated he was left with a hard, horn-like mass. By adding castor oil and camphor to the solution he made the material less brittle – more plastic.

Parkes was an enthusiastic inventor and he made great claims for his *Parkesine* and exhibited it at the International Exhibition of 1862. Unfortunately his material was not as good as he claimed, and his company went bankrupt.

Figure 8.19
Objects made from Parkesine.

It was an American, John Hyatt, who finally made a success of Parkes's discovery when trying to make artificial ivory for billiard balls. Hyatt realized the importance of the camphor in the product. He developed *celluloid*, the first commercially successful plastic.

Celluloid is perhaps most famous for its use as movie film. Unfortunately it is highly flammable and becomes increasingly dangerous with age, so that it may burst into flames while on the projector. Much historic film made on celluloid is likely to become too dangerous to view unless it is copied onto modern film.

In the early 1900s, another plastic based on cellulose was developed. It is usually called cellulose acetate. It is made by treating cellulose with ethanoic (acetic) acid, instead of nitric acid. There are two common forms of cellulose acetate, depending on the amount of ethanoic acid used.

Figure 8.20
A sample of celluloid film.

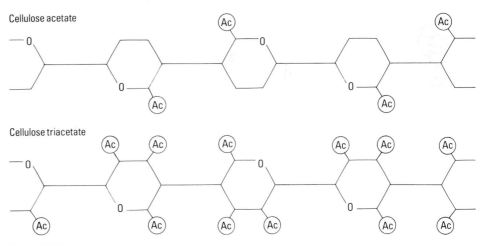

Figure 8.21
Cellulose acetate and triacetate.

Figures 8.22
A use of cellulose acetate – a face mould.

During the First World War, this plastic was used as "dope" to tighten the fabric on aircraft wings. After the war, demand dropped and so other outlets were sought. Cellulose acetate became widely used for such things as screwdriver handles, pens, toys and steering wheels. It is also used to make photographic film. It is much safer than celluloid and does not deteriorate with age.

We can imitate the silkworm and shape plastics into fibres. This can be done with cellulose acetate by dissolving it in a solvent which evaporates quickly. The usual solvent is propanone (acetone) which is more familiar as nail varnish remover. The solution is forced through hundreds of tiny holes in a man-made spinneret. The propanone evaporates in warm air, leaving a solid

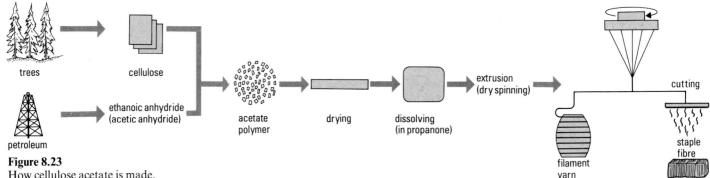

Figure 8.23
How cellulose acetate is made.

Figure 8.24
"Winter luxury in Tricel".

fibre which is wound onto a bobbin. In the early days the fabrics made from cellulose acetate were called "artificial silk". Modern forms of cellulose acetate include the fabrics Tricel and Dicel.

Synthetic polymers

We live in an age of polymers. Figure 8.25 will bring home to you the number and variety of synthetic polymers in modern homes.

3 Look at figure 8.25 and make a list of some important uses of each of these polymers: polythene, polypropylene, polystyrene, PVC, and polyester.

4 Draw up a list of all the advantages of plastics. (Look at the list of advantages of glasses and ceramics in section **C6.3** to give you an idea of the sorts of properties you might include.) Give examples of the way in which we take advantage of these properties with the help of your own experience and figure 8.25.

5 Plastics have some disadvantages too. Make a list of all the disadvantages you can think of. Give examples of the way these disadvantages can affect you.

BOX 2   Problem
Are plastic films best for keeping food fresh?

Polythene bags, cling film, greaseproof paper and metal foil are all used as food wrappings. Which is the best choice? You are asked to plan an investigation to find out.

Here are some of the questions you may need to think about when planning your experiments:

● What happens when food goes stale? Why does it stay fresh longer if it is wrapped up?
● Which are the popular wrapping films for food?
● What are the useful properties of the wrapping films? Which is the most important property when food freshness is being considered?
● How can this property be tested and measured? Will it be better to test the film using a real example of food (such as bread)? Or can the investigation be carried out more reliably without using an actual food?
● How can the experiments be designed to make sure they are a fair test?
● What apparatus and measuring instruments will you need for your experiments?
● How long will the experiments have to run to give reliable results?

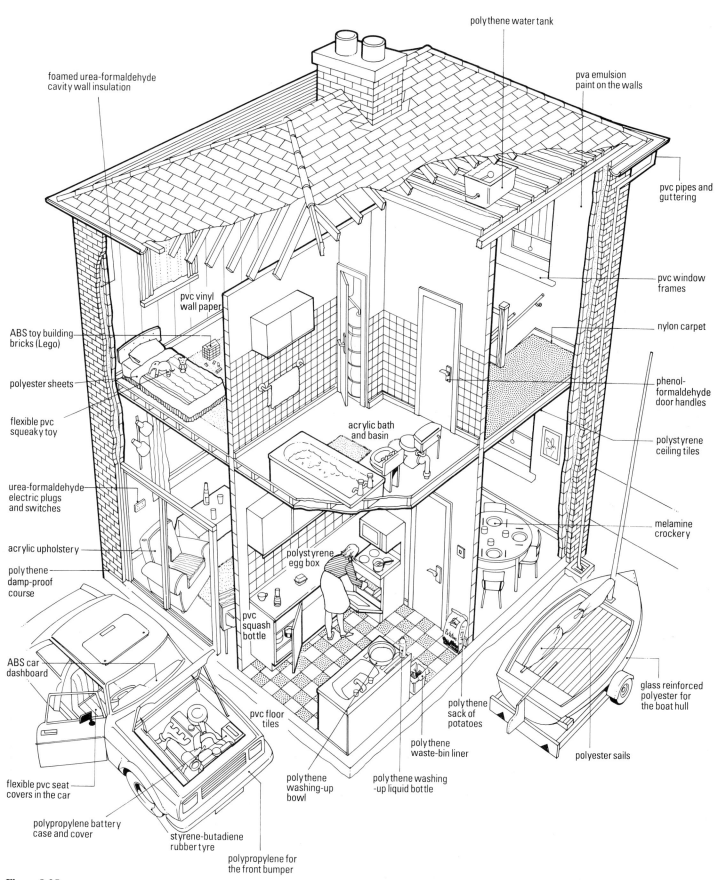

polythene water tank

pva emulsion paint on the walls

foamed urea-formaldehyde cavity wall insulation

pvc pipes and guttering

pvc window frames

nylon carpet

pvc vinyl wall paper

ABS toy building bricks (Lego)

phenol-formaldehyde door handles

polyester sheets

flexible pvc squeaky toy

acrylic bath and basin

polystyrene ceiling tiles

urea-formaldehyde electric plugs and switches

melamine crockery

acrylic upholstery

polystyrene egg box

polythene damp-proof course

polystyrene ceiling tiles

ABS car dashboard

pvc squash bottle

glass reinforced polyester for the boat hull

flexible pvc seat covers in the car

pvc floor tiles

polythene sack of potatoes

polythene waste-bin liner

polyester sails

polypropylene battery case and cover

styrene-butadiene rubber tyre

polythene washing-up bowl

polythene washing-up liquid bottle

polypropylene for the front bumper

Figure 8.25
Some uses of polymers in the home.

C8.3   What is the structure of plastics?

Why can one plastic be used to make a saucepan handle when another will melt if placed near a flame? Why are some polymers plastic while others are elastic? How can there be two different types of polythene with different densities? Why are only some polymers suitable for spinning into fibres? Why do some fibres get stronger if you stretch them as they form? Why are some polymers fully transparent while others are translucent or opaque? In this theoretical section we will not try to answer all these questions, but they are some of the problems which can be solved by investigating the structure of polymers.

It is important to remember that on an atomic scale polymer molecules are very long, with between 1000 and 50 000 monomer units joined in a single chain. The diagrams in section **C**8.1 only show **very**, **very** short fragments of polymer molecules. To emphasize the great length of the chains, we will now use an even simpler picture of polymer molecules when trying to imagine how these molecules are arranged. In the diagrams in this section the molecules are represented by lines.

Thermosoftening and thermosetting plastics

Thermosoftening plastics are like chocolate. When you warm them they get soft and can be moulded. When cool they set hard. Reshaping these plastics by heating and moulding can be repeated many times.

Thermosetting plastics are like an egg. Once you have hard boiled an egg, it sets and cannot be turned back to a liquid by further heating. Once thermosetting plastics have been heated and moulded they cannot be remelted.

Examples of these two types of plastic are shown in figure 8.26.

Thermoplastics

They can be repeatedly softened by heating and hardened by cooling because the polymer chains are not cross-linked.

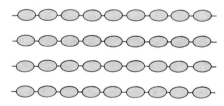

Examples:
polythene, polypropylene, pvc, polystyrene, acrylic, nylon.

Thermosets

They can be heated and moulded only once. When heated the chains become permanently cross-linked.

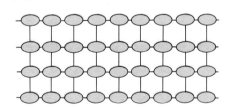

Examples:
phenol-formaldehyde, urea-formaldehyde, melamine-formaldehyde, epoxy resin, polyester resin.

Figure 8.26
Thermosoftening and thermosetting plastics.

If you could see the molecules in a thermosoftening plastic, they might look something like the tangled mass in figure 8.27. In parts the molecules are in a muddle, but in other parts they may line up in a regular way, more like the molecules in a crystal.

Section **C**5.3 explains that the bonds holding the atoms together **within** molecules are strong, while the forces **between** molecules are weak. This is also true for polymer molecules.

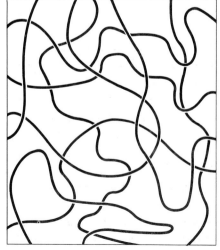

Figure 8.27
Molecules in a thermosoftening polymer.

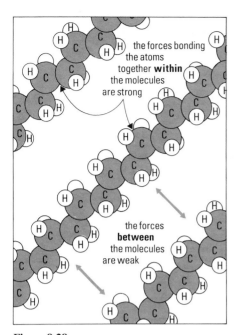

Figure 8.28
The difference between forces within and between polymer molecules.

In thermosoftening plastics, it is not too difficult to get the molecules to slide past each other, because the forces between the molecules are weak. When thermosoftening plastics are warmed up, the molecules gain more energy and can move past each other more easily. The softened plastic can then be shaped in a mould. It hardens again on cooling.

Next time you feel the handles of a heavily laden polythene carrier bag giving way, you can imagine the molecules slipping past each other as the plastic stretches. It is only because the polythene molecules are so long compared with other molecules that the bag has any strength at all. Figure 8.29 may help you to see why this is so.

Methane. The molecules are small. Methane is a gas because the weak forces between the molecules cannot hold them together at room temperature.

C_6H_{14}

Hexane is one of the compounds in petrol. The forces between the molecules are the same as in methane but, acting over longer molecules, they keep them together as a liquid at room temperature.

$C_{20}H_{42}$

Eicosane is one of the compounds in candle wax. Weak forces acting between these longer molecules keep this substance solid at room temperature. Even so the wax is brittle and melts easily.

a hydrocarbon with up to 100 000 carbon atoms in each chain

Polythene is a plastic. On the scale of this diagram one of the polymer molecules should be drawn about 900 metres long. The forces between the molecules are similar to those in other hydrocarbons, but the molecules are so long that they hold together quite strongly at room temperature. Even so the plastic softens and melts between 100 and 150 °C.

Figure 8.29
How strength depends on the length of the molecules.

Thermosetting plastics are manufactured in two stages. The first stage produces a resin consisting of long chain molecules. During the second stage, the resin is heated and moulded. The molecules react further and chemical cross-links are formed between the chains. This produces a continuous network of linked molecules more like a giant structure. The polymer sets hard and cannot be melted, because the cross-links are permanent.

Do-it-yourself car exhaust repair kits contain a bandage which is soaked in a thermosetting resin. The bandage is wrapped around the damaged part of the pipe. When the engine is started, the polymer is cross-linked as the exhaust pipe heats up.

Some thermosetting polymers can be made to cross-link and harden at room temperature by using catalysts. One example is the polyester resin used in the kits to repair car body work. The polymer molecules in the resin react and become permanently cross-linked when mixed with the catalyst.

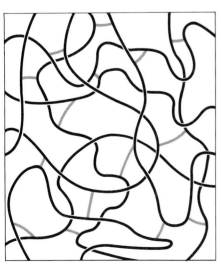

Figure 8.30
Cross-linking in a thermosetting plastic.

Figure 8.31
Polypropylene granules.

C8.4   Who does what in the plastics industry?

This section is about people as well as processes. You can find out more about what it is like to work in the plastics industry from careers literature. Compare this section with experience you may have had in making models at home or working with plastics in a CDT course.

Companies which make polymers are part of the chemical industry. Carrying out polymerization reactions on an industrial scale is much more difficult than making small amounts of polymer in a laboratory, and the products must be of constant good quality.

Plastics are supplied in a variety of forms: powders, small chips, granules or sticky liquids. These go to fabricators who change the polymers into finished products by moulding, extruding, and other processes. The engineering firms which make the special equipment used in these processes are also part of the plastics industry.

The usual way of making finished articles from polymers is some form of heating and moulding under pressure. The methods used depend on whether the polymer is thermosoftening or thermosetting.

Plastics processing is an important industry in Britain. There are some big companies, but also a very large number of small ones. Many companies specialize in only one or two of the processes discussed here. You may find more variety in bigger organizations. Whatever the company, you will usually meet with the same sort of workers.

● *Machine operators*
They start and stop the machine, feed it with raw materials, and take off the finished goods. They may trim, inspect and pack them.

● *Machine minders*
You don't make all machines start and stop by hand. Some are automatic. The operator's job may be just watching, and checking all is well. That is often called "machine minding".

● *Machine setters*
The same machine may make different sizes and shapes of goods. All the working parts have to be set correctly, to suit the measurements of a product, before a batch can be made. A machine setter does this, and also makes adjustments, if things don't seem to be working right.

● *Supervisors and managers*
Most supervisors know a lot about the operators' and setters' work. They probably did their jobs themselves, before they were promoted. Their main task is to see that work is done in the most efficient way, and that the highest quality of goods is produced. To do this, they have to be in charge of other workers, and have authority to say what should be done. Supervisors may progress to be in charge of a whole section of the works; they will then probably be called "managers". Managers have a say in decisions about how the company is run and what is made.

Vacuum forming

Rosalind makes the covers for indoor electric display signs that advertise the

Figure 8.32
Rosalind at work.

names of shops and garages. They are made from a thermoplastic called "polystyrene".

Rosalind works a vacuum-forming machine.

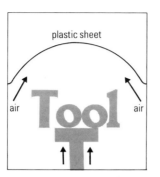

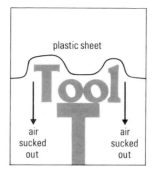

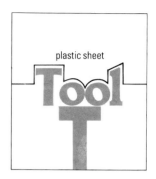

In the machine, a sheet of plastic is heated till it gets soft. Then it is blown up with air, like a balloon. A 'tool' is inserted into this bubble. That is a block of wood or metal shaped exactly like the finished sign.

Then a vacuum is turned on. All the air is drawn out of the bubble and it collapses. The plastic is sucked onto the surface of the tool.

It fits like a skin, taking on the tool's shape. As it cools, it sets.

Figure 8.33
Vacuum forming.

Looks complicated? Rosalind doesn't agree . . . .

"The machine does most of the work, while I watch. Some presses are fully automatic. On others, you need to switch the heater, blower, vacuum, and fan on and off.

"I suppose you could get bored, but I don't. Each job's different – the colour, the design, and the time it takes to mould. If I did get fed up, I could move to another department. I used to clean and pack the finished signs, before I moved onto vac. forming.

"I enjoy factory work, because I like being with other people. I once did a Saturday job in a shop, but I felt cut off from the other workers. Everyone's so friendly here. If there's a problem with a job, someone will help, so you have no worries. And we all go out together, to football, tennis, or to dances. I often do overtime, not so much for the money, as to be here. It's at home that I get bored."

Extrusion

When you work a mincer, you turn a screw that alters the shape and consistency of meat, by forcing it through little holes. A plastics extruder works in basically the same way. You put plastic powder or granules in one end, and then heat it. The plastic is shaped by forcing it through a metal nozzle, called the "die". While it is still soft, it passes through extra "formers", which adjust the shape. Then it is cooled, to harden it.

Plastic pipes, rods and "profiles" are often extruded. A profile is a shaped rod – for example a curtain rail. Extrusions like these are produced in continuous lengths.

An extruder takes a long time to heat up from cold. During that time, it's producing waste. You couldn't start and stop it every day, or the company would never make a profit. Once a week is the limit. That means there has to be a night shift, and possibly week-end working.

Figure 8.34
Colin working the extruder.

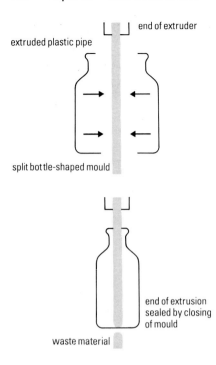

end of extruder

extruded plastic pipe

split bottle-shaped mould

end of extrusion sealed by closing of mould

waste material

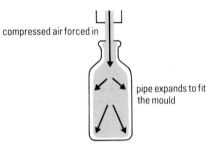

compressed air forced in

pipe expands to fit the mould

finished bottle cut off from extruded pipe

mould opens

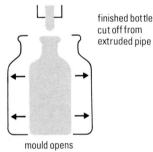

Figure 8.35
How blow moulding is done.

Colin's company works straight through from Monday to Friday. There are three shifts each day, and Colin changes shift every week.

He finds his job a challenge . . . .

"It's not just a case of putting stuff in one end of a machine, and watching it come out the other. You train for a month here even to make pipes. They're easier than profiles, which can be complicated shapes.

"When you first start up, you have to form an extrusion gradually to the proper shape and size, by adjusting the way the machine is set, and getting the temperature right. It has to be supported as it comes out of the die and formers, because it is still warm, and can sag out of shape. I also control the water-cooling system, and set the automatic saw. I don't touch the wires or mechanics of the machine, though – that's the engineer's job.

"Speed's important. While you're setting up, you're not producing, and we get a bonus for high production. Every shift has to pull its weight.

"Once everything's set, you still can't doze off. You keep checking the measurements of the extrusion with a micrometer, in case they vary. When there is trouble, you have to pinpoint what's going wrong. Experience comes into that. It's good when you spot a fault you've seen before, and you can remember what you did to put it right. But there's also satisfaction in coming across something new."

Blow moulding

Extrusion is often the main part of longer processes. One of the easiest to understand is blow moulding. That is the way you make a plastic bottle – any shape and size, from a tiny perfume phial to a gallon bleach container, from a bottle of engine oil to one that holds a saline drip for hospital patients.

A blow moulding machine is an adapted extruder. You clamp a split, bottle-shaped mould round an extruded plastic pipe, sealing off the end. Then you blow compressed air into the pipe, which expands like a balloon, till it fills the mould. Cut it off at the other end, let it cool, open up the mould, and there's your bottle.

Machine operators look after one or more machines, depending on whether they have to start and stop them, or whether the process is fully automatic. When the mould opens, they remove the finished bottle. Usually, there is a ridge of waste plastic round it, where the two halves of the mould joined. They must cut that off with a knife and inspect the bottle for faults. Some operators learn to do this at amazing speed. They may get a bonus for high production.

Injection moulding

An injection moulding press is like a giant syringe. It has a long barrel, where plastic granules are heated up till they melt. A rotating screw then forces the hot material through a small hole into shaped moulds. When the plastic has cooled and set, the moulds open. The operator takes out the mouldings, trims and packs them.

Like an extruder, an injection moulding press needs several hours to warm up, before it will produce good results. So injection moulders can expect to work shifts.

Magid always works days, as his company has a separate night-shift. He started as a press operator two years ago . . . .

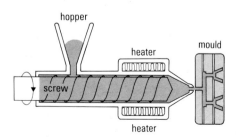

hopper

heater

mould

screw

heater

Figure 8.36
How an injection moulding machine works.

Figure 8.37
Magid.

"I was glad when I got the chance to train for setting. There's more to it than operating. It can be quite an art to get the press working right.

"The presses here are fairly small. Setting them isn't difficult work, once you've got the hang of it, but it needs a lot of concentration. It can take me between two and seven hours to set a machine. It depends how complicated the shape of the tool is. Then the pressure and heat have to be adjusted to suit different materials.

"I have to see that my presses turn out good work. I go to Quality Control if I'm doubtful about standards. If they say a customer will accept something, that's OK. If they give me the thumbs down, then it's back to work – re-setting the machine till I make an improvement.

"I recently became a chargehand here. I'm still a setter, but now I'm in charge of workers, as well as machines. I share out the work, and move the operators to whichever machine suits a particular job."

You can injection mould plenty of different shapes. Magid's firm makes office equipment, car accessories and bottle caps, and also some household insecticide and air freshener packs. They don't just mould the containers: the scented bit in the middle can be made of plastic too.

Compression moulding

Victor works in compression moulding. That uses the old sealing-wax principle – heat up the wax, then impress it hard with a tool that's shaped exactly how you want your seal to look. Victor works fast. He puts hot plastic pellets in his machine. Down comes the tool under high pressure, and bingo – ashtrays, light-bulb sockets or electric plugs. Dead easy. Or is it? . . . .

Figure 8.38
Victor.

"Well, it's not **that** simple. For a start, you wouldn't make anything out of wax that had to withstand heat, would you? For the same reason, you don't use thermoplastics either. Just try marketing the amazing melting ashtray. We use thermosetting plastics. They soften with heat, so they can be moulded, but further heating causes a chemical reaction. That makes them hard, and they can't be melted again. It's called 'curing'. The timing and temperature of the process, and the weight of the pellets vary with each job and each type of plastic.

"The chargehand sets up the machine, but I'm responsible for the temperature of the pre-heater. That's where I heat the pellets before they go into the press. You have to learn how to adjust the heat for each job by experience. I thought at first I'd never pick it up. I made so much waste. I'm confident now, though.

"I put the pellets in the mould, and fix into the press. I close the press to form the moulding, but after a short while, it has to be opened slightly, so the plastic can 'breathe'. That means that any gases caused by the chemical reactions are allowed to escape. When the moulding is cured, I take it out of the press, and check it. I clean off waste material, and clear any holes or slots with a drill or punch. Sometimes metal inserts are moulded into the plastic while it's being cured in the press."

Some presses at Victor's firm are fully automatic, including the pre-heater. The machine even loads itself. One operator runs several machines. He just sees that the hoppers are full of pellets, and that the finished goods are all right.

Figure 8.39

Among the bigger firms that use compression moulding are record companies. Although some singles are injection-moulded nowadays, LPs are still mass-produced by compression moulding, after the master disc has been cut. The moulding has to be very accurate, to get a good quality of sound reproduction when the record is played.

Summary

1 Some of the important terms and processes used in this chapter are listed below. Write a brief explanation of each and, where possible, give an example to show what you mean.

Polymer, addition polymerization, condensation polymerization, natural fibre, synthetic fibre, thermosoftening plastic, thermosetting plastic, cross-linking, extrusion.

2 Draw up a table comparing the properties of polymers with other materials such as metals, ceramics and glasses. Look at figures 8.40 and 8.41 for ideas. Discuss the advantages and disadvantages of the materials you compare.

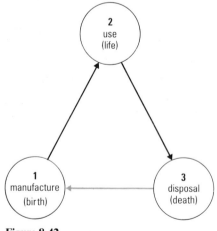

Figure 8.40

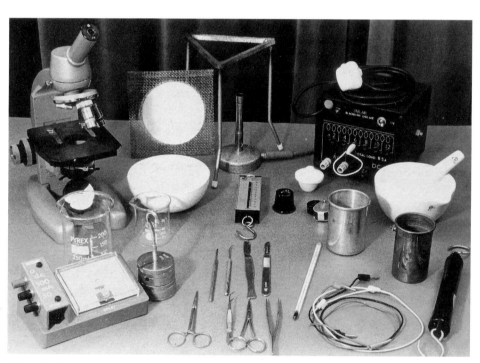

Figure 8.41

Figure 8.42
Three stages of the life of a plastic article.

3 There are problems involved in the use of plastics. Imagine the "life" of a plastic bag, or a plastic mac, or any other plastic article as having three phases as shown in figure 8.42. Make a list showing some of the disadvantages, at each stage in their "life", of using plastic objects.

Topic C3

Chemicals in our homes

Chapter **C9**   **Foams, emulsions, sols and gels**   164
Chapter **C10**   **Keeping clean**   178
Chapter **C11**   **Dyes and dyeing**   201
Chapter **C12**   **Chemicals in the medicine cupboard**   213

In this topic you are going to take a fresh look at some of the things you take for granted in your home: pure water, food, cosmetics, soaps and detergents, coloured fabrics and medicines. What are they? Where do they come from? What do they do?

You will learn more about the importance of chemical analysis. How can we check the information on the labels of supermarket products? In the laboratory you will learn how to analyse one or two products. You will find out for yourself how chemists can check that the products we buy are what the manufacturers say they are.

As you study this topic you will also be introduced to some new chemical theory. This will help you to understand why it can be difficult to get clean using soap in hard water areas. It will give you a better understanding of acids and what happens when an acid is neutralized by an alkali.

Cosmetics are made from chemicals.

Chapter C9

Foams, emulsions, sols and gels

*You may be wondering why foams, emulsions, sols and gels are grouped together in the title of this chapter. As you read it, you will discover that they are all examples of colloids. You may not have met the term "sol" before but you will know about "aerosols" if you have read Chapter **P**2.*

*Milk is an everyday example of a colloid. Worksheet **C**9A will give you some practical experience of the properties of colloids.*

In this chapter you will have to think carefully about some of the language used. You will meet a number of new technical terms. You will also meet words which have a more precise meaning in science than they do in everyday life.

C9.1 What are colloids?

Try to imagine what the world would be like if there were no clouds, no brilliant sunsets, no mist on hills or fog on city streets. Think how different your diet would be if there were no butter or milk, no ice cream and jellies, no salad cream, no sauces, no bread and no cake. How would you be affected if there were no cosmetics, and no medical creams? What would your home look like if there were no paint? How would you manage if you could not clean up your hands and clothes by washing with soap or detergents?

Figure 9.1
Most dairy products are colloids. Butter is a solid emulsion.

Figure 9.2
Emulsion paint is a colloid.

Figure 9.3
Cosmetic and medical emulsions.

Figure 9.4
Pumice is a natural, solid foam.

Figure 9.5
Morning mist in a Hertfordshire village.

One way or another, many of the things which we take for granted involve colloids.

Colloids muddle up the simple idea that everything is either a solid, a liquid or a gas. A colloid is made by mixing two things which cannot mix! This sounds stupid until you look at some examples and think about the meaning of the word "mix".

Figure 9.4 shows a piece of pumice which is a volcanic rock. In pumice the air is spread through the rock trapped as little bubbles. This is an example of a *solid foam*. You can see that the air bubbles are thoroughly blended with the stone. But if you think about the atoms and molecules you will see that the two parts are separate. The air molecules have not mixed with the atoms and ions in the rock.

Figure 9.5 shows a mist in a village in Hertfordshire. The mist consists of minute droplets of water suspended in the air. The air and water are all muddled up together but the water molecules in the water drops are not mixed with the air molecules. Mist is an example of an *aerosol*.

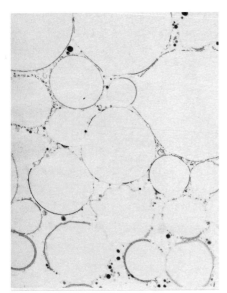

Figure 9.6
Fat droplets in milk under the microscope
(× 5000).

Milk is a more complicated colloidal system. Figure 9.6 shows milk under the microscope. What you see are fat droplets spread through a watery liquid. The average diameter of the fat droplets is 3 to 4 µm (see question **3** on page 168). Larger fat droplets may cluster together and separate out as a cream layer when milk is left to stand.

Under even more powerful magnification, you can see that there are other particles in milk too. Some of these are included in figure 9.7 which shows the small protein particles of casein. The average size of the casein particles is 0.4 µm.

Here we are giving the word "particle" its everyday meaning. The particles are little bits of a solid or small droplets of a liquid. The same word "particle" is also used in science to mean "atom, molecule, or ion" which is the way the word is used in Chapters **P**2, **P**10, and **C**5.

Emulsion paint is an example of a man-made colloidal system. A white emulsion paint may have three types of particles suspended in water. You can see pictures of the particles magnified 4000 times in figure 9.8. In a white paint all three sets of particles are mixed up with water but not dissolved in it. The particles are scattered about in the water, so we can say that they are *dispersed*.

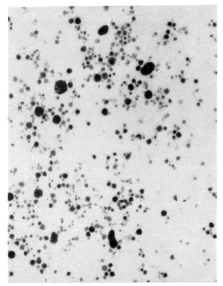

Figure 9.7
Particles of casein in milk (× 15 000).

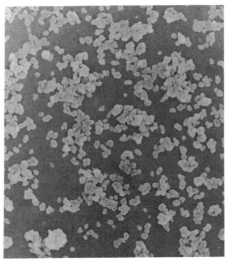

a

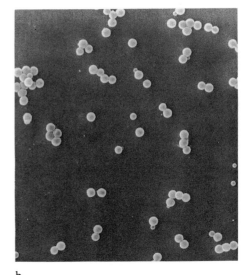

b

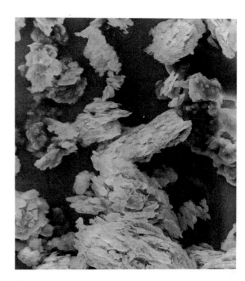

Figure 9.8a, b, and **c**
Titanium dioxide, latex and china clay particles in emulsion paint under the microscope (× 4000).

c

With emulsion paint we have another example of a word which has a different meaning in science from its meaning in everyday use. In chemistry the word *emulsion* is only used for colloids in which one liquid is dispersed in another liquid as fine droplets. Salad cream is an emulsion of vegetable oil in vinegar. Milk is an emulsion of fat droplets in water.

In paint the dispersed particles are solid. So paint is not truly an emulsion. Paint is an example of the type of colloid called a *sol*.

Photographic "emulsions" are also examples of colloids which are not really emulsions. The light-sensitive layer of a film consists of fine particles of silver bromide trapped in a gelatine *gel* (figure 9.9). Figure 9.10 shows that the light-sensitive layer is supported on a film of cellulose acetate.

Figure 9.9
Electronmicrograph of undeveloped silver bromide crystals in a photographic emulsion (× 7000).

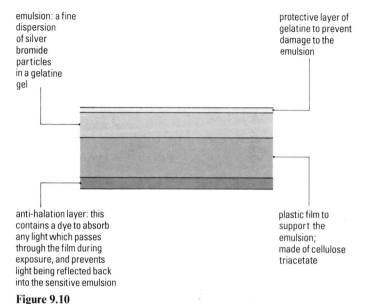

emulsion: a fine dispersion of silver bromide particles in a gelatine gel

protective layer of gelatine to prevent damage to the emulsion

anti-halation layer: this contains a dye to absorb any light which passes through the film during exposure, and prevents light being reflected back into the sensitive emulsion

plastic film to support the emulsion; made of cellulose triacetate

Figure 9.10
Cross-section of a photographic film.

A gelatine gel is made with warm water. Gelatine is a protein. Proteins are natural polymers, and the molecules of a protein are very large (see section **C3.3**). The large molecules disperse in water to form a sol (figure 9.11). As the gelatine-in-water sol cools, the gelatine molecules are attracted to each other and form a continuous network. In this way, a gel is formed in which water droplets are trapped in the protein.

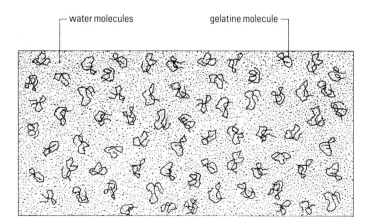

water molecules   gelatine molecule

Figure 9.11
The molecules in a gelatine sol.

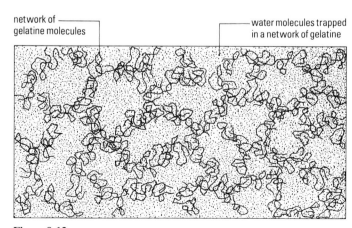

network of gelatine molecules

water molecules trapped in a network of gelatine

Figure 9.12
The molecules in a gelatine gel.

There are eight different types of colloid made by dispersing fine particles of one substance in another. These are listed in figure 9.13.

Continuous phase	Disperse phase	Type	Example
gas	gas	–	–
gas	liquid	aerosol	mist
gas	solid	aerosol	smoke
liquid	gas	foam	whipped cream
liquid	liquid	emulsion	hand cream
liquid	solid	sol	paint
solid	gas	solid foam	pumice
solid	liquid	solid emulsion	butter
solid	solid	solid sol	pearl

Figure 9.13
Types of colloidal systems.

Every colloid has at least two parts. One part is the *continuous phase* like the water in milk or the air in mist. The other part is split up into minute particles like the fat in milk or the water droplets in mist. These scattered particles make up the *disperse phase*.

The word *phase* is often used in the sciences but it can have different meanings. In biology the stages in the life cycle of animal are sometimes called phases. In a similar way astronomers talk about the phases of the moon. These are both examples of the use of the word phase to mean a period of time.

In chemistry and material science, the word is used in another way to describe different substances which are present together but separated by a clear boundary. In fibre glass there is a glass phase (the fibres) bound together by a separate plastic phase (the polymer resin). In a colloid there are particles of a disperse phase scattered through a continuous phase.

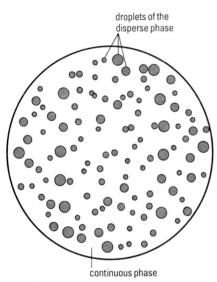

droplets of the disperse phase

continuous phase

Figure 9.14
Continuous phase and disperse phase in a colloid (drops of water in oil).

1 Why is it impossible to make a colloid from two gases?

2 Classify the following colloids according to the types listed in figure 9.13: meringue, fog, salad cream, clouds, the "head" on a glass of beer, a jelly baby, muddy river water, whipped cream, the spray from a pressure can of insecticide, a sponge, a crunchy toffee bar, bread, chocolate mousse, hand cream, lipstick, mascara, a rubber pillow, toothpaste, a bubble bath, cold custard, non-drip paint.

3 In each of figures 9.8 a, b and c and 9.9, measure the distance across some of the particles. Use the magnification given in the captions to work out the sizes in micrometres (μm). 1 μm $= \frac{1}{1\,000\,000}$ m. Compare your answer with the size of atoms. The diameter of an atom is about 0.1 nm $= \frac{1}{10\,000}$ μm.

4a Could you draw a continuous line from the top of figure 9.14 to the bottom without touching an oil droplet?
b Could you draw a continuous line from top to bottom, staying in the oil and not touching any water?

5 When you first open a bottle of ketchup and turn it upside-down, nothing pours out onto the plate. To get some sauce you have to bang on the bottom of the bottle. Suggest an explanation for this. (Figures 9.11 and 9.12 may give you a clue.)

BOX 1   Problem
Why won't the pineapple gel?

APPLY   PLAN

If you try to make a fruit jelly using fresh pineapple and gelatine you will find that no gel forms. It is believed that this is because there is an enzyme in the fruit which breaks down protein molecules.

6 Why might an enzyme which breaks down protein molecules prevent a gelatine gel forming?

7 Suggest a series of experiments which you could do to test the theory that it is an enzyme in pineapple which prevents the gel forming. (Remind yourself of the properties of enzymes in Chapters **C3** and Biology **B5**.)

8 How could you treat fresh pineapple in a kitchen to make it possible to use it in jelly?

C9.2   How can we recognize a colloid?

*You may have heard of the use of dialysis to treat kidney patients (see section **B**12.5). In this section you can find out how the process works.*

Light scattering

Egg white is transparent (figure 9.15). So is the air. Now look at figure 9.16 which shows what happens when egg white is whisked to produce a foam of air bubbles trapped in the egg. The foam looks white, and you cannot see through it.

Figure 9.15
Egg white in a glass bowl before whipping.

Figure 9.16
Whisked egg white.

This illustrates an important property of colloids: they scatter light. That is, when a light ray passes through a colloid it is reflected or refracted every time it meets one of the droplets or particles.

Milk looks white because of the scattering of light rays by colloidal particles of protein and fat. Mist and fog are dangerous on the roads because motorists cannot see through them. The light scattering is particularly

Figure 9.17
The effects of a sugar solution and a solution of egg white on a light beam – note the difference.

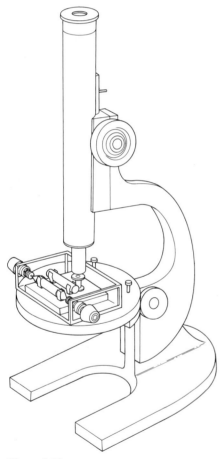

Figure 9.18
How to demonstrate Brownian motion with smoke particles.

obvious at night when car headlight beams show up brightly as the rays are diffused by the water droplets.

One of the test-tubes in figure 9.17 contains sugar solution, the other contains a diluted egg white sol. The light beam shows up the difference between a solution and a sol. The colloidal particles in the sol scatter light. The dissolved sugar molecules do not.

Light scattering gives rubies and emeralds their colour. The colour of some stained glass is caused by colloidal particles in the glass.

Blue light is scattered by small particles much more than red light. When the sun is low on the horizon, the light which passes through to someone watching the setting sun looks red because much of the blue light has been scattered. Sunsets are often particularly beautiful after a volcanic eruption when there is much more dust in the air than usual.

Brownian motion

Smoke is a colloid. Smoke particles are too small to see directly with an ordinary microscope, but their movement can be observed using the apparatus shown in figure 9.18. When you look down the microscope you see bright points of light wherever there is a smoke particle to scatter the light up into the lenses.

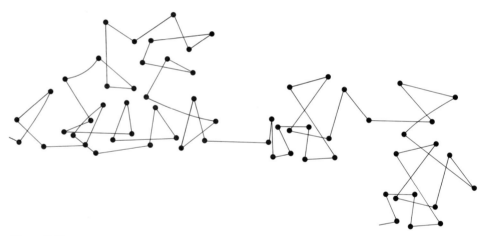

Figure 9.19
The random motion of colloidal particles – much enlarged.

Even if there are no air currents, the smoke particles are not still. They can be seen to jiggle about. This movement was first seen by Robert Brown (1773 – 1858) when he examined pollen grains in water. The random movements are called Brownian motion.

We now think that Brownian motion happens because fast moving but invisible molecules collide with the colloidal particles. This is one of the most direct pieces of evidence for the particles-in-motion (kinetic) theory.

Dialysis

The colloidal particles in a sol are much bigger than the molecules and ions which will dissolve in water; but they are smaller than the particles in a *suspension*. Look at figure 9.20. The tube on the left contains a solution of sugar in water. There is no sign of any particles and nothing settles. The tube on the right contains a milk of magnesia suspension which was shaken a

Figure 9.20
Sugar solution, colloidal clay and milk of magnesia.

minute or two before the photograph was taken. In this short time the particles have already begun to settle at the bottom. The middle tube contains colloidal clay which has been standing for some time. It looks slightly cloudy because the particles scatter light, but there is no sign of the particles settling.

The difference in size between dissolved molecules and colloidal particles is demonstrated by the experiment described in section C3.2. Turn back and look at figure 3.16. This shows starch and glucose in water inside Visking tubing. The starch molecules are very big, so big that starch "solution" is really a sol. You can see that this is so if you hold some starch "solution" in a light beam. It scatters light like other colloids.

Starch molecules are small enough to pass through the small holes in filter paper, unlike the particles in a suspension which get caught.

However, starch molecules are too large to pass through the small holes in Visking tubing. Glucose molecules are much smaller and can get through. In this way small dissolved molecules can be separated from colloidal particles. The process is called *dialysis*.

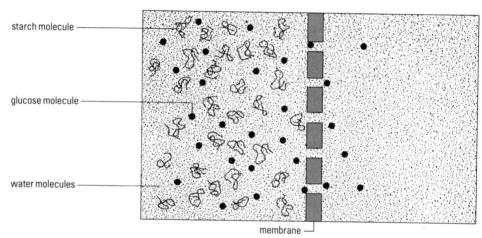

Figure 9.21
Dialysis separates small molecules from colloidal particles.

Dialysis is used to prolong the life of people suffering from kidney failure. The kidneys are an efficient mechanism for separating the soluble waste products of the body from the blood. One of these waste products is urea which is normally excreted in urine. Kidney failure leads to a rapid build-up of urea in the bloodstream, and this can cause death unless some other means is found to remove it.

C9.3   How do we make colloids?

Mixing ingredients to make something new is essential to cookery. It is also important in chemistry. In the food industry, additives are included to preserve food and to make it taste better. You can compare the food additives mentioned in this section with the ones mentioned in Biology Chapter B6.

Worksheet C9C gives you a chance to make a cosmetic emulsion. You can compare the quality of your product with the cosmetics you can buy.

Some colloids are quite easy to make, as you will know if you have ever made a jelly with gelatine or a white sauce with flour. Sometimes it is not so easy. If you shake up oil and vinegar to make a salad dressing they quickly separate.

Figure 9.22
Three jars containing oil and vinegar. One has not been shaken, one has just been shaken, and the third has been shaken and then allowed to stand for a minute or so.

Shaking breaks up the oil into little droplets. Unfortunately the droplets soon join up to form bigger droplets and this continues until the two liquids are separated into two layers again. This separation occurs because of *surface tension* (see figure 9.23). Surface tension causes liquids to try to have the minimum surface area. When two droplets collide and combine to form a single larger droplet the total surface area is less. (If you answer question **9** you will discover that the surface area is much greater when something is broken up into small particles.)

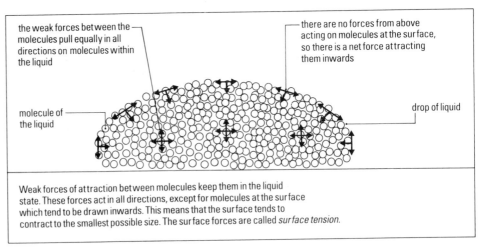

Figure 9.23
Why surface tension occurs.

Within an emulsion there is a very large surface area. The small droplets tend to join together to make bigger ones. To stop this happening and make the emulsion stable, something has to be added which will lower the surface

Figure 9.24
A soap film.

Figure 9.25
Pond skaters take advantage of the unusually high surface tension on water.

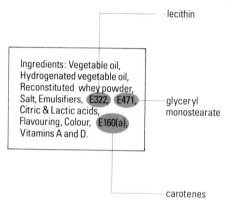

Figure 9.26
E numbers on a tub of soft margarine.

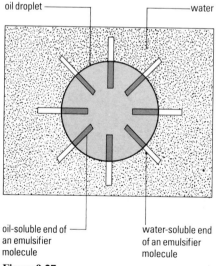

Figure 9.27
How emulsifiers stabilize an emulsion.

tension at the boundary between the two liquids. The substances which are used in this way are called *emulsifiers*.

Mayonnaise is an emulsion of vegetable oil in water (in the vinegar). The emulsion is made to last by including egg yolk in the recipe. The yolk contains an emulsifier called lecithin. Food manufacturers usually use lecithin obtained from soya bean oil because it is cheaper. Another emulsifier which is often used by food manufacturers is glyceryl monostearate (GMS). This is a compound which is formed naturally in your body when you digest fats. It can be made of glycerol and stearic acid. Emulsifiers like these are called "additives".

You can see which additives are present in the food you eat if you know their E numbers and look at the labels (see figure 9.26). The additives used to make emulsions have E numbers in the range E322 to E494.

A molecule of an emulsifier has the special property that one end is attracted to water while the other end is attracted to oil. When an emulsifier is added to a mixture of an oil and water, the molecules make for the boundary between the two liquids so that they can be partly in the water and partly in the oil. This lowers the surface tension between the two liquids and stabilizes the emulsion.

Stabilizers also help to make food emulsions. They increase the thickness of the mixture and this stops the oil droplets coming together. Natural polymers are used including starch. Other carbohydrates with long chain molecules are used too, including agar (E406) and carrageenan (E407) which are extracted from seaweeds. Alternatively gelatine may be added or natural gums such as locust bean gum (E410) or gum tragacanth (E413). You will notice that the emulsifiers and stabilizers added to food mainly come from natural sources.

Figure 9.28
Seaweed – source of carrageenan.

Many cosmetics are also emulsions. These are of two types, as explained in figure 9.29. Foundation creams, hand cleaning creams, and brushless shaving creams are all usually oil-in-water emulsions. Cold creams and cleansing creams are water-in-oil emulsions.

The diagram in figure 9.30 shows how a cleansing cream can be made. In this example, the oil phase is a mixture of beeswax, liquid paraffin and stearic acid. The liquid paraffin is not the same as that used as a fuel. It is a mixture of hydrocarbons which is refined enough to be used as a medicine. The water phase is a solution of borax. Borax reacts with stearic acid and beeswax to produce the emulsifiers needed to stabilize the emulsion.

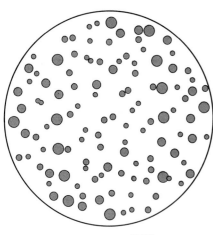

An oil-in-water (O/W) emulsion

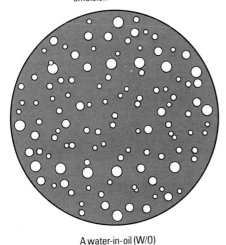

A water-in-oil (W/O) emulsion

Figure 9.29
The difference between oil-in-water and water-in-oil emulsions.

9a What is the surface area of a cube of side 1 cm?

b Suppose that the cube of side 1 cm is broken up into many smaller cubes of side $\frac{1}{10\,000}$ cm. How many of these cubes are formed?

c What is the surface area of one of the small cubes formed in **b**?

d What is the total surface area of all the cubes formed in **b**?

10 Look at food labels in your kitchen and make a list of all the foods which contain emulsifiers or stabilizers which you can identify with the help of the E numbers given in this section.

11 Use reference books to find out the natural sources of these food additives: alginic acid (E400), agar (E406), guar gum (E412), gum tragacanth (E413), pectin (E440a), sodium carboxymethylcellulose (E466).

12 What are the advantages of turning the ingredients used to make cosmetics into an emulsion? (See figure 9.31.)

13 Oil slicks at sea are sometimes dispersed by spraying them with powerful emulsifiers. Should the emulsifier be chosen to make a water-in-oil emulsion or an oil-in-water emulsion?

14 You may have noticed that energy is needed to create colloids. The methods for making them involve shaking or stirring, grinding or whisking. Why do you think that this is so? (Think about the forces between molecules, and look at figure 9.23.)

solution of borax in water at 75 °C

mixture of beeswax, stearic acid and liquid paraffin at 75 °C

Figure 9.30
Making a cosmetic cream.

Figure 9.31
The ingredients used to make the cosmetic cream in figure 9.30.

C9.4   How can we destroy colloids?

*When you have studied this section you will find it easier to understand some of the stages in the treatments used to purify water and dispose of sewage described in Chapter **C10**.*

*An understanding of colloids can help to solve some pollution problems. So there are links between this chapter and Biology Chapters **B17** and **B18** which deal with human influences on the environment.*

*This section also illustrates the usefulness of an understanding of ions and the forces between charged particles. The theory of ions is described in Chapters **C5** and **C18**. The forces between charged particles are investigated in Chapter **P16**.*

Fog on motorways, smoke from factory chimneys, impurities in our water supply, and solids in sewage are examples of colloids which we may want to try to get rid of. To destroy a colloid we have to make the dispersed particles join together and separate out from the continuous phase.

Figure 9.32 illustrates another example. If you are living through a drought it can be frustrating to see clouds passing overhead carrying masses of water as a stable colloid. In recent years much research has been done worldwide in attempts to control the weather by modifying the behaviour of clouds.

Figure 9.33 provides a clue to understanding how the two phases of some colloids can be made to separate. The picture shows the point where the

Figure 9.32
The hole in this thick cloud layer was produced by seeding it.

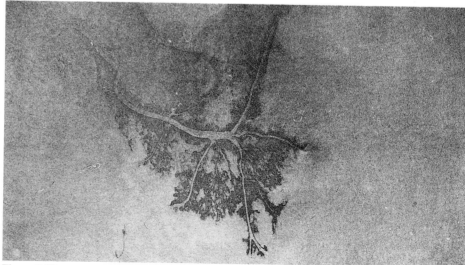

Figure 9.33
The delta at the mouth of the Mississippi.

BOX 2 **Experiment**
Colloidal particles and electric charges

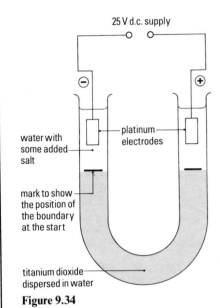

25 V d.c. supply

⊖   ⊕

platinum electrodes

water with some added salt

mark to show the position of the boundary at the start

titanium dioxide dispersed in water

Figure 9.34

Figure 9.34 shows an apparatus which can be used to see what happens to colloidal particles in an electric field.

Titanium dioxide is dispersed in water as in emulsion paint (see figure 9.8a). Marks on the outside of the U-tube show the positions of the boundaries between the white titanium dioxide sol and the water at the beginning of the experiment. Then the electrodes are connected to a d.c. supply of electricity.

Slowly, the boundary on the left moves down and the boundary on the right moves up. So the titanium dioxide particles move away from the negative electrode towards the positive electrode. All the particles move in the same direction.

15 What evidence is there from this experiment to show that the particles of titanium dioxide are electrically charged and that they all carry the same charge?

16 Is the charge on the particles negative or positive?

17 How do you think that the charges on the colloidal particles help to prevent them from coagulating?

18 The titanium dioxide particles coagulate, separate, and sink soon after a solution of magnesium sulphate is added to the sol. Suggest a reason for this. (A solution of magnesium sulphate contains magnesium ions, Mg^{2+}, and sulphate ions, SO_4^{2-}.) Can you offer an explanation for the fact that adding ions to a colloid can make the particles coagulate?

Mississippi river meets the sea. The river carries with it clay and other particles as a colloidal sol. When it flows into the sea the sol is destroyed. The particles clump together, settle out, and are deposited as silt. This happens at the mouth of all rivers, but in strongly tidal areas the silt is carried away by the waves and no delta forms.

Lime is used on clay soils. The particles of clay join up to form larger particles when lime is added. This makes the clay easier to work and improves the drainage. The process of making small particles join up to form bigger ones is called *coagulation*.

Aluminium chloride is included as an ingredient of deodorants. It coagulates the proteins in sweat so that they block the pores of the sweat glands. This prevents B.O. (body odour) which is produced by bacteria decomposing proteins on the skin.

Sodium chloride in sea water, calcium hydroxide in lime, and aluminium chloride are all ionic compounds. Many colloids are affected by added ions. The ions cause the dispersed particles to join together and separate from the continuous phase.

19 Write down the symbols of the ions in each of the following compounds with the help of the Data section:
sodium chloride, calcium hydroxide, and aluminium chloride.

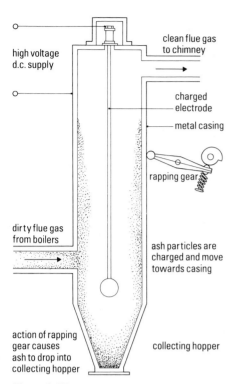

Figure 9.35
How an electrostatic precipitator works.

high voltage
d.c. supply

clean flue gas
to chimney

charged
electrode

metal casing

rapping gear

dirty flue gas
from boilers

ash particles are
charged and move
towards casing

action of rapping
gear causes
ash to drop into
collecting hopper

collecting hopper

Figure 9.36
Car body being painted by an electrostatic process.

Why does adding ions affect colloids? Ions are electrically charged atoms. Does this mean that the particles in colloids are charged too? The experiment described on box 2 investigates these questions.

It shows that some colloidal particles are electrically charged and can be attracted or repelled by a charged electrode. There are a number of important industrial applications of this fact. For instance, electrostatic precipitators are used to prevent smoke and dust particles escaping from factory chimneys.

The same principle is used to make sure that car bodies are thoroughly coated with paint. The paint droplets in the aerosol spray are charged. The metal parts of the car body are given an opposite charge. This ensures that paint is attracted into all the crevices, giving a complete coating.

Natural rubber latex is a sol in which the rubber particles are negatively charged. The rubber can be coagulated on the surface of a positively charged mould. This technique is used to make rubber goods such as surgical gloves and contraceptive sheaths (condoms). It is important to control the thickness of the rubber and to ensure that there are no holes. Any gap in the deposit of rubber is soon filled because of the electrical attraction between the rubber particles and the mould.

20 Clouds are an aerosol of water droplets in air. Suggest an explanation for the observation that heavy rain often follows flashes of lightning.

21 Suggest a connection between these two statements.
"Stream water in limestone regions is often very clear."
"The water flowing from limestone hills contains calcium ions in solution."

22 Fog and mist cause fatal accidents on motorways. How would you set about trying to develop a method for getting rid of these hazards?

23 Suggest a reason for the fact that alum, which contains Al^{3+} ions, is much more effective for coagulating some colloids than salt, which contains Na^+ ions.

Summary

1 Copy and complete the table in figure 9.37, giving at least one example of each of the types of colloid shown in figure 9.13. For each example, state what the two phases consist of.

Example of a colloid	Type of colloid	What does the continuous phase consist of?	What does the disperse phase consist of?

Figure 9.37

2 Describe:
a two ways of preventing colloidal particles coagulating; and
b two ways of making colloidal particles coagulate.

Chapter **C**10

Keeping clean

This chapter shows how chemistry helps to keep us clean and healthy. Soap only became widely available once chemists had discovered ways of manufacturing alkalis as described in Chapter C4. In this chapter you can find out how soaps and other detergents work. You will carry out more detailed studies of washing and dry cleaning if you are taking a course in Home Economics.

Alkalis are manufactured from sodium chloride and chlorine is made at the same time. One of the uses of chlorine is to treat drinking water to make sure that it is free of harmful bacteria. As you read about the chemistry of water treatment and the control of water pollution in this chapter you can relate the information to any work you have done on the water cycle in geography.

C10.1 Why do we get so dirty?

If you go for a run, mend your bicycle, or work in the garden you expect to get dirty. But even if you do none of these things you will still have to wash yourself and your clothes. So where does all the dirt come from?

Figure 10.1
Rugby is a muddy business!

We live in a cloud of dust. There are tiny particles of dirt suspended in the air. You do not usually notice the dust except when a ray of sunlight shines

Dry cleaning

The symbols in the circle indicate the type of dry cleaning solvent which can be used on the fabric. If you use a coin-operated machine make sure that the symbol on the fabric label is the same as that on the machine.

(A) These goods can be dry cleaned in any solvent.

(P) These goods can be dry cleaned in perchloroethylene, white spirit, solvent 113 and solvent 11.

(F) These goods can be dry cleaned in white spirit and solvent 113.

(⊗) These goods should not be dry cleaned.

Figure 10.2
Codes for dry cleaning fabrics.

into a room or when it scatters light from the beam of a projector. The dust settles on furniture and gets caught on your clothes.

Dirt particles stick to your clothes held there by grease and sweat. Added to this are the flakes of skin which rub off and catch in the fibres of cloth. There may also be stains from food or drink and accidental spills of paint or cosmetics.

So grease and dust are the main causes of soiling. An obvious way to clean clothes is to use something which will dissolve the grease. If the grease is removed perhaps the dust will go with it. This is the idea behind "dry cleaning".

The word "dry" is not really correct, but is used because the grease is dissolved in a solvent which is not water. The main solvent used commercially is called perchloroethylene (see question **2**). It is gradually being replaced by trichlorotrifluoroethane (solvent 113).

Dry cleaning can be effective and is essential for some fabrics, but it needs expensive equipment and so is not practicable at home. Dry cleaning is carried out in special shops and is not a cheap way of cleaning clothes.

1a What does the word *solvent* mean?
b Which solvents do we use at home when cooking, cleaning, decorating, and gardening, and what are they used for?

2 Perchloroethylene molecules are related to eth**ene** (see section **C**2.5). All the hydrogen atoms in ethene are replaced by chlorine atoms which, like hydrogen atoms, form one bond with other atoms. Draw a diagram of a perchloroethylene molecule.

3 In a molecule of solvent 113, the hydrogen atoms of eth**ane** (see section **C**2.2) have been replaced by chlorine and fluorine atoms. Two of the chlorine atoms are attached to one carbon atom and one to the other atom. Draw a diagram of a solvent 113 molecule.

4 Which of the substances in the following list are soluble in water? Which of them would you expect to be soluble in a dry-cleaning solvent? Which of them are insoluble in both?
sugar, vaseline, salt, lard, chalk, lipstick, baking powder, clay.

C10.2   Why do we need detergents?

*We spend a lot of money on chemicals to keep us clean. Is the money well spent? This section shows you that a packet of washing powder contains more chemicals than you might expect. There is a good reason for including each of the ingredients. Your knowledge of molecules and ions (Chapter **C**5) as well as an understanding of colloids (Chapter **C**9) will help you to make sense of how soaps and detergents work.*

*Some washing powders include enzymes. Sections **B**5.3 and **B**5.4 in your Biology book describe the properties of enzymes. They will help you to explain why enzymes can be used to remove some stains.*

Water is cheap and easily available, but on its own it is not very good at getting things clean. Try washing your hands or clothes just in water and see how long it takes.

There are two main problems. The first one seems very odd, but the fact is that water is not very good at wetting things! This sounds ridiculous, but you can see from figure 10.3 that it is true. We often find that water will not wet materials by spreading out and soaking into them. The second problem is that water cannot dissolve grease or dust.

Figure 10.3
The water on the left contains detergent; the water on the right does not.

To get clothes clean with water we have to

- get the water to wet the fibres
- separate the grease and dust from the fibres
- suspend the dirt in the water so that it can be rinsed away.

Anything which will do these three things is a detergent. The word "detergent" means "something which cleans".

One of the successes of modern chemistry has been to discover a variety of substances which can be used to make effective detergents.

There are two types of detergents:

- soap detergents (usually just called "soap") which are made from animal fats or vegetable oils
- soapless detergents (usually just called "detergents") which are made using chemicals from oil

Figure 10.4 shows a model of a typical soap with its chemical formula. This is sodium stearate. Figure 10.5 shows a model of a typical soapless detergent also with its formula.

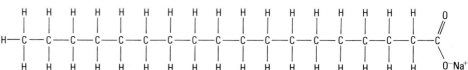

Figure 10.4
The structure of soap.

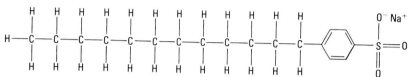

Figure 10.5
The structure of a soapless detergent.

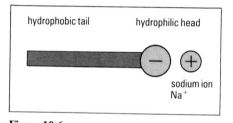

Figure 10.6
Simplified way of showing a detergent molecule.

You can see from figures 10.4 and 10.5 that there are two parts to a detergent molecule. There is a long part made of carbon and hydrogen atoms. This hydrocarbon chain is chemically very similar to the hydrocarbons in oil (see section **C2.2**) and will mix with grease or oil. It is grease-loving, water-hating, and said to be *hydrophobic*.

The shorter part of each detergent molecule has oxygen atoms in it and is ionic. Ionic salts generally dissolve in water. The ionic end of a detergent molecule is water-loving, so it is said to be *hydrophilic*.

You can read about another example of molecules which have one end which likes water and the other end which likes oil in section **C9.3**. We can explain how these emulsifiers and detergents work in terms of their molecular structure. The structure diagrams in figures 10.4 and 10.5 are complicated and it takes a long time to draw them; so from now on we will represent detergent molecules as in figure 10.6.

When a detergent is dissolved in water the ions separate and can be considered to be independent.

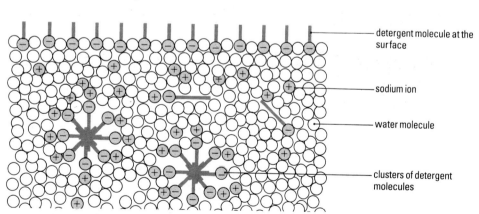

Figure 10.7
Two ways in which detergents behave in water.

You can see from figure 10.7 that the hydrophobic parts of detergent molecules have two ways of getting away from the water they "hate". Some molecules cluster together with their hydrocarbon chains inside the group. These clusters are as large as colloidal particles, so a detergent solution is really a sol. If you hold some soap solution in a projector beam you will see that it scatters the light.

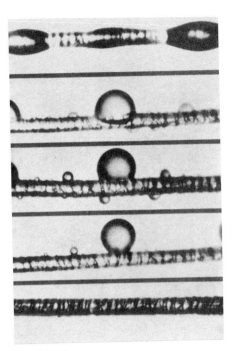

Figure 10.8
The removal of grease from a fibre.

Other detergent molecules in water make for the surface, and settle with their hydrophobic tails sticking out. This is very important, because it has the effect of lowering the surface tension of the water. This can then spread out and wet cloth better as shown in figure 10.3.

The pictures in figure 10.8 show stages in the removal of grease from a fibre. At first the grease is smeared along the fibre. When a detergent is added, the grease gradually rolls up into a ball and breaks away from the fibre. Figure 10.9 shows how we can imagine that detergent molecules help this process to happen.

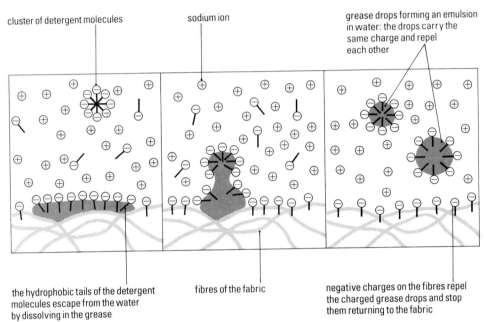

cluster of detergent molecules   sodium ion   grease drops forming an emulsion in water: the drops carry the same charge and repel each other

the hydrophobic tails of the detergent molecules escape from the water by dissolving in the grease

fibres of the fabric

negative charges on the fibres repel the charged grease drops and stop them returning to the fabric

Figure 10.9
How detergents separate grease from cloth. (Note that the water molecules are not shown.)

You can see from figure 10.9 that the grease forms an emulsion in water with the detergent acting as the emulsifying agent. Colloidal particles are small enough to pass through filter paper and they will also pass through the fibres in clothes. This means that when the washing is wrung out or spun dry, the water filters through the clothes, carrying the dirt with it. Rinsing with fresh water makes sure that no dirt is put back on the clothes when they are dried.

There are many different types of washing powder on sale and they do not all have the same washing properties. Some are more alkaline (pH above 7) and are good at removing grease. Some are neutral, to allow them to be used with delicate fibres or special fabric finishes. Washing powders are quite complicated chemical mixtures as you can see from figure 10.10.

The action of the *detergent* in a washing powder has been explained in this section. The *builders* include a form of sodium phosphate. They both help to soften the water (see section **C**10.4) and to remove clay particles. The presence of phosphates in washing powders creates a pollution problem which is described in section **C**10.5.

The *enzymes* break down proteins in the same way as the digestive enzymes in your small intestine (see Biology Chapter **B**5). The *bleach* added is in the form of sodium perborate which is most effective above 60 °C. The *blueing ingredient* and the *brightening agents* are designed to stop "whites" becoming yellow as they get old.

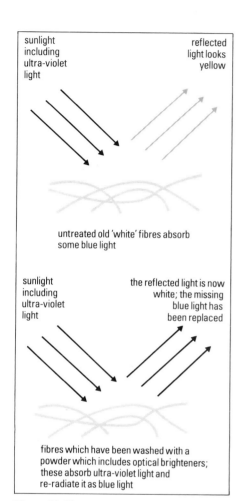

Figure 10.11
The action of optical brighteners.

Figure 10.12
Water droplets on a leaf.

synthetic detergents and soap
increase 'wetting power', loosen bond between fabric and dirt, and keep dirt in suspension

enzymes
break down protein stains (*e.g.* blood, gravy)

blueing ingredient
gives a 'blue-white' hue to white fabrics

metal protector
protects aluminium parts of washing machines

builder
(complex chemicals) helps cleaning process, softens water, and suspends dirt

brightening agents
make 'whites' and coloureds brighter

perfume
gives clothes a fresh, clean fragrance

oxygen bleach
removes stains such as black tea/coffee; gives added whiteness only at high temperatures and during long periods of soaking (so that at lower temperatures and short washes coloureds are not bleached)

Figure 10.10
Some of the ingredients in a brand of washing powder.

The brightening agents absorb ultra-violet light, which you cannot see, and then re-radiate the energy as blue light. This is the effect which makes clothes glow when ultra-violet lights are switched on at some parties. Old fabrics gradually turn yellow. Blue light added to the yellow light reflected from old fabrics makes them look white. You will understand this better when you have read the section on colour and the spectrum in Chapter **P**15 of your Physics book.

The common detergents not only stabilize emulsions; they also stabilize foams, which is why you usually get suds on the surface of the water when you wash. In an automatic washing machine the suds can stop the pump working; so washing powders for automatics are made using detergents which are just as good at removing and emulsifying grease, but do not produce foams.

5 Why do you think that it is generally easier to wash away grease in hot water than in cold water?

6 Why is it usual to agitate the clothes while they are being washed?

7 Why do the instructions for removing protein stains with enzyme washing powders say that you should soak the clothes at about 40 °C? Why must the water not be hotter than this? (You can remind yourself of the properties of enzymes by reading sections **B**5.3 and **B**5.4 of your Biology book.)

8 Why do the instructions for removing fruit juice stains from white cotton using a washing powder suggest that the material should be given a very hot wash?

9 Washing powders generally work better in hotter water, and the brands sold in the UK are ineffective below 30 °C. Heating the water uses more energy than any other part of the laundry process. Look at the table of instruction on the side of a packet of washing powder and suggest a temperature which is likely to be both economic and effective when washing most of the clothes used in your household.

10 If you have watched the preparation of a solution of a garden weedkiller or pesticide, you may have noticed that a foam appears on the water. This is because a detergent is included as one of the ingredients. With the help of figure 10.12, suggest a reason for including a detergent.

C10.3   How do we get pure water?

A safe water supply makes a huge contribution to our health. Typhoid, cholera, dysentry and infective hepatitis can all be spread if untreated sewage contaminates drinking water. Before this was recognized there were epidemics in big cities. Over 50000 people died in London alone during the cholera epidemic of 1831. In this section you can study the way in which chemistry is involved in water treatment.

*Fluoride is added to drinking water in some parts of Britain and other countries to reduce tooth decay. The reasons for this are given in Biology Worksheet **B4**E. The issue of fluoridation is still being debated vigorously in the press, particularly in local newspapers in areas where fluoridation is in operation, or being proposed. In this section you can find out about the chemicals used and the amounts which are added.*

The water cycle

You are probably already familiar with the water cycle. (See Biology Chapter **B**15.) Fresh water evaporates from the oceans, leaving impurities behind. The vapour is carried across the Earth's surface by winds. When the air is forced to rise as it meets land masses, or other air currents, the vapour is cooled and droplets of water form. These eventually fall as rain or snow.

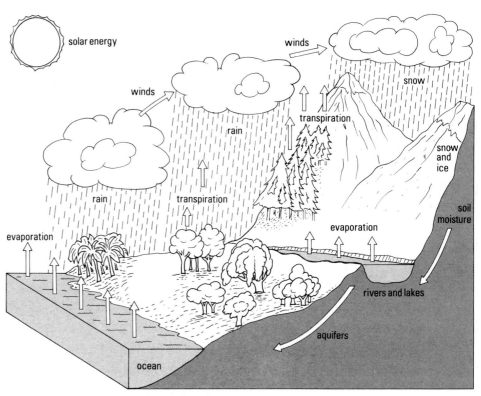

Figure 10.13
The water cycle.

The water flows back to the sea, evaporating as it travels. We can collect the water we need from rivers and lakes. Some water is pumped up from underground *aquifers*. Aquifers are huge natural reservoirs under the ground. The water is absorbed in porous rocks.

The purity of water constantly changes. As it falls through the air, the rain dissolves carbon dioxide and oxygen. Near the coast, sea spray adds salts to rain. In industrial regions, rain absorbs the sulphur dioxide and oxides of nitrogen produced by burning fuels. (See section C13.4 and Biology Chapter **B**17.)

Most of the oxides of carbon, sulphur and nitrogen are acidic. They react with water to form acids. Carbon dioxide and water combine to form carbonic acid, H_2CO_3. So the pH of rain water is usually below 7.

More carbon dioxide dissolves in water as it soaks through the soil. Water which contains dissolved carbon dioxide may react with rocks dissolving some of the minerals. This is important in areas where the hills are made of chalk or limestone (see section C10.4).

Sulphur dioxide combines with water and oxygen to make sulphuric acid. Nitrogen dioxide produces nitric acid in a similar way. Rain water which contains sulphuric acid or nitric acid is much more acidic than rain which just contains carbonic acid. This is the origin of *acid rain*. (See Biology Chapter **B**17.)

BOX 1   What makes water acid?

I
INTERPRET

Here are some true statements about acids and acid solutions:

- All acids have hydrogen in their formulae.
- Solutions of acids in water conduct electricity.
- When acid solutions conduct electricity, the product at the cathode is always hydrogen.

11 Look up the formulae of these acids in tables 4 and 5 in the Data section to check that they all have hydrogen in their formulae: methanoic acid, ethanoic acid, nitric acid, sulphuric acid.

12 Draw a diagram of an apparatus which you could use to show that acid solutions conduct electricity and to collect the gas formed at the cathode. (See table 7 in the Data section.)

13 Which chemical test is used to detect hydrogen gas?

The results of experiments suggest that all acidic solutions in water contain hydrogen ions, H^+. This can explain why all acids have similar chemical properties when they dissolve in water.

14 In what ways are the reactions of acids similar? Illustrate your answer by giving examples of the reactions of acids with:
a indicators, such as litmus
b metals, such as magnesium
c carbonates, such as calcium carbonate.

Water in rivers and lakes may become polluted by natural processes such as the decay of dead animals or plants. It may also be polluted as a result of domestic, agricultural or industrial activities. (See section C10.5 and Chapter **B**17.) Eventually most of the water flows back into the sea, carrying all the impurities with it. The sea is therefore a vast store of chemicals.

	Percentage of the Earth's total water
Oceans	over 97.00
Ice caps	2.10
Ground water	0.58
Glaciers	0.015
Freshwater lakes	0.009
Saline lakes	0.007
Soil moisture	0.005
Atmosphere	0.001
Rivers	0.0001

Figure 10.14
How the Earth's water is distributed. (See also Biology Chapter **B**15.)

15 Are most non-metal oxides acidic? Find out whether or not this is true with the help of reference books. Are there any exceptions? (The Periodic Table in the Data section shows which of the elements are non-metals.)

16 Why do we find that water in the soil contains more carbon dioxide than rain? Where does the carbon dioxide come from? (See Biology Chapter **B3**.)

17 Write balanced symbol equations for these reactions:
a the reaction of carbon dioxide with water to form carbonic acid
b the reaction of sulphur dioxide with water and oxygen to form sulphuric acid.

Figure 10.14 shows that we depend on a minute part of the total amount of water on Earth for our supply of fresh water. In Britain this is controlled by the area Water Authorities which are responsible for water purification, drainage, and sewerage. (See figure 10.17 overleaf.)

Water treatment

All the water supplied to our homes is pure enough for drinking and cooking even though we use most of it for other purposes (see figure 10.15).

Purpose	Volume of water used per person per day (litres)
Flushing toilets	40
Washing and bathing	37
Washing clothes	20
Dishwashing and cleaning	12
Cooking and drinking	6
Outside uses	5
Total	120

Figure 10.15
How we use water in our homes.

Figure 10.16 demonstrates that chemistry makes an important contribution to the supply of fresh drinking water. The flow diagram shows the way in which water from a river or reservoir is processed in some treatment works.

Aeration dissolves oxygen in the water and helps to remove dissolved iron compounds. High concentrations of iron compounds are not dangerous but they can be very annoying. Rice and vegetables may turn brown if cooked in water containing iron. Tea has an inky colour and tastes bitter.

Another problem caused by iron compounds in the water is that clothes may be covered in rusty stains after washing. Washing powders make the water alkaline, and this can produce a precipitate of iron(III) hydroxide which is deposited on the clothes.

What does the "(III)" mean in the name of iron(III) hydroxide? What is a precipitate? Why do alkalis precipitate iron(III) hydroxide? These questions are answered in boxes 2 and 3 overleaf.

Carbon in the form of charcoal is a very effective material for absorbing unwanted smells, tastes and colours. It has been used in gas masks to absorb

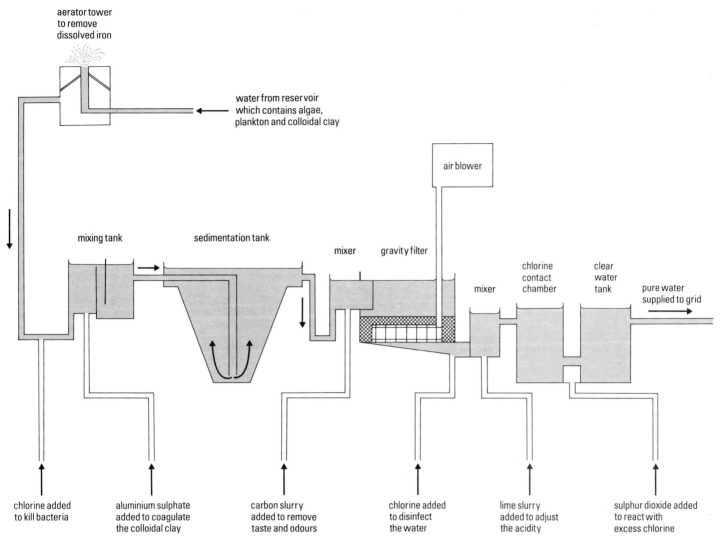

aerator tower
to remove
dissolved iron

water from reservoir
which contains algae,
plankton and colloidal clay

air blower

mixing tank

sedimentation tank

mixer

gravity filter

chlorine
contact
chamber

clear
water
tank

mixer

pure water
supplied to grid

chlorine added
to kill bacteria

aluminium sulphate
added to coagulate
the colloidal clay

carbon slurry
added to remove
taste and odours

chlorine added
to disinfect
the water

lime slurry
added to adjust
the acidity

sulphur dioxide added
to react with
excess chlorine

Figure 10.16
Flow diagram for the water treatment
process.

poisonous gases. You will find it in cooker hoods to capture kitchen smells. In water treatment it is also used to get rid of unpleasant tastes and odours. Then the charcoal is removed with other solids in the gravity filter.

Rapid gravity filters consist of graded gravel and sand. Water passes through at a fast rate. At regular intervals the filters are cleaned with the help of a blast of compressed air which loosens the dirt. The dirt is then washed away by a reverse, upward flow of clean water.

The lime slurry in figure 10.16 is milk of lime (see figure 4.29). A slurry is a suspension of particles in water. In milk of lime, the suspended particles are calcium hydroxide which is only slightly soluble in water. Calcium hydroxide is an alkali and will neutralize acids. The pH of the water leaving the treatment works has to be carefully controlled. If it is too alkaline (pH above 8.5) it may start to dissolve the zinc from galvanized steel water tanks.

The use of chlorine to disinfect water was first used in Britain in 1904 after a typhoid epidemic in Lincoln caused by contaminated drinking water. Chlorine kills bacteria in the water. Less than 0.1 mg/L is ineffective. Over 1.0 mg/L gives the water an unpleasant taste. The amount of chlorine in the water as it leaves the works is carefully adjusted to about 0.7 mg/L in winter. In summer the level is 0.9 mg/L to allow for the fact that some chlorine will be lost if the water is warmer.

If there is too much chlorine, the excess can be removed by adding sulphur dioxide.

BOX 2 Roman numbers in chemical names

One of the properties of the transition metals in the middle of the Periodic Table is that many of them can form more than one ion. Iron forms Fe^{3+} ions and Fe^{2+} ions. Copper forms Cu^{2+} ions and Cu^{+} ions.

These metals can therefore form two chlorides and two oxides, and so on. Roman numbers are included in the names to show which ions are present in the compounds.

Iron forms iron(III) chloride, $FeCl_3$, with Fe^{3+} ions and iron(II) chloride, $FeCl_2$ with Fe^{2+} ions.

Copper forms copper(II) oxide, CuO, with Cu^{2+} ions and copper(I) oxide, Cu_2O, with Cu^{+} ions.

Notice that the roman numbers in the chemical names tell you the **size of the charge** on the metal ion. They **do not** tell you the numbers which appear in the formulae.

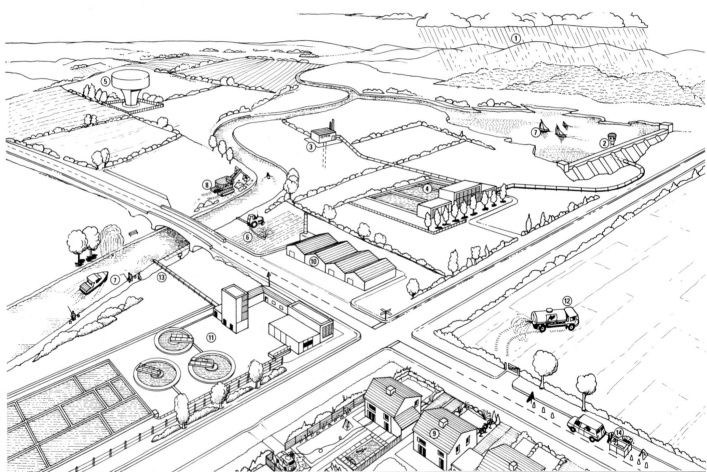

Key
1 rainwater: our natural water supply

2 reservoir: used to store rain and river water and for recreational purposes

3 borehole: to obtain water naturally stored below ground

4 water treatment works: to ensure water supplies are suitable for human consumption

5 water storage tower: to store treated water and provide pressure to deliver the water

6 farmer: extracting river water for irrigation

7 boating and fishing: popular leisure activities

8 dredging: to improve drainage

9 domestic use of water: drinking, cooking, washing, watering, swimming

10 industrial use of water: washing, cooking, purifying, diluting

11 sewage treatment works: receives all dirty water for treatment

12 sewage sludge: sprayed onto the farmland – a good soil conditioner

13 treated water returning to the water course

14 the water board repairers: keeping the system working

Figure 10.17
The activities of a water authority.

Figure 10.18
Part of a water treatment works: the sand filters.

BOX 3 Precipitation reactions

Sometimes a solid forms when two solutions are mixed. The solid drops out of the mixture so it is called a *precipitate*.

A precipitate forms when a solution of an alkali, such as sodium hydroxide, NaOH, is added to a solution of an iron compound such as iron(III) chloride, $FeCl_3$. Both sodium hydroxide and iron(III) chloride are ionic and soluble in water.

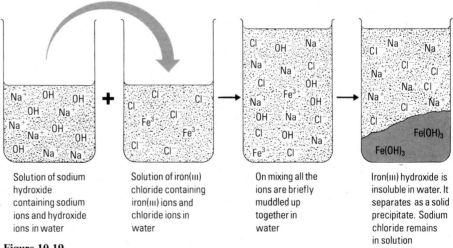

| Solution of sodium hydroxide containing sodium ions and hydroxide ions in water | Solution of iron(III) chloride containing iron(III) ions and chloride ions in water | On mixing all the ions are briefly muddled up together in water | Iron(III) hydroxide is insoluble in water. It separates as a solid precipitate. Sodium chloride remains in solution |

Figure 10.19
The formation of a precipitate.

When solutions of two ionic compounds are mixed, new combinations of ions are possible. A precipitate forms if one of the new combinations forms an insoluble compound.

When sodium hydroxide is added to iron(III) chloride, the new combinations are sodium chloride (which is soluble) and iron(III) hydroxide (which is insoluble). A precipitate of iron(III) hydroxide forms.

18 Find out where the water supplied to your home comes from. How is it treated before it arrives? Where does it go after you have finished with it?

19 Why do you think that there are often fears of outbreaks of disease after cities have been damaged by natural disasters such as earthquakes?

20 The colloidal clay particles in the water from rivers are negatively charged. Use your knowledge of colloids from section **C9.4** to explain why aluminium sulphate is used to make the clay settle out in the sedimentation tanks.

21 How is chlorine produced for use in industry? (See section **C4.3**.)

22 What are the names of the compounds with these formulae:
Cu_2S, $CuCl$, $CuCl_2$, $CuSO_4$, FeO, Fe_2O_3, $FeSO_4$, $MnCl_2$, MnO_2?

23 Alkalis neutralize acids, forming salts. Write word and symbol equations for the reaction of calcium hydroxide with hydrochloric acid. (Refer to table 5 in the Data section for help with the formulae.)

24 Concentrations are sometimes quoted in parts per million by mass (p.p.m.). Show that a concentration of 1 mg/L is the same as 1 p.p.m. Remember that the density of water is 1 g/ml and that there are 1000 ml in one litre (L).

25 Sulphur dioxide reacts with chlorine and water to form a mixture of sulphuric and hydrochloric acids.
a Write word and symbol equations for the reaction including state symbols. (Refer to table 5 in the Data section for help with the formulae.)
b Use the equation to calculate the mass of sulphur dioxide which must be added per litre of water to reduce the chlorine concentration from 1.42 mg/L to 0.71 mg/L. (Use the method explained in section **C5.7**.)

26 If a water sample contains 0.71 mg/L of chlorine, Cl_2, what is the concentration of chlorine in mol/L? (See box 4 and compare the answer with the concentrations of some common laboratory reagents calculated in the next question.)

27 Standard laboratory reagents can be prepared by dissolving weighed amounts of solid in water and then adding more water until the total volume of the solution is as required. Calculate the concentrations (in mol/L) of the solutions made up according to the quantities shown in figure 10.20. Use the method described in box 4.

Solid in solution: name and formula	Mass of solid in g	Volume of solution in ml
Sodium carbonate Na_2CO_3	530	5000
Potassium hydroxide KOH	224	2000
Copper(II) sulphate $CuSO_4 \cdot 5H_2O$	31.25	250
Silver nitrate $AgNO_3$	8.50	500

Figure 10.20
Quantities used to make some standard laboratory reagents.

Chlorine molecules are used to kill bacteria, and the process is not

BOX 4   Concentrations

In chemistry, amounts of substances are measured in moles. This is explained in section **C5.7**. It is important to know the concentrations of chemical reagents. Bottles of solutions are labelled to show the amount of substance (in moles) dissolved per litre of the solution. Laboratory acids and alkalis are often supplied as 2 mol/L solutions. This means that there are two moles of the dissolved substance in each litre of the solution.

Example
What is the concentration of an alkali which contains 320 g of sodium hydroxide in 4000 ml of the solution?

Method
The concentration is measured in mol/L. This means that it is calculated by dividing the amount of dissolved substance (mol) by the volume of the solution in litres (L).

Step 1: Calculate the amount of substance (mol).

- Amount of substance (mol) $= \dfrac{\text{Mass of sample (g)}}{\text{Molar mass (g/mol)}}$

Step 2: Calculate the volume of the solution (L).

Step 3: Calculate the concentration of the solution (mol/L)

- Concentration (mol/L) $= \dfrac{\text{Amount of substance (mol)}}{\text{Volume of solution (L)}}$

Answer
Step 1: The molar mass of sodium hydroxide $= 40$ g/mol

The amount of sodium hydroxide in solution $= \dfrac{320 \text{ g}}{40 \text{ g/mol}}$

$= 8$ mol

Step 2: There are 1000 ml in 1 L,

so the volume of the solution $= \left(\dfrac{4000}{1000}\right)$ L

$= 4$ L

Step 3: The concentration of the alkali $= \dfrac{8 \text{ mol}}{4 \text{ L}}$

$= 2$ mol/L

controversial because everyone accepts the need for drinking water free of disease. However, chlorine is not the only halogen involved in water treatment. Much more controversial is the fluoridation of water supplies. This is done in some areas because there is evidence that it helps to limit tooth decay, especially in children. (See Biology Worksheet **B4E**.)

One of the substances used to add fluoride to water is disodium hexafluorosilicate, Na_2SiF_6. This releases fluoride ions, F^-, when it dissolves in water. It is the ions which are the active ingredient.

Note that it is not the highly reactive fluorine which is added to water but one of its compounds. Sodium fluoride is as different from fluorine as sodium chloride is from chlorine (see section **C**5.4).

The concentration of the added fluoride is about 1 mg/L. This compares with concentrations of up to 8 mg/L found in natural water in some parts of the country. At much higher concentrations than this, fluoride ions can inhibit enzymes and are poisonous.

---

28 What is the formula mass of disodium hexafluorosilicate?

29 What percentage by mass of disodium hexafluorosilicate consists of fluoride ions? (See the worked example in section 4.6, box 7.)

30 What mass of disodium hexafluorosilicate must be dissolved in one litre of water to give 1 mg/L of fluoride ions?

31 If there is 1 mg/L of fluoride ions, what is the fluoride ion concentration in mol/L ?

C10.4   Hard or soft?

You may already know something about the weathering of limestone from your work in geography. Here you can study the chemistry of the process and find out why water is "hard" in areas with chalk or limestone rocks.

*Worksheet **C**10B shows you how to compare the methods of water softening which we can use at home.*

*While studying this section, you may be asked to plan an investigation to solve one of the problems of using hard water. You may then be able to find out which is the best way of getting rid of the scale which furs up kettles and blocks pipes. Alternatively you may try to find out whether getting rid of scale helps to save energy. You may have made surveys of energy savings while studying Physics Chapter **P**9, and if so can compare your findings with what you discover here.*

Much of the water used in Britain has flowed through areas where the rocks consist of limestone or chalk. In these regions the water is hard. If you belly-flop onto hard water, it feels no different from a pool full of soft water but you notice the difference when you try to wash your hands with soap. If the water is hard you need a lot of soap to get a lather, and scum forms round the edge of the basin. If the water is soft you can get a good lather with little soap and no scum forms.

How does water become hard?

The results of the experiment in box 5 will help you to decide what it is that causes hardness in water.

Marble, chalk and limestone are all composed of calcium carbonate which is insoluble in water. However these minerals are affected by water which contains carbon dioxide. Some of the carbon dioxide dissolves in the rain as it

BOX 5 Experiment
Which ions make water hard?

Several salts were dissolved in distilled water. All the concentrations were the same. A measured sample of one solution was put into a test-tube and mixed with soap solution. The tube was corked and the mixture shaken. This procedure was repeated with each of the solutions with the results shown in figures 10.21 and 10.22.

Substance in solution	Ions present in solution		Height of lather in mm
	Positive	**Negative**	
Sodium sulphate	Na^+	SO_4^{2-}	21
Sodium nitrate	Na^+	NO_3^-	18
Potassium chloride	K^+	Cl^-	23
Potassium sulphate	K^+	SO_4^{2-}	20
Calcium chloride	Ca^{2+}	Cl^-	1
Calcium nitrate	Ca^{2+}	NO_3^-	1
Magnesium sulphate	Mg^{2+}	SO_4^{2-}	1
Magnesium chloride	Mg^{2+}	Cl^-	1

Figure 10.21
Two test-tubes: one shows the result of shaking a solution of sodium chloride with soap solution; the other shows the result of shaking calcium chloride with soap solution.

Figure 10.22
Results of the experiment.

32 Which of the salts tested make water hard?

33 Which of the salts tested do not affect the hardness of water?

34 Does it seem to be the positive metal ions or the negative non-metal ions which make water hard? Show how you decide on your answer.

35 Do all metal ions make water hard or only some of them?

36 Which ions make water hard according to this evidence?

falls through the air. More dissolves as the water soaks through the soil. Carbon dioxide reacts with water to form carbonic acid, H_2CO_3.

Carbonic acid reacts with calcium carbonate to form calcium hydrogencarbonate which is soluble in water.

$$CaCO_3(s) + H_2CO_3(aq) \longrightarrow Ca(HCO_3)_2(aq)$$

Calcium hydrogencarbonate is ionized in water as calcium ions, Ca^{2+}, and hydrogencarbonate ions, HCO_3^-. In this way calcium ions get into the water. (The way in which the formula of calcium hydrogencarbonate is written is explained in Chapter **C5**, box 3.)

The effects of chemical weathering of limestone can be seen in many parts of Britain above and below the surface of the earth. (For more information about the weathering of rocks see section **C4.1**.)

Figure 10.23
Limestone pavement above Malham Cove, North Yorkshire. The surface of limestone is often criss-crossed by small cracks. The rain runs into these and dissolves the limestone more quickly here, making larger cracks.

Figure 10.24
Limestone cave ("the Temple", Gibraltar).

The mineral gypsum (calcium sulphate) is slightly soluble in water and so will make water hard even if there is no carbon dioxide present. Figure 10.25 compares the solubility of calcium sulphate and calcium carbonate in water.

Chemical	Temperature in °C	Solubility in g per 100 g water
Calcium sulphate	0	0.24
	100	0.22
Calcium carbonate	0	0.0015
	100	0.0019

Figure 10.25
Solubilities of calcium sulphate and calcium carbonate.

37 How can you demonstrate that a solution of carbon dioxide in water is acidic?

38 Show that the equation for the reaction of calcium carbonate with carbonic acid is balanced.

39 With the help of the results of the experiment described in this section (box 5), explain how water becomes hard when it flows through the mineral magnesite (magnesium carbonate).

40 What happens to the solubility of calcium sulphate as the temperature is raised? In what way is this unusual?

41 Why does hard water form a scum with soap?

The structure of a typical soap is shown in figure 10.4. The soap in the picture is a sodium salt called sodium stearate. This is soluble in water. However the calcium salt is insoluble and so when soap is added to hard water which contains calcium ions a precipitate of calcium stearate forms. This appears as a greasy scum. The formation of scum wastes soap and makes it more difficult to make things clean. Figure 10.26 uses the simplified picture of a soap molecule to show how scum forms.

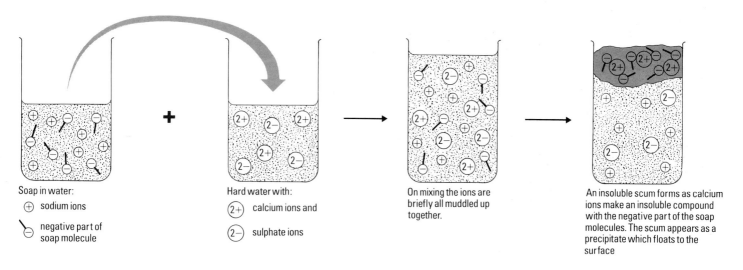

Soap in water:
⊕ sodium ions
⎯ negative part of soap molecule

Hard water with:
(2+) calcium ions and
(2−) sulphate ions

On mixing the ions are briefly all muddled up together.

An insoluble scum forms as calcium ions make an insoluble compound with the negative part of the soap molecules. The scum appears as a precipitate which floats to the surface

Figure 10.26
The formation of scum with hard water.

Magnesium stearate is also insoluble in water, and so a scum forms if soap is added to a solution containing magnesium ions.

Soapless detergents do not produce a scum when added to hard water because they do not form insoluble compounds with calcium or magnesium ions.

What is scale?

If you live in a hard water area and look inside a kettle, you will probably find that the heating element and sides are coated with an off-white solid. This is scale. Scale forms when the water is heated. It furs up boilers and hot water pipes as well as kettles.

Figure 10.27
Scale on the inside of a furred up electric kettle.

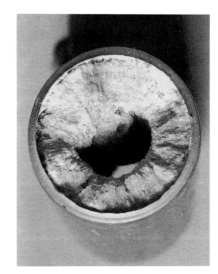

Figure 10.28
Furred up pipe.

Scale forms because the reactions between calcium carbonate, carbon dioxide, and water are reversed when the temperature is raised. The limestone which was dissolved by water trickling through rocks is reformed inside kettles and boilers.

$$Ca(HCO_3)_2(aq) \longrightarrow CaCO_3(s) + H_2CO_3(aq) \longrightarrow CaCO_3(s) + H_2O(l) + CO_2(g)$$

Boiling the water removes the calcium ions and softens the water. This is an expensive way of removing hardness, and it does not get rid of hardness caused by dissolved calcium sulphate.

BOX 6   Problem
Which is the best descaler?

Imagine that you are a scientist working for a manufacturer of electric kettles and other appliances. You have been given the job of selecting a suitable descaler to be sold with the products of your company.

Figure 10.29
Using a commercial descaler.

Plan the experiments which you will carry out in the laboratory to solve this problem. What other information will you need and where will you get it? The following points of information and questions may help you.

- Descalers are acids which dissolve calcium carbonate. Acids used by other manufacturers include methanoic (formic) acid, phosphoric acid and aminosulphuric acid. You also have access to hydrochloric, sulphuric and nitric acids.
- The company makes kettles from aluminium, stainless steel and plastic.
- How much descaler should be included in a single pack sold to the public? What should the pack be made of?
- What instructions should be given on the packet to make sure that the scale is removed quickly and efficiently?
- Can you be sure that it is safe to allow the chemical to be used at home? What warnings should be included on the packet?
- Can you recommend a suitable name for the product?
- What is the cost of the chemical in one pack?

BOX 7 Problem
Does descaling kettles save energy?

This is an investigation which you could carry out if you live in a hard water area and have a kettle which is affected by scale. You can try the investigation with an electric kettle, or with a kettle heated on an electric hot plate.

Plan your investigation in four stages:

- Measure the time taken to boil the kettle before descaling.
- Descale according to the instructions and note the quantity and cost of the descaler used.
- Measure the time taken to boil the kettle after descaling.
- Calculate the energy saved (if any) using the information given about the power of the kettle, or electric hot-plate, and the method given in section **P**17.2 of your Physics book.
- Calculate the cost of the energy saved and compare it with the cost of the descaler. Would you recommend the regular descaling of kettles on the basis of the energy saved?

Consider how you will have to carry out each step to make sure that you are making fair comparisons and that the results are reliable. Are there any other arguments for or against descaling kettles?

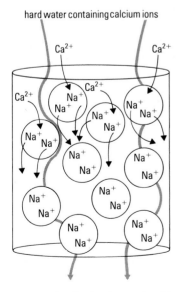

soft water in which sodium ions have replaced the cal‑cium ions in hard water

Figure 10.30
An ion exchange resin in action.

How can hardness be removed?

Boiling is an expensive way of softening water and it does not remove all the hardness. There are better ways. The method used in washing powders has already been mentioned. One of the ingredients is a form of sodium phosphate which combines so strongly with calcium ions that they are not free to react with soap. The chemical used is available commercially as Calgon. This technique stops the formation of scum, but it does not remove the calcium ions from the water so it cannot entirely prevent the formation of scale.

One of the better methods of softening water is to use an *ion exchange resin*. This is what is used in the water softeners which are built into many washing machines and can be used to soften all the water used at home. The resin consists of a mass of tiny plastic beads. They are made of polystyrene which has been treated so that it is negatively charged and will attract metal ions. When ready to use, the ions on the resin are sodium ions (figure 10.30). As the hard water passes through a column of the beads the calcium ions are exchanged for the sodium ions held on the resin.

Eventually all the sodium ions are replaced by calcium ions, at which point the resin becomes useless. Fortunately it is possible to reverse the process by pouring a concentrated solution of sodium chloride (salt) through the resin

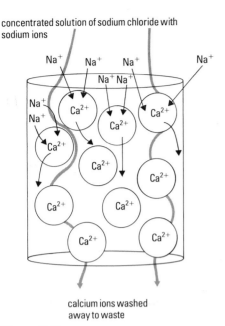

concentrated solution of sodium chloride with sodium ions

Na⁺ Na⁺ Na⁺   Na⁺

Na⁺ Na⁺

Na⁺
Na⁺

Ca^{2+}

calcium ions washed away to waste

Figure 10.31
Regenerating an ion exchange resin.

Ingredients: Salt, Anti-caking agents:

Magnesium carbonate,

Sodium hexacyanoferrate(II).

Figure 10.32
The label on a packet of salt.

(figure 10.31). Sodium ions are put back, the calcium ions are washed away, and the resin is ready to use again.

This explains why it is necessary to put salt into water softeners. But why is it necessary to use a special form of salt called granular salt? The main reason is that granular salt is pure. Table salt is not, as you can see if you look at the label on the packet (figure 10.32). The salt we use at table and for cooking contains chemicals which help to stop it getting moist and caked up so that it will not pour. The added chemicals would gradually choke up a water softener and stop it working.

---

42 When ion exchange water softeners are installed, they are used to soften water used for heating and washing but not drinking water. How do water softeners change the concentrations of calcium and sodium ions in water? Why might these effects be bad for our health if we drink a lot of softened water? (See Biology Chapter **B6**.)

43 Another way of softening water is to use bath salts or bath crystals. These both work by adding carbonate ions, CO_3^{2-}, to the water. What happens when carbonate ions are added to a solution containing calcium ions? Why does this soften the water?

C10.5   Keeping water pollution under control

*You can read about pollution in more detail in Biology Chapter **B**17. This section mentions some of the chemical aspects of water pollution. It shows that chemical analysis is very important in identifying and controlling pollution.*

Polluted water is harmful to animals and plant life. It is unfit to drink, dangerous to swim in, and unpleasant for boating. Severe pollution can kill off all life in a river. Figure 10.33 illustrates some of the ways in which water can become polluted.

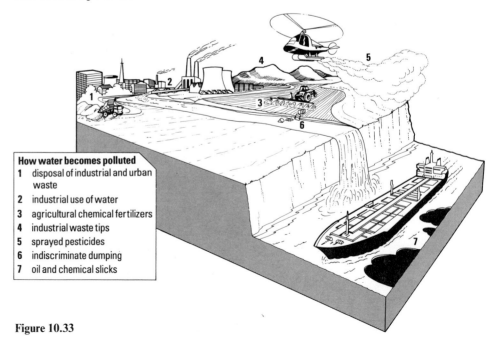

How water becomes polluted
1 disposal of industrial and urban waste
2 industrial use of water
3 agricultural chemical fertilizers
4 industrial waste tips
5 sprayed pesticides
6 indiscriminate dumping
7 oil and chemical slicks

Figure 10.33

Pollution can arise in three main ways:

- by increasing the amount of organic matter
- by enrichment with chemicals which acts as fertilizers
- by adding substances which are poisonous to fish and other animals

All natural waters are subject to some degree of pollution from decaying vegetation and impurities washed out of the soil. Bacteria and fungi feed on organic material in water and use up oxygen as they do so.

Severe pollution can arise when the amount of organic material in a river is increased so much that the bacteria remove oxygen from the water faster than it can be replaced at the surface. The concentration of dissolved oxygen falls and this reduces the variety of animal life.

When most of the dissolved oxygen has been removed, oxygen is then taken out of nitrates, NO_3^-, and sulphates, SO_4^{2-}, in the water. Sulphates turn in to sulphides, including hydrogen sulphide. This is responsible for the foul smell of polluted stagnant water.

The extent of pollution by organic materials is measured by its biochemical oxygen demand (BOD). The test determines the amount of oxygen used by bacteria as they feed on the organic material in the water.

Sewage can make a big contribution to the BOD, and so can many farm wastes such as vegetable washings, farmyard manure, waste milk and water draining from silage. They are all richly organic and so farmers have a responsibility to ensure that these farm wastes do not enter streams and rivers. It is better to store them ready to be put back on the land at a suitable time to fertilize and condition the soil.

Figure 10.34
Correct practice when handling agricultural waste.

Industrial effluents can also add to the BOD of rivers.

Nitrates and phosphates in the water act as fertilizers which stimulate too much growth of plants including algae. One of the main sources is washing powders, which include phosphates used to soften the water. In addition, up to 40 per cent of the fertilizers spread on farm land may be washed into rivers. This enrichment of water is often called *eutrophication*. (See Biology Chapter **B**17.) It is a particular problem in lakes such as the Norfolk Broads.

Rivers can be poisoned as a result of carelessness. It is highly irresponsible to dispose of waste engine oil by pouring it down a drain.

Undiluted fungicides, herbicides and insecticides are often extremely toxic to fish. They are also a hazard for animals and human life. In some parts of Britain nearly half of the incidents which kill fish are caused by agricultural pollution.

Industrial processes can be dangerous too. For example electroplating uses cyanide solutions with ions of metals such as copper and chromium. These are highly toxic.

Chemists have an important role to play in monitoring and controlling pollution. All the Water Authorities test water samples at regular intervals. They measure suspended solids, dissolved oxygen, and the biochemical oxygen demand as well as ammonia, nitrates, and chlorides. Further tests measure the hardness of the water and the concentrations of sulphate ions, phosphate ions, and detergents.

Figure 10.35
Checking water samples.

Summary

1 Dirt and stains can be removed by washing, bleaching or dry-cleaning. What are the advantages and disadvantages of these three methods of keeping clothes clean?

2 Make a summary of your knowledge of hardness of water by answering these questions:

- What is hard water?
- How does water become hard?
- What are the disadvantages of hard water?
- How can hard water be softened?

Chapter C11 Dyes and dyeing

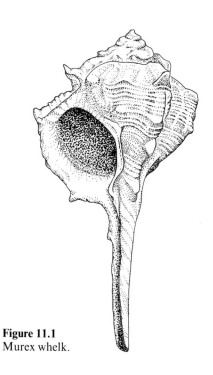

Figure 11.1
Murex whelk.

Figure 11.2
Woad.

Chemistry has brought colour into our lives. In this chapter you can read about some of the natural dyes which were used for thousands of years until William Perkin discovered the mauve dye for which he is famous. Since then a great industry has grown up to make new and brighter colours starting with chemicals from coal and oil. You will find that more recently a knowledge of chemical structure and bonding has made it possible to develop brighter and faster dyes.

While studying this chapter you will have the opportunity to try dyeing fabrics with both natural and synthetic dyes. You will be able to use the alum you made earlier in the course and you may be asked to test dyes to see if they are fast to washing and light.

Many natural dyes can be extracted from plants so there is a link between this chapter and Chapter C3. The green pigment in plants has a vital part to play in photosynthesis, as explained in Biology Chapter B3. Physics Chapter P15 explains why different materials have different colours

C11.1 What is a dye?

We use dyes to colour our clothes. We also use them to colour paper, hair, leather and food. If a coloured chemical is to be a dye it must somehow be able to attach itself to the material to be coloured. A good dye is a *fast* dye, which means that it is held strongly to the material so that it will not wash out or fade in the light.

In ancient times, dyes were extracted from plants and animals. The Egyptians and Assyrians, the Greeks and Romans were skilled in the art. The city of Tyre was famous for its purple dye which was extracted from molluscs.

Tyrian purple was extremely expensive because up to a quarter of a million molluscs were required to obtain about 30 g of the dyestuff. In the days of Imperial Rome its use was reserved for the Emperor's family. Lucretius (98–55 BC), a Roman poet, wrote as follows about the dye:

"The purple dye of the shell fish so unites with the body of the wool alone, that it cannot in any case be severed – not if the whole sea were willed to wash it out with all its waters." (*De rerum natura.*)

Plants were the biggest source of natural dyes. The two dyes which were most widely used were indigo and alizarin. The discovery of Egyptian mummy cloths dyed with indigo has shown that it was known over 5000 years ago. The same blue dye was obtained from the woad plant. Woad coloured the warpaint of British tribes and continued to be used to dye cloth blue in Britain until it was replaced, in the sixteenth century, by the superior indigo imported from India.

Indigo plantations in India were the basis of a big industry producing up to eight thousand tonnes per year of the dye. Then a method for making the

Figure 11.3
Workers loading a vat with indigo in nineteenth-century India.

dyestuff synthetically was discovered in 1897. Within a few years of the discovery, over a million acres of land which had been used for growing indigo could be used instead to grow food. At the same time, many people were thrown out of work.

Indigo is insoluble in water and has to be changed to a water soluble form before it can be used to dye cloth. Originally this was done in large vats and so indigo (and other similar dyes) are still called *vat dyes*.

The soluble form of indigo is yellow. It turns blue again when it reacts with oxygen in the air. This is another example of the type of chemical reaction called *oxidation* (see section **C4.5**, box 5).

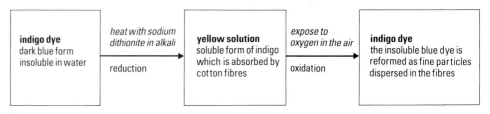

Figure 11.4
Vat dyeing with indigo.

---

1 What are the properties of a good dye?

2 The reaction involved in turning indigo into its soluble yellow form is the opposite of oxidation. What name is given to this type of reaction?

Alizarin has been used as a red dye for cotton for thousands of years. It used to be obtained from the roots of the madder plant which was widely grown in southern Europe and parts of Asia. Hundreds of thousands of acres of land were used for this purpose until about 1870, when a method for making the dye synthetically from coal tar was discovered.

So you can see that these discoveries made by chemists changed the pattern of agriculture in Europe and Asia. Indigo is still used to dye denim for blue jeans, but it is now obtained synthetically.

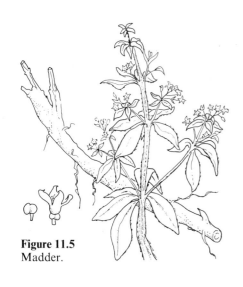

Figure 11.5
Madder.

Name and source	Colour and type	Other details
Indigo (*Indigofera tinctoria*) from leaves of plants native to India. Used there before 3000 BC.	Blue, vat dye. Fairly fast.	The colouring matter is called indigotin. After harvesting, the plant leaves were cut, left for a day, and dried by treading. Then they were mixed with urine and heated in the sun. Urine provided ammonia which made the mixture alkaline and helped fermentation. Afterwards the mixture was boiled and fermented for a considerable time. The juice was yellowish at first but turned blue on exposure to oxygen in the air.
Woad (*Isatis tinctoria*) from leaves of a plant native to Eastern Europe. Used in Britain before Roman times.	Blue, vat dye. The dye is indigotin as in indigo leaves but it is less concentrated.	It was more difficult to extract indigotin from woad than from the indigo plant. When the all-sea route from India was opened up in about 1500, indigo began to replace woad in Western Europe. English and Scots dyers opposed its use for a considerable time.
Madder (*Rubia tinctorum*) from the root of a hardy shrub.	A mordant dye. Fast, brilliant red with alum; orange if mordanted with tin; purple with chrome; and brownish with alum and tin.	The colouring matter is alizarin contained in the yellow pith of the root. Plants were gathered after about two years' growth and the roots were washed, dried, ground, and stored. This was the cheapest of the good red dyes and was widely used for ceremonial robes. Mixed dyeing with indigo or woad produced a cheap substitute for Tyrian purple.
Saffron (*Crocus sativus*) from the stigmas and styles of an autumn-flowering crocus.	A shade of orange–yellow.	Pliny, in the first century AD, speaks of wild saffron from Sicily as being superior to that grown in Italy. The Romans perfumed their baths and streets with it and it was (and still is) used for colouring food. The Phoenicians gave saffron-dyed cloth as part of the tribute to Assyrian conquerors. The plant was grown in Essex in mediaeval times – hence the name Saffron Walden.
Archil (*e.g. Rochella tinctoria*) from lichen on rocks in the East Mediterranean	Red/purple dye usually mordanted with alum.	The dried lichen was extracted with potash in the presence of air. It produced a poor – but cheap – substitute for purple or crimson. A similar extract is used for Scottish tweeds and tartans. Lichen dyes are what give tweeds their characteristic smell. The dye includes litmus which is commonly used in laboratories as an indicator.

Figure 11.6
Natural dyes from plants.

Name and source	Colour and type	Other details
Tyrian purple molluscs of several species known from at least 1500 BC.	A vat dye. Colour varies according to the species of mollusc: from scarlet to purple.	The colouring matter – dibromoindigotin – was found in a gland at the back of the neck of the mollusc. The molluscs were caught in lobster pots and killed suddenly to preserve the secretion which decomposed after death. They were opened, crushed and left in salt for 3 days. Then they were extracted with water, impurities removed and the solution concentrated. Initially the secretion from the molluscs is colourless; the colour develops in the presence of air and light.
Cochineal (*Coccus cacti*) a scale insect breeding on a species of cactus in Mexico.	A mordant dye. Scarlet/crimson. Often mordanted with tin.	The dye was valued for ceremonial dress and hunting coats. It was, and is, used as a colouring for food. Its use became widespread in Europe after the arrival of the Spaniards in Mexico (1518). Napoleon was concerned to find a good substitute for it when the British blockade cut off his supplies.

Figure 11.7
Natural dyes from animals.

Another famous natural dye is cochineal, which is extracted from a Mexican insect, *Coccus cacti*. This was the dye used for the uniforms of British red-coats throughout the seventeenth, eighteenth and nineteenth centuries.

Henna is also a natural dye which is known to have been cultivated in Egypt and Syria from about 2500 BC. It was, and is, used for dyeing hair, hands and feet.

Dyes can be obtained from many plants found in the British Isles. The colours of the traditional tweeds and tartans of Scotland show what skilful dyers can achieve with natural materials.

Many natural dyes lack brightness. They are often not very fast and large amounts of plant or animal material are needed to provide significant quantities of the dye. Now the only natural dye of widespread commercial importance is extracted from South American logwood. It is used to dye nylon black.

Figure 11.8
Dyeing silk in the dyeworks of the Gobelin tapestries in eighteenth-century France.

C11.2   What is a mordant?

In Chapter C4 you can study the history of the alum industry. This section shows why alum was a valuable chemical. Also in this section you will find a theory to account for the properties of alkaline solutions which follows up the explanations about precipitates and acidity in Chapter C10. Consult your teacher to discover whether you are expected to study these topics in detail.

Fabrics are treated with mordants before they are dyed. Without mordants most natural dyes are not fast to light or washing. The term *mordant* is based on an old French word meaning "to bite". This gives us a picture of the mordant sinking its teeth into the dye so that it cannot get away. Chemically, the "teeth" are the bonds which hold the dye to the mordant in the cloth.

The most commonly used mordant was alum. The alum industry around Whitby described in Chapter **C4** grew up to supply the dyeing industry with this chemical. It had to be pure otherwise the colours of dyes were affected.

Mordants help to make dyes fast. They can also change the colour of dyes. The Roman writer, Pliny (AD 23–79), who compiled an encyclopaedia of popular science, seems to have been fascinated by the craft of dyeing and in this passage he described the use of mordants which he called "medicaments".

"In Egypt they dye cloths in a remarkable way. The white material is treated not with the colours, but with medicaments which absorb the colours. This done, the materials appear unchanged, but when immersed in a cauldron of boiling dye and immediately removed, they are coloured. It is remarkable that though the dye in the cauldron is of one colour only, the materials when taken out are of various colours according to the quality of the medicaments applied."

The full procedure for mordanting with alum involves simmering the cloth in a solution of alum and tartaric acid for several hours. If you try mordant dyeing you can precipitate the mordant in the fibres of the cloth more quickly by first dipping the fabric into ammonia solution and then into an alum solution.

BOX 1   What makes water alkaline?

Sections **C4.3** and **C4.4** describe the manufacture and uses of two important industrial alkalis: sodium hydroxide and calcium hydroxide. We know that metal hydroxides produce alkaline solutions if they dissolve in water. They are ionic compounds and their solutions in water all contain hydroxide ions, OH^-.

The theory is that it is hydroxide ions which make water alkaline. So how can ammonia, NH_3, be an alkali? This can be explained if ammonia molecules react with water molecules as shown:

$$NH_3(aq) + H_2O(l) \longrightarrow NH_4^+(aq) + OH^-(aq)$$

| ammonia molecule | water molecule | ammonium ion | hydroxide ion |

Alum solution contains aluminium ions, Al^{3+}. When ammonia solution and alum solution are mixed, a white precipitate forms. This precipitate is aluminium hydroxide which is insoluble in water.

$$Al^{3+}(aq) + 3OH^-(aq) \longrightarrow Al(OH)_3(s)$$

| from the alum solution | from the ammonia solution | white solid mordant formed in the fibres of the cloth |

---

3 Traditional dyers' recipes often called for the use of rainwater and dyeworks were usually built in areas where there was a natural supply of pure, soft water. Why do you think that it was important that dyers should have pure soft water for their work?

4 Another metal hydroxide which can act as a mordant is iron(II) hydroxide. Write an equation to show the formation of iron(II) hydroxide when iron(II) ions from an iron salt combine with hydroxide ions from an alkali. (See table 6 in the Data section.)

Figure 11.9
A photograph of William Perkin when he was fourteen years old. It was taken by himself.

C11.3   How were synthetic dyes discovered?

This section, and section C11.4, are about "people and problems". These sections describe some of the people who made the discoveries and the problems they were trying to solve. You are not required to remember the historical details.

Scientific discoveries may arise in a number of ways. Chance plays a part in many discoveries but, as Pasteur said, "Chance favours the prepared mind". In this section you can read about an unexpected discovery. Compare this with the discovery of the magnetic field surrounding an electric current described in Physics Chapter P18, and the development of the theory of evolution discussed in Biology Chapter B23.

The first synthetic dye was made by William Perkin in 1856 when he was only eighteen years old. At the time he was a student at the Royal College of Chemistry in London working as an assistant to the distinguished chemist, August Hofmann.

Perkin was interested in the idea of making natural organic substances artificially. He decided to try to make quinine which was extracted from cinchona bark and used to treat malaria. But he failed to make any quinine – this is not surprising because the chemical formula which he was using as the basis for his plans is now known to have been incorrect.

At his first attempt Perkin produced a dirty reddish brown precipitate, so he decided to try a simpler reaction to see if he could work out what was happening. He heated a mixture of aniline sulphate and potassium dichromate. This time he got a black precipitate. He did not throw it away but decided to investigate further in his rough laboratory at home during the Easter holiday.

Perkin found that when he added alcohol to the black stuff he could extract a purple coloured material from it. He then found that the coloured substance could be used as a dye and that it was fast to light. At that time aniline was obtained from coal, and so this mauve dye became known as the first of the "coal tar dyes".

This is how Perkin continued the story in the Hofmann memorial lecture which he gave in 1893 and which was published in the *Journal of the Chemical Society* 1896, Volume 69:

"After the vacation, experiments were continued in the evenings when I had returned from the Royal College of Chemistry . . . . I then got an introduction to Messrs Pullar of Perth [a firm of dyers] and sent them some specimens of dyed silk. On 11 June 1856, I received the following reply:

'If your discovery does not make the goods too expensive, it is decidedly one of the most valuable that has come out for a very long time. This colour is one which has been very much wanted in all classes of goods, and could not be obtained fast on silks, and only at great expense on cotton yarns. I enclose you a pattern of the best lilac we have on cotton – it is dyed only by one house in the United Kingdom, but even this is not quite fast, and does not stand the test that yours does, and fades by exposure to air . . . .'

"This first report was very satisfactory; the 'if' with which it commenced was, however, a doubtful point.

"During the summer vacation, however the preparation of the colouring matter on a very small technical scale was undertaken; my brother (the late T. D. Perkin) assisted me in the operations, and, after preparing a few ounces of the product, the results were thought sufficiently promising to make it desirable to patent the process for the preparation of this colouring matter. This was done on 26 August (Patent no. 1984)...."

The dyeing firm, Pullars of Perth, were still not sure of the value of the new dyes and Hofmann tried to discourage Perkin from pursuing this line of business.

"... Still, having faith in the results I had obtained, I left the College of Chemistry and continued my experiments....

"...In January 1857 Mr R. Pullar, however, advised me to see Mr Thos. Keith, a silk dyer of Bethnal Green, London, and, after making a few experiments with the colouring matter, and exposing the specimens he dyed to light for some time, he was much pleased with the result and encouraged me to go on with its production.

"I was then joined in the undertaking by my father – who was a builder, and had sufficient faith in the project to risk the necessary capital – and also by my brother, who also had a good knowledge of building, and, as he had taken part in the preliminary experiments on the preparation of the dye, his assistance proved most valuable, especially as he was possessed of good business capabilities. Plans were prepared and a site obtained at Greenford Green, near Harrow, and in June 1857, the building of the works was commenced."

Figure 11.10
A sample of Perkin's original mauveine.

Figure 11.11
Sketch by Perkin of his first factory.

Perkin's new dye was called mauveine and it was immensely successful. Clothing dyed with it was the height of fashion. Queen Victoria wore a mauve dress at the International Exhibition of 1862. A postage stamp was dyed with the colour: the penny mauve. If Punch is to be believed, London policeman were to be heard directing loiterers to "get a mauve on" at that time, and there were frequent references to the dye in that magazine for about ten years.

Perkin's discovery started a great rush to discover new dyes. Perkin joined the race, but in 1868 he submitted his application for a patent for the manufacture of synthetic alizarin twenty four hours after the identical process had been patented in Germany. A series of new discoveries followed including

COLOURING EXTRAORDINARY.

THERE is no accounting for tastes as to female beauty, and MAUVE is so much the rage, that we are hardly surprised to learn from the following advertisement in the *Times* of last Wednesday that some of our fair friends have devised means of transferring the fashionable colour from their clothes to their complexions :—

FOUND, on the 30th ult., a handsome LADY'S PARASOL, left there by two ladies, of mauve colour, lined inside with white, which may be had at ARTHUR GRANGER'S Stationery Warehouse, 308, High Holborn, W.C.

At the same time we should be rather inclined to consider that "two ladies, of mauve colour, lined inside with white," deserve to be classed as at once "plain and coloured," instead of "handsome," as in the polite language of the advertisement.

SCENE—COMMERCIAL ROOM.

INCIPIENT COMMERCIAL TO CRUSTY OLD TRAVELLER. *" You're always in the Fashion, I see. Last time I had the pleasure of seeing you, Mauve was the prevailing Colour, and your Nose was Mauve. Now Magenta is all the go, and it's changed to Magenta."*

Figure 11.12
Thoughts about mauveine from *Punch* (1859 and 1861).

Magenta, Methyl violet and Malachite green. Rapid commercial development took place in Germany. Within a few years German dye manufacturers came to dominate the industry. By 1874 Perkin found that he could no longer compete and so he sold his factory.

Figure 11.13
Some examples of synthetic dyes for use with cotton.

> **5** Survey the curtains, carpets and upholstery in your home. Also look at any coloured fabrics outside. Which of the fabrics seem to have faded in use? Is the fading due to washing or to exposure to light? Which colours seem to be less fast than others?

Type and examples	Characteristics
Sulphur dyes Sulphur blacks ICI Thional range	First introduced in 1873. Very fast to washing but colours are dull. There are no reds. They are easy to use and cheap. Used for cotton, viscose and nylon.
Direct dyes Congo red Cuprofix navy blue	First discovered in 1884. These are dyes which can be applied directly without mordants. Low price but only moderately fast to washing and light. They remain popular because they are cheap and easy to use.
Vat dyes Indigo ICI Soledon range	Fully synthetic vat dyes were first introduced in 1901. Complicated to use because the dye is insoluble in water and must be converted to a soluble form before dyeing. Few reds. Excellent fastness to washing. Used for cotton and modified cellulose fibres.
Azo dyes A very large number made by linking combinations of 50 bases with 52 coupling agents.	Discovered in 1859 but only used extensively for cotton from 1912. The final stage of making the dye takes place in the fibres by linking two chemicals to produce an insoluble coloured compound. The linking group of atoms is called an "azo group" – hence the name of the dyes. They are cheap, and very fast to washing. There are many reds but no greens. Several azo dyes are permitted for use as food colours including tartrazine (E102). Methyl orange indicator is also an azo dye.
Reactive dyes ICI Procion range CIBA Cibacron dyes	First available from 1956. These dyes react with the fibres in the fabric, forming strong chemical bonds. They are very fast to washing and usually fast to light. Medium price, easy to use, and a wide range of colours.

C11.4   Why are modern dyes so fast?

*Many natural dyes are held to cloth by the forces **between** molecules which are weak. The forces which hold atoms together **within** molecules are much stronger, as explained in section C5.3. This section shows how an understanding of these differences led to the discovery of very fast modern dyes. Unlike Perkin's discovery this was a planned rather than a chance discovery.*

Another major breakthrough in the dye industry occurred in Britain during the early 1950s. At that time, the two main textile fibres were wool and cotton. Research scientists working for the dyestuffs division of the large chemical company ICI were synthesizing about 3000 new substances a year hoping to find new colouring materials which could be manufactured and sold profitably. Very few of the tested substances were developed because to be successful a new dye has to pass many tests.

Potential for large sales	There must appear to be a market for the new dye for one of the textiles used on a large scale.
New properties	It must have clear advantages over existing dyes.
Patentable	It must be possible to patent the process so that the dye can be exploited free of competition for long enough to recover the research and development costs.
Low cost	Price must compare favourably with existing dyes.
Easy to make	It must preferably be possible to make the new dye with existing plant, machinery and raw materials.
Safe	It must be possible to manufacture and use the dye without endangering the health of those who work on the processes.
Easy to use	It must be possible to use the dye on existing dyeing and printing plants.
Fast	The dye must not fade perceptibly in the light or when washed.
Wide range of colours	The wider the range of bright and dull colours the better.

Figure 11.14
Characteristics of a successful new dye.

At the time the methods for dyeing textiles depended either on forming an insoluble dye in the fibres or on weak forces holding the dye molecule to the fibre molecules. The challenge was to discover a new range of dyes which would react with the molecules in textile fibres to form strong chemical bonds.

At first the aim of the research group was to make new dyes for wool. Dr W. E. Stephen prepared a number of new compounds by combining existing dye molecules with a reactive molecule which he hoped would combine with wool. Unfortunately the results of dyeing tests were disappointing.

The breakthrough came when I. D. Rattee, a senior technologist, realized that some of the new dyes might be used to dye cotton under more alkaline conditions which could not be used with wool without damaging it. This idea was shown to be correct on a small scale in October of 1953.

As you can see from figure 11.15, the reactive part of the experimental dye combines with —OH groups in the fibres. If you look back to figure 3.13 showing the structure of glucose, you will see that there are several —OH groups in the molecule. Cellulose is a polymer of glucose so there are —OH groups all along its long-chain molecules.

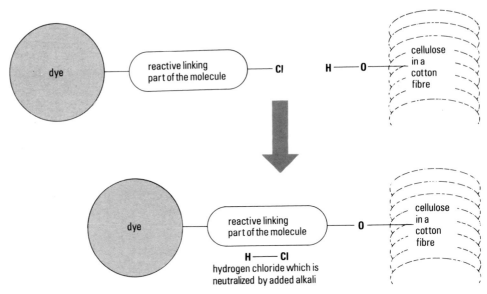

Figure 11.15
The principle of fibre-reactive dyes.

By April of 1954 a red, yellow and a blue dye had been made and tested on a laboratory scale. The next stage was to show that the dyes could be used successfully in the dyeing trade and then to scale up the manufacturing processes.

The reactivity of the new dyes was a problem. The dyes work by reacting with the —OH group of atoms in cellulose molecules. But water, H—O—H also contains this group. Dyes are usually dissolved in water when they are applied to cloth. So the active part of the dye was in danger of reacting with water and losing its activity. This would waste much of the dye. Solving this problem was the key to the manufacture and use of the new dyes.

Dr Stephen showed that the problem could be overcome by carefully controlling the pH of the solutions at all stages both when manufacturing the dye and when preparing to use it. However, the problem was not considered solved until samples of the dyes had been stored for up to a year without serious loss of chemical activity.

The next problem was to develop new processes for large scale commercial dyeing suited to this new range of dyes – now called Procion dyes. The first trials were not without their problems. Even when they were solved the sales teams of ICI had to establish that there was a big enough market for the new dyes to justify the expense of setting up manufacturing. Eventually the dyes were launched in 1956, just one hundred years after Perkin discovered mauveine.

The value of the dyes increased when it was realized that they could be applied at room temperature. This saved the cost of the energy needed to heat up the dye baths as required by most other dyes.

As a result of the success of these dyes, ICI invested in building two modern dye works to make them, one at Trafford Park in Manchester, the other at Grangemouth in Scotland. During the first twenty-one years of commercial

Figure 11.16
The old Huddersfield works where Procion dyes were first manufactured.

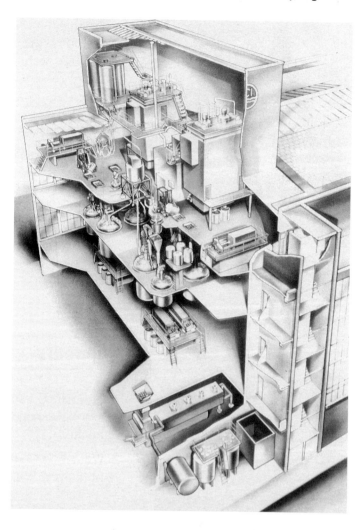

Figure 11.17
Trafford Park works.

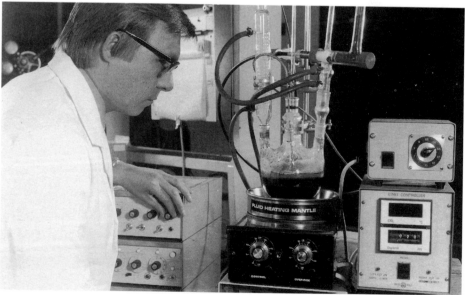

Figure 11.18
Scientist at work in ICI Organics Division research department.

activity large sums of money were needed to provide plant, machinery and buildings. It was not until 1971–2 that money earned from the sale of dyes exceeded the money spent.

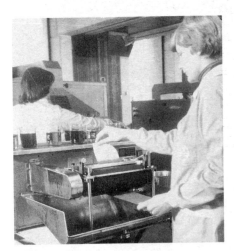

Figure 11.19
Testing dyes in a laboratory.

Throughout this period, and still today, research continues to develop the range of Procion dyes. A blend of cotton with polyester is now a popular fabric. New techniques are needed to dye two different types of fibre in a single process.

6 Why are Procion dyes called ''reactive dyes''?

7 In what way is the reaction between a Procion dye and cellulose in cotton similar to the reactions used to make condensation polymers? (See section **C**8.1.)

8 Look at figure 11.14. To what extent can you account for the success of **a** mauveine and **b** Procion dyes in terms of the information in the table?

9 Draw a diagram in the style of figure 11.15 to explain the fact that Procion dyes lose their activity if they react with water, H—O—H.

Summary

1 Compare the stories of the discovery and development of mauveine and of reactive, Procion dyes. In what ways are they similar? In what ways are they different?

2 What are the advantages of modern synthetic dyes when compared with natural dyes? What are the advantages of natural dyes?

Chapter **C**12	# Chemicals in the medicine cupboard

A typical medicine cupboard in the home will include a variety of pills and ointments as well as bandages and other first-aid materials. This chapter concentrates on mild painkillers and antacids.

*Aspirin is taken as an example to show how drugs can be discovered in plants, and then modified chemically to make them safer and more effective. Many drugs are dangerous and can be abused. This is discussed in Biology Chapter **B**11.*

The worksheets which go with this chapter will introduce you to some of the skills of chemical analysis. You will practise a technique for measuring amounts of substances and use it to analyse antacid tablets.

C12.1 What is the difference between a drug and a medicine?

We usually take medicines hoping that they will prevent or cure diseases. We may also take them to make the symptoms of illness more bearable, so you will often find painkillers and cough pastilles in a medicine cupboard.

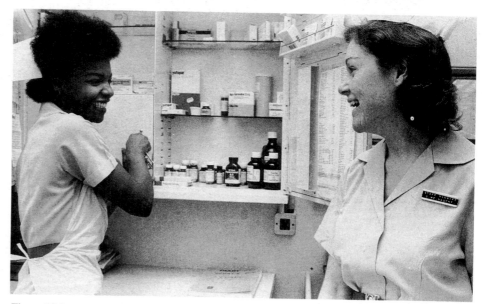

Figure 12.1
A hospital medicine cupboard.

Most medicines are made up from a mixture of ingredients. One or more of these ingredients may be a drug. A drug is a chemical substance which in some way changes the way in which our bodies work. Most medicines contain drugs but not all drugs are medicines. We do not think of ourselves as taking medicine when we drink tea or coffee.

Figure 12.2
The amount of caffeine in a cup of tea.

Caffeine is a drug present in tea and coffee. A cup of tea contains 50 to 100 mg of caffeine. Like other drugs it affects the body in several ways. There is enough of the drug in two cups of tea or coffee to stimulate the brain and make you feel less sleepy. Some people do not drink tea or coffee before going to bed at night because they find that it keeps them awake. If you work at a keyboard, or are involved in similar practical skills, you may find that caffeine helps you to work faster and more accurately. Caffeine is also a *diuretic* which means that it stimulates the production of urine.

Figure 12.3
Some alcoholic drinks.

Wine, beer and spirits all contain alcohol which is another drug which we do not usually think of as a medicine. It can be a dangerous drug. Drinking alcohol is controlled by law: there are laws about the age at which you can drink, the times at which you can drink and about drinking and driving.

The study and science of drugs is called *pharmacology*. A person who has studied the behaviour of drugs and has been trained to dispense them is called a *pharmacist*. If you get a prescription from your doctor you will probably take it to a chemists' shop where a pharmacist in the dispensary will prepare your medicines.

Notice the two uses we have of the word *chemist*. A chemist can be another word for a pharmacist but more generally a chemist is someone who has been trained in the study of chemistry. Any qualified pharmacist will have to be a chemist in this sense too.

Some medicines are mild enough to be sold without prescription. These include pain killers such as aspirin, antacids to treat mild stomach upsets, and a variety of preparations to make coughs and colds more bearable.

Figure 12.4
Pharmacist making up a prescription in a hospital dispensary.

In any medicine cupboard you are also likely to find a variety of antiseptics and disinfectants. This is a reminder that perhaps the biggest contribution of science to the prevention of disease has been to explain the importance of good hygiene. Above all, efficient sewage treatment and the supply of safe drinking water (see section C10.3) have dramatically reduced the danger of serious illness. We are reminded of the dangers of carelessness about hygiene when there are outbreaks of food poisoning.

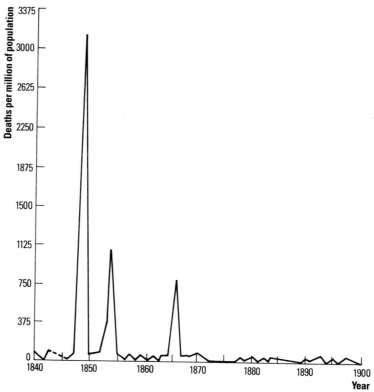

Figure 12.5
The decline of cholera during the nineteenth century.

C12.2 What is an analgesic?

Analgesics relieve pain. They may be mild like aspirin, or powerful like morphine from the opium poppy. Most drugs have unwanted side effects and the problem with the strong painkillers is that they are highly addictive.

You can buy mild painkillers without having a doctor's prescription. Over fifty competing brands are on sale, but the active drug in all of them is one, or more, of these four compounds: aspirin, ibuprofen, paracetamol, and codeine. (You may also find caffeine in some pain relievers but, if so, there is likely to be less than you would get by drinking a cup of tea or coffee.)

Codeine is the most powerful of these drugs, it is chemically related to morphine and only very small quantities can be included in medicines sold without prescription.

Aspirin and ibuprofen reduce fever and inflammation as well as relieving pain. The word *fever* is used to describe illnesses which involve a body temperature above normal. An *inflammation* is a reaction to infection or injury. Inflamed parts of the body are likely to be red, swollen, and painful.

Paracetamol is similar to aspirin in its effects but it does not reduce inflammation. It is now recommended for children under twelve years old instead of aspirin. This is because it seems that aspirin can sometimes contribute to the development of a rare illness called Reye's syndrome.

A *syndrome* is a combination of symptoms which are the signs of a particular disease. The symptoms of Reye's syndrome affect the liver and brain and can be fatal. The syndrome affects about five in a million children in Britain each year.

| Paracetamol BP 500 mg |
| Caffeine BP 32 mg |

| Aspirin BP 325 mg |
| Caffeine BP 22 mg |

Aspirin BP	250 mg
Paracetamol Ph Eur	250 mg
Codeine Phosphate Ph Eur	6.80 mg

Figure 12.6
The ingredients of some pain killers.

The bitter drug extracted from the bark of Cinchona trees is quinine, which was for a long time the only effective treatment for malaria. Malaria used to be a common disease in Britain especially in marshy places. Nowadays synthetic drugs are used to treat malaria, but quinine is still used to prevent muscle cramp.

Figure 12.7
The cinchona tree.

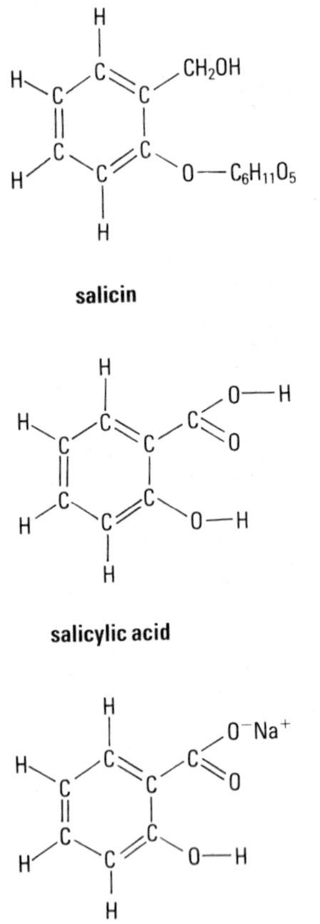

salicin

salicylic acid

sodium salicylate

Figure 12.8
Chemical structures of salicin, salicylic acid and sodium salicylate.

C12.3   How was aspirin discovered?

The story of aspirin is typical of many drugs. It was first discovered in plant material. Later the active chemical was separated and identified by chemists. Once its chemical structure was known it could be modified to improve its properties.

*You are not required to remember the details of the history of aspirin and you will not be expected to know the chemical formulae. Compare the discovery of aspirin with the discovery of mauveine in Chapter **C11**.*

Aspirin is almost certainly the drug which is most widely used as a medicine. Its history began with a paper called "An account of the success of the bark of willow in the cures of agues" (meaning fevers). This was read to the Royal Society of London in 1763 by Edward Stone, a clergyman of Chipping Norton in Oxfordshire. Stone first thought of using willow bark in the treatment of malaria and other fevers because it tasted bitter like the bark of the cinchona tree. Cinchona bark was imported from Peru to provide the only effective treatment for malaria.

Stone held the common belief of the time that a disease and its cure could be found together. Outbreaks of malaria were common in marshy districts where willow trees grew.

Stone's extract of willow certainly gave relief from fevers but it did not cure malaria because it does not contain quinine which is the active drug from cinchona bark. Later the chemical in willow bark was shown to be a compound of salicylic acid and glucose. Salicylic acid is named after the willow tree in which it was discovered. *Salix* is the Latin name for willow.

In the hundred years following Stone's discovery, salicylic acid was widely used to treat fevers and rheumatism. In many ways it was a "wonder drug" but unfortunately it was unpleasant to take because it caused irritation of the mouth, gullet and stomach. The acid can be made less irritating by neutralizing it with alkali and turning it into a sodium salt. However the sodium salt has an unpleasant taste.

The breakthrough came when Felix Hofmann discovered a simple way of modifying the structure of salicylic acid. He was interested in the problem because his father suffered from rheumatism but could not take sodium salicylate. Hofmann converted salicylic acid to acetylsalicylic acid which was soon shown to be effective as a pain killer without the unpleasant taste of sodium salicylate. It became known as aspirin.

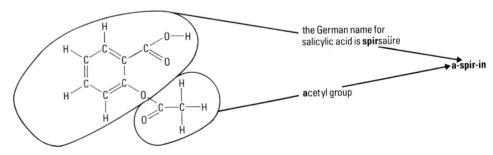

the German name for salicylic acid is **spir**saüre

acetyl group

a-**spir**-in

Figure 12.9
Aspirin was the commercial name given to this compound when it was introduced as a medicine by the German firm Bayer in 1899.

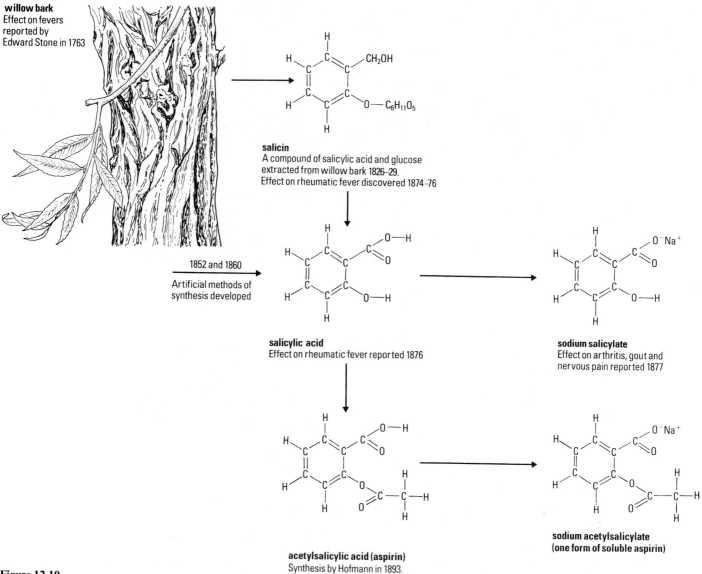

willow bark
Effect on fevers reported by Edward Stone in 1763

salicin
A compound of salicylic acid and glucose extracted from willow bark 1826–29. Effect on rheumatic fever discovered 1874–76

1852 and 1860
Artificial methods of synthesis developed

salicylic acid
Effect on rheumatic fever reported 1876

sodium salicylate
Effect on arthritis, gout and nervous pain reported 1877

acetylsalicylic acid (aspirin)
Synthesis by Hofmann in 1893. Effects on rheumatism and pain discovered 1893–1900

sodium acetylsalicylate
(one form of soluble aspirin)

Figure 12.10
A history of the discovery of aspirin.

1 Hofmann tested acetylsalicylic acid on his father. What precautions are taken now before a new drug can be prescribed by doctors?

2 Look at the structure of salicylic acid in figure 12.8. How many bonds are formed by each carbon atom, each hydrogen atom and each oxygen atom in the structure? (Compare your answers with the table in figure 5.14.)

3 Figure 12.11 is an unfinished version of a simplified equation for the reaction of aspirin with sodium hydroxide. Copy and complete the equation.

the rest of the aspirin molecule   + ...   ⟶   the rest of the aspirin molecule   + ...

Figure 12.11

Aspirin can also damage the lining of the stomach, so you may see warnings on packets of the medicine saying that it should not be taken by anyone suffering from a stomach ulcer.

SOLUBLE ASPIRIN TABLETS   24 Tablets BP 300 mg

DOSE unless otherwise directed by a doctor.   **ADULTS AND CHILDREN OVER 12** one to three tablets

WARNING:
Do not give to children under 12 years old unless your doctor tells you to.
Dose should not be repeated more frequently than 4-hourly intervals and
not more than 4 times in any 24-hour period.
If symptoms persist for more than three days consult your doctor.
The tablets should be dissolved in water before being taken.
DO NOT EXCEED THE STATED DOSE
DO NOT TAKE IF YOU HAVE
A STOMACH ULCER
KEEP OUT OF THE REACH OF CHILDREN

Figure 12.12
What it says on a packet of aspirin.

Most modern aspirin tablets are designed to dissolve quickly to limit the time that an acidic lump can touch the stomach wall. Some tablets form the soluble calcium salt as they dissolve in the stomach. Other tablets are designed to produce the soluble sodium salt in a glass of water before the medicine is taken.

BOX 1   Problem
Patterns of solubility

INTERPRET

The sodium salt of aspirin is much more soluble in water than aspirin itself. The sodium salt of aspirin is ionic. Aspirin is molecular. (See figure 12.10.)

4 Is this a common pattern? Are ionic compounds generally more soluble in water than molecular compounds? Use reference books and tables of data to find evidence to help answer these questions.

Aspirin can react with moisture and turn back to salicylic acid if it is not stored correctly. This is a good reason for not using a bathroom cabinet as a medicine cupboard. Nowadays the tablets are often individually wrapped to prevent this happening. If you have any doubts about an old bottle of aspirin you should throw it away and not use the tablets.

C12.4   What is an antacid?

Section C10.3 describes how acids are formed when the oxides of non-metal elements dissolve in water. In this section you will find that the oxides of metal elements have opposite properties. They form alkalis if they dissolve in water.

The box on page 185 in Chapter C10 shows that it is hydrogen ions which make water acid. The box on page 205 in Chapter C11 explains that it is hydroxide ions which make water alkaline. In

Calcium carbonate BP   534 mg
Magnesium hydroxide BP 160 mg

Calcium carbonate 680 mg
Light magnesium carbonate Ph Eur 80 mg

Figure 12.13
The ingredients on some antacid packs.

this section the ionic theory is taken further to explain what happens when an acid is neutralized by an alkali. Your teacher will tell you if you have to study the details of this ionic theory.

This section includes a box to explain how to do calculations involving amounts of substances in solution. Consult your teacher to find out whether you are expected to learn how to do these calculations.

The word *antacid* is short for anti-acid which means "opposite to acid". Antacids are sold as medicines which will cure indigestion by neutralizing acid in the stomach. If you look at the lists of ingredients on the packets of antacid tablets, you will see that they contain chemicals such as: magnesium hydroxide, aluminium hydroxide, magnesium carbonate, calcium carbonate and sodium hydrogencarbonate. (Note that the common name for sodium hydrogencarbonate is sodium bicarbonate or bicarbonate of soda.)

Magnesium hydroxide is the active ingredient in Milk of Magnesia tablets. It is formed when magnesium oxide reacts with water.

Magnesium hydroxide is slightly soluble in water forming an alkaline solution. It can neutralize the acid in the stomach which is hydrochloric acid. (See Biology Chapter **B**5.)

$$2HCl(aq) + Mg(OH)_2(s) \longrightarrow MgCl_2(aq) + 2H_2O(l)$$

Remember that the "2" outside the bracket in the formula of magnesium hydroxide applies to both atoms inside the brackets. The "2" in front of the formula for hydrochloric acid and the "2" in front of the formula for water are included to balance the equation.

Magnesium chloride is an example of a salt. Salts are ionic compounds made up of metal positive ions and non-metal negative ions. They are formed when acids are neutralized.

BOX 2   What happens when an acid is neutralized by an alkali?

Acids are compounds which produce hydrogen ions when they dissolve in water. A solution of hydrochloric acid contains hydrogen ions, H^+, and chloride ions, Cl^-. A solution of nitric acid contains hydrogen ions, H^+, and nitrate ions, NO_3^-.

Alkalis are compounds which produce hydroxide ions when they dissolve in water. A solution of sodium hydroxide contains sodium ions, Na^+, and hydroxide ions, OH^-. A solution of magnesium hydroxide contains magnesium ions, Mg^{2+}, and hydroxide ions, OH^-. In magnesium hydroxide solution there are two hydroxide ions for each magnesium ion.

The equation shows what happens when hydrochloric acid is neutralized by sodium hydroxide solution.

$$\underbrace{H^+(aq)+Cl^-(aq)}_{\substack{\text{dilute hydrochloric}\\\text{acid}}} + \underbrace{Na^+(aq)+OH^-(aq)}_{\substack{\text{sodium hydroxide}\\\text{solution}}} \longrightarrow \underbrace{H_2O(l)}_{\text{water}} + \underbrace{Na^+(aq)+Cl^-(aq)}_{\substack{\text{sodium chloride}\\\text{solution}}}$$

During this neutralization process, the sodium ions and the chloride ions do not change. They are present in solution before and after the reaction.

The ions which react are the hydrogen ions from the acid and the hydroxide ions from the alkali. These ions combine to form neutral water molecules. So if equal amounts of acid and alkali are mixed, the result is a solution of sodium chloride in water which is neutral.

Acids are also neutralized when they react with carbonates but carbon dioxide is formed as well as a salt and water.

acid + metal carbonate $\longrightarrow$ salt + carbon dioxide + water

An example of this reaction is the use of magnesium carbonate to neutralize hydrochloric acid. You may see reference to "heavy" and "light" magnesium carbonate on antacid packets. Chemically both forms are the same. In the heavy form the shape and size of the crystals allows the powder to be more densely packed than in the light form.

$$2HCl(aq) + MgCO_3(s) \longrightarrow MgCl_2(aq) + CO_2(g) + H_2O(l)$$

5 Antacid preparations are said to cure: indigestion, dyspepsia, heartburn, acid stomach, and flatulence. With the help of a dictionary, find out what these terms mean. Which of the terms are used by doctors?

6 Write a balanced symbol equation for the reaction of magnesium oxide with water. (Turn to the box on page 17 to remind yourself of the rules for writing balanced equations.)

7 Write balanced symbol equations for the neutralization of hydrochloric acid by:
a aluminium hydroxide
b calcium carbonate
c sodium hydrogencarbonate.
(Table 5 in the Data section will help you with the formulae.)

8 Which antacids produce a gas when they neutralize hydrochloric acid? How do you think you might be affected when taking an antacid which produces gas when it reacts?

9 Rewrite the equation for the neutralization of hydrochloric acid by magnesium hydroxide to show what happens to the ions present. (Box 2 will help you.)

The chemicals used to make medicines have to be pure and meet strict standards. The standards are laid down in the British Pharmacopoeia and chemicals which are up to standard are labelled BP.

The active ingredients in a tablet are held together by a harmless binder which makes up most of its bulk. Sometimes flavourings are added as well to make the medicine taste better.

When you eat a meal, the cells in the lining of your stomach produce gastric

Figure 12.14
The specifications for sodium hydrogen carbonate: "reagent grade" and AnalaR.

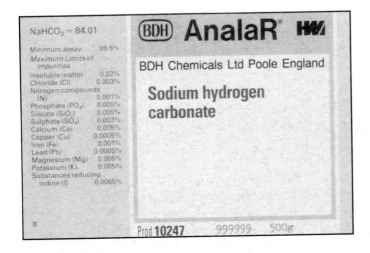

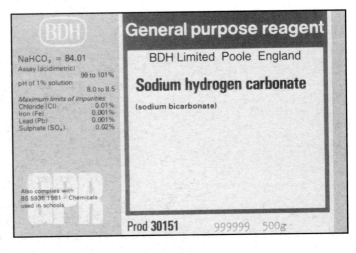

juice. This consists of a mixture of hydrochloric acid with mucus and enzymes. After a meal the stomach produces about 200 ml gastric juice per hour. In a day your stomach may form between one and one and a half litres of gastric juices. (See Biology Chapter **B**5.)

Gastric juice is very acid (pH $\approx$ 1) but at first it is partly neutralized by food. The stomach contents gradually become more acid falling to about pH 2 over a period of about an hour.

---

10 Use the method in box 3 to calculate the volume of gastric juice neutralized by:

a 500 mg calcium carbonate

b 63 mg magnesium carbonate

c 63 mg sodium hydrogencarbonate

d one antacid tablet which contains the ingredients in a, b and c mixed together.

BOX 3   How much gastric juice can an antacid tablet neutralize?

A typical antacid tablet contains 290 mg of magnesium hydroxide. The concentration of hydrochloric acid in gastric juice is about 0.1 mol/L.

● Amount of substance(mol) $= \dfrac{\text{Mass of substance (g)}}{\text{Molar mass (g/mol)}}$

The formula mass of magnesium hydroxide, $Mg(OH)_2 = 24\,u + 2 \times (16\,u + 1\,u)$
$= 58\,u$

The molar mass of magnesium hydroxide $= 58\,g/mol$

The mass of magnesium hydroxide in one tablet $= \dfrac{290}{1000}\,g = 0.29\,g$

The amount of magnesium hydroxide in one tablet $= \dfrac{0.29}{58}\,mol$

$= 0.005\,mol$

The equation for the neutralization reaction is:

$$Mg(OH)_2(s) + 2HCl(aq) \longrightarrow MgCl_2(aq) + 2H_2O(l)$$
$$\text{1 mol} \qquad\quad \text{2 mol}$$

Thus 1 mol of magnesium hydroxide neutralizes 2 mol of hydrochloric acid. The amount of hydrochloric acid neutralized by one tablet $= 2 \times 0.005\,mol$
$= 0.01\,mol$

The concentration of the hydrochloric acid is 0.1 mol/L.

● Amount of substance (mol) = Volume (L) × Concentration (mol/L)

This formula can be rearranged to give:

● Volume of solution (L) $= \dfrac{\text{Amount of substance (mol)}}{\text{Concentration (mol/L)}}$

Thus volume of gastric juice neutralized by one tablet $= \dfrac{0.01\,mol}{0.1\,mol/L} = 0.1\,L$
$= 100\,ml$

C12.5 What is chemotherapy?

Chemotherapy involves the use of chemicals to cure disease. Nowadays you are most likely to hear about chemotherapy in connection with the treatment of cancer. The drugs are powerful and are only used under the supervision of a doctor.

The main aim of research into chemotherapy is to develop chemicals which attack the diseased cells but do not harm the healthy body. This is very difficult to achieve, and many of the drugs used have unpleasant side-effects.

The discovery of antibiotics was one of the greatest successes of chemotherapy. Antibiotics are very effective for treating diseases caused by bacteria. Unfortunately there has been much less success in finding treatments for diseases caused by viruses.

Figure 12.15
Advances in the treatment of cancer and virus diseases are made through the work of people like Jane C. Wright.

Jane C. Wright   Physician and Chemotherapist

1   Dr Jane C. Wright, who was born in 1919, is one of the world's leading specialists in cancer research, particularly in its treatment by chemotherapy. This is a method in which various drugs are used to try to kill or slow down the spread of cancer.

A native of New York City, Dr Wright graduated when she was 23, won a four-year scholarship to the New York Medical Centre, and was awarded the degree of MD (Doctor of Medicine) in 1945.

She worked in various hospitals and in 1952 became Director of the Harlem Hospital Cancer Research Foundation. Three years later she was made a Professor of Research at New York University and also Director of Cancer Chemotherapy at the Medical Centre there.

2   One of the most unusual and fascinating adventures in Dr Wright's career was a safari in Africa where she and medical colleagues took a mobile medical unit through the villages, giving skilled medical treatment to the local people as they went. The unit included medical-surgical supplies, X-ray equipment, electricity, and running water, and it served as an operating room, clinic and laboratory on the safari.

3   On 1st July 1967, Dr Wright made history when she became the first black woman to become a Dean of a medical school – it was at the college where, as a young woman, she had studied for her MD.

Summary

This chapter has included a number of technical terms, some medical and some chemical. Write sentences which show the meaning of the words listed below.

Medical terms:
drug, medicine, pharmacist, analgesic, antiseptic, antacid, chemotherapy, antibiotic, bacterium, virus.

Chemical terms:
acid, alkali, oxide, hydroxide, neutralize, salt.

Topic C4   Energy changes in chemistry

Chapter **C**13   **Fuels and fires**   224

Chapter **C**14   **Batteries**   248

*Energy is one of the big ideas in science. You have probably studied energy transfers in Physics and Biology. In Topics **C** 1 and **C** 2 of this book you can see that large energy changes are involved in manufacturing useful products from raw materials. This topic will help you to understand better the link between energy transfers and chemical reactions.*

    Much of what we do depends on burning things. Oil, gas and coal are burned on a huge scale to generate electricity, to keep us warm and to power many forms of transport. The discovery of gas and oil in the North Sea provided a big boost to the British economy. Yet in your lifetime most of the fuels under the North Sea will have been used up. The use of fuels also creates environmental problems including the acid rain which can kill off the fish in lakes and damage forests. This topic will help you to think about these issues.

    Batteries are a convenient source of electricity. What are they made of? How do they work? Why are they expensive? Why are torch batteries different from car batteries? These are more questions for you to think about as you study this topic.

Fire!

Chapter **C**13

Fuels and fires

In Chapter **C**2 you can read about how we make chemicals from oil. Here you can find out more about oil as a source of fuels. You will compare fuels which are made from oil with natural gas and coal.

Physics Chapter **P**8 shows that most engines depend on a rise in temperature brought about by burning a fuel. We keep warm by using the energy which is released when fuels burn and you can study how to do this efficiently in Physics Chapter **P**9.

Directly or indirectly we depend on the Sun for all of our energy. Biology Chapter **B**3 explains that we, like all living organisms on Earth, depend on photosynthesis for every activity of our lives. When reading this chapter you will realize that most of the fuels we now burn were formed by photosynthesis millions of years ago.

In Physics Chapter **P**12 you have the opportunity to think about all our energy resources. Is there an energy crisis? What are the alternatives to fuels? Can we find a safe way of using nuclear fusion, or fission, to supply the useful energy we need?

C13.1   What is burning?

Figure 13.1 shows a rocket being launched. Inside the rocket engine the fuel burns in oxygen. The fuel is hydrogen and the oxygen comes from tanks of liquid oxygen in the rocket. The rocket has to carry its own oxygen with it, because it is being launched into space where there is no air. (You can read more about rockets and how they work in section P6.5 of Physics Chapter **P**6.)

Since we live on the surface of the Earth, we are surrounded by the air which contains oxygen. Normally when things burn they react with oxygen in the air.

Burning is a chemical reaction. The reaction is generally slow at room temperature, so we usually have to heat things before they will start to burn.

The fire triangle shown in figure 13.2 reminds us of the three things needed to make a fire. We need:

● fuel
● a source of oxygen, which is usually the air, and
● heating to raise the fuel to a high enough temperature.

Figure 13.1
The launch of a Saturn V rocket.

Figure 13.2
The fire triangle.

FIRE

1 Most things will not burn until they have been heated. How are the following heated so that they start to burn?
a the wood of a match
b the methane gas supplied to a Bunsen burner
c the paper used to start a bonfire
d the petrol burning in a car engine
e trees burning in a forest fire.

BOX 1   Experiment
What are the products of burning fuels?

Figure 13.3 shows an apparatus which can be used to investigate the products which are formed when a fuel burns. In the experiment described here, the fuel is ethanol (alcohol).

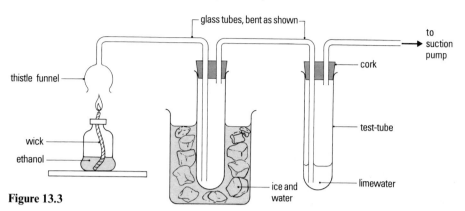

Figure 13.3

In a short time the limewater turns milky white and a colourless liquid condenses in the test-tube cooled in ice. A little soot forms on the glass funnel.

After about 20 minutes, there is enough liquid in the cold test-tube for it to be possible to measure its boiling-point. It boils at 100 °C.

2 What does the limewater test tell you about the products of burning?

3 What is the liquid which boils at 100 °C? Suggest another test which could be used to identify this liquid.

4 How could you check, by experiment, that the two compounds identified by these tests come from burning the fuel and not from the air?

5 Which element is likely to make up most of the black soot? Suggest where this soot comes from.

6 Look up the formula of ethanol in table 4 of the Data section. Write a balanced equation to explain the results of this experiment.

7 How would the results of this experiment differ if the experiment is repeated with **a** hydrogen and **b** hexane instead of ethanol?

Figure 13.4
Burning candle.

8 Why do most fuels continue to burn once they have been heated to a high enough temperature to start combustion?

9 Figure 13.4 shows a candle burning.
a Which part of a candle is the fuel which burns?
b Why must a candle have both wax and a wick?

10 Methane gas *burns* when heated in air (see section C1.3). Limestone *decomposes* when it gets hot (see section C4.4). Use these examples to explain the difference between burning and decomposing.

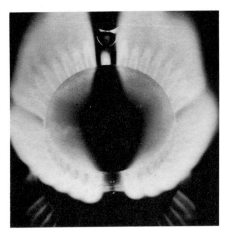

Figure 13.5
Gas is used for home heating and cooking.

Another word for burning is *combustion*. A car engine is called an internal combustion engine because the fuel burns inside the cylinder of the engine. See Physics Chapter **P**8, section **P**8.4.

C13.2   What is a good fuel?

*In this section you can read about the fuels we use for home heating, cooking, transport and for generating electricity. We also need energy for living and growing. The fuels for the "living engine" are described in Chapter **B**9 of your Biology book.*

In the laboratory you will be asked to decide what you think makes a good fuel and carry out tests on possible fuels (see Worksheets C13B, C and D).

Fuels are burned to provide us with energy. Often we use this energy directly to keep us warm or to cook our food.

We also use the energy from burning fuels to drive cars, lorries, tractors and other vehicles. In most power stations, fuels are burned in boilers designed to produce steam at high pressure. High pressure steam is needed for the turbines which turn the dynamos. The dynamos transfer energy to electricity, as explained in Physics Chapters **P**11 and **P**18.

---

11 The energy from burning fuels is also required in the manufacture of materials.
Explain why energy is needed to make:
a petrol from oil (see section **C**2.3)
b ethene from ethane (see section **C**2.4)
c sodium hydroxide from salt (see section **C**4.3)
d aluminium from bauxite (see section **C**4.5)
e glass from limestone, sand and sodium carbonate (see section **C**6.1)
f pottery from clay (see section **C**6.5)
g brass from copper and zinc (see section **C**7.3)
h a plastic bottle from polythene (see section **C**8.5).

Fuels are chemicals which react with an oxidizing agent (see section **C**4.5). Usually the oxidizing agent is oxygen itself. Energy is released during the reaction and new chemicals are formed.

Any substance which reacts with oxygen, or with another oxidizing agent, could be used as a fuel.

The overall changes which take place when a fuel burns are shown in figure 13.6.

The best known fuels include petrol, diesel, coal and natural gas. All these fuels are burned in oxygen.

Scientists have developed other fuels for special purposes. For example, a chemical called hydrazine is used as a rocket fuel. It is not burned in oxygen, but it reacts instead with concentrated nitric acid.

Not every chemical which burns is a good fuel. The fuel must release plenty of energy when it is burned, but there are many other important points to consider. For most people it is probably convenience and cost which seem important. We prefer fuels which are safe to use and which do not produce unpleasant gases and smoke when they burn.

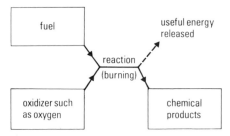

Figure 13.6
What happens when a fuel burns.

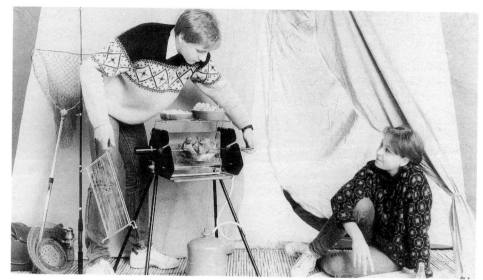

Figure 13.7
Gas barbecue.

Figure 13.8
Modern heater burning smokeless solid fuel.

Figure 13.9
Motor mowers often run on petrol.

12 Campers used to cook on wood fires and on paraffin (primus) stoves. Why do many campers now use gas burners?

13 In smokeless zones it is illegal to heat homes by burning coal in open fires. What are the problems with using coal in an ordinary fireplace? What types of solid fuel are now available which get round the problems?

14 What dangers are involved in using petrol as a fuel? What safety precautions are necessary when storing this fuel?

15 Make a survey of the fuels which you, and your family, use every day for such things as room heating, cooking, and transport. Try to explain why you use these fuels instead of the alternatives. (Note that electricity is **not** a fuel.)

16 Explain why it is incorrect to call electricity a fuel.

If cost alone is considered, then the best fuel is the one which gives the most energy when it is burned for the smallest amount of money.

BOX 2   Experiment
Measuring the energy
transferred when fuels burn

The calorimeter is designed to
make sure that as much as
possible of the energy released
when the fuel burns is
transferred to the water. The
experiment is carried out in
two stages.

Stage 1
A measured quantity of fuel is
burned and the temperature
rise of the water is recorded.

Stage 2
The heater is switched on. The
heater is kept on until the
temperature rise of the water
in the calorimeter is the same
as it was with the fuel. The
time taken for the heater to
produce this measured
temperature rise is
recorded.

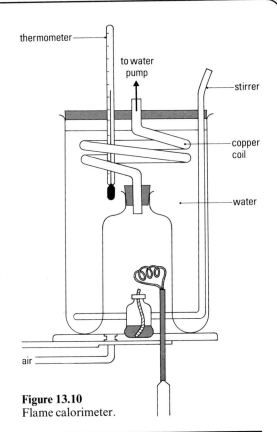

Figure 13.10
Flame calorimeter.

17a Explain the purpose of the coiled length of copper tube in the calorimeter
shown in figure 13.10.
b Why is the tube made of copper and not glass?
c Why must the water in the calorimeter be stirred well before any temperature
measurements are made?
d Some of the energy transferred from the burning fuel, or from the electric
heater, is used to heat up the apparatus as well as the calorimeter. How is it that
this does not affect the accuracy of the results?
e What measurements have to be made to find the mass of fuel burned during the
experiment?

18 The calorimeter shown in figure 13.10 was used to measure the energy output
from burning ethanol. Burning 0.5 g of ethanol produced a temperature rise of
5.5 °C. The same temperature rise was produced by switching on the 25-watt
heater for 600 seconds. Use this information to calculate the energy released when
1 g of ethanol burns, by answering these questions:
a How much energy was transferred to the calorimeter from the electric heater in
600 seconds? (See sections P16.1 and P17.2.)
b How much energy was transferred to the calorimeter by burning 0.5 g of
ethanol?
c How much energy is released when 1 g of ethanol burns according to this
experiment? (Convert your answer to kilojoules per gram, kJ/g. Remember that
1000 J = 1 kJ.)

19 Why does the apparatus shown in figure 13.10 give more accurate results than
the apparatus shown in figure 13.11, which you may have used in the laboratory?

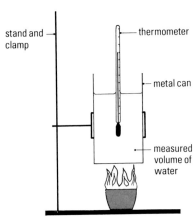

Figure 13.11
Apparatus for measuring the energy
output from burning fuels.

The amount of energy released when a fuel is burned can be accurately measured. The measurement is made using a calorimeter. Figure 13.10 illustrates one type of calorimeter. Figure 13.11 shows the equipment you may use if you do the experiment yourself.

The energy values of many different fuels have now been measured. The results compare the number of joules of energy which can be obtained from a gram of each fuel. Some values are shown in figure 13.12.

For comparison, the energy which is available from some foods is also shown in figure 13.12. This energy is released during respiration, and is necessary to keep us alive (see Biology Chapter B9).

Fuel	Energy value in kJ/g
Wood	21
Coal	34
Heating oil	46
Natural gas (methane)	52
Propane	51
Butane (calor gas)	50
Petrol	42
Sugar	17
Butter	31

Figure 13.12
Fuel values.

Fuel	Cost in pence/kg
Wood	2.5
Coal	14
Heating oil	20
Natural gas	19
Propane	61
Butane	69
Petrol	55
Sugar	50
Butter	200

Figure 13.13
Fuel prices in 1986.

It can be seen at once that fuels like natural gas (methane) and petrol are the best sources of energy per gram. However, this does not necessarily mean that they are the best value for money. The costs of some fuels are given in figure 13.13.

20a For each fuel in figure 13.13 work out how much energy you could obtain for 1p using the energy values given in figure 13.12.
b Which fuel is the best value for money?
c Suggest why we do not all use this fuel.
d Suggest why propane and butane are much more expensive than methane. (See section **C2.3**.)

21 The price of a fuel can change quite quickly. List the reasons:
a why a fuel might become more expensive, and
b why a fuel might become less expensive.

22 Imagine that the price of coal falls rapidly, so that it becomes much cheaper than natural gas. If your house was heated by natural gas, explain how you would decide whether or not to convert to using coal.

C13.3   What are fossil fuels?

In this section you can read about processes which took place millions of years ago and which took millions of years to happen. This is hard to imagine. You may find it helpful to look at sections C4.1 and Biology B23.2 which suggest ways of thinking about these huge periods of time.

Figure 13.14
Fossilized ferns are found in coal.

You may be able to link the information in this section to your studies of the formation, discovery, extraction, transport and uses of fuels in geography.

Most of our energy is obtained by burning coal, oil or gas. These fuels are sometimes called *fossil fuels* because the energy they release on burning was stored up millions of years ago. This stored energy came from the sun and was used to make chemicals in the leaves of plants by the process of photosynthesis.

Coal

The coal now being mined in Britain was formed about 300 million years ago. At that time much of northern England and the North Sea was a large, tropical swamp. Huge ferns, trees, and other plants grew in the swamp. Over many years these plants grew, reproduced and died. In this way a layer of dead and rotting plants formed.

Over millions of years sand and mud was washed down by rivers so that it covered the decaying plants. In time the sediments over the future coal seams reached depths of over 3000 metres. Heat and pressure gradually converted the plant remains into coal. Fossilized plants found in lumps of coal are part of the evidence for this theory of how coal was formed.

Coal consists almost entirely of the element carbon. This is why the period of geological time when coal was formed is called the *carboniferous* period. (See section **C4.4**.)

The largest reserves of coal are found in the USSR, the USA and in China, and these countries are the largest producers. The UK mines about 3 per cent of the total world production. Many old mines are now closing as the coal runs out and becomes more expensive to extract. At the same time new mines are being opened.

Figure 13.15
A high-powered shearer cutting its way through a rich seam of coal.

Mining has always been dangerous because of the risk of fires, explosions, flooding and roof falls. Humphry Davy (see figure 4.26) contributed to the

Figure 13.16
Sampling the air in a mine.

development of the safety lamp which is still used in the pits as a safety device. Changes in the flame of the lamp can give a warning of a dangerous build-up of explosive gas.

Today scientists are still working to improve safety in mines. Fires may result when a coal seam becomes overheated by absorbing oxygen. We now know that the gases given off from the coal can act as a warning of a dangerous rise in temperature. The amount of gas involved is small, so sensitive methods of analysis such as chromatography have to be used. In some mines, samples of the mine atmosphere are fed through fine tubes to instruments at the surface for analysis.

Coal provides more than 20 per cent of the world's energy. About 65 per cent of the coal mined in the world is burned in power stations.

Figure 13.17
Coal mine and power station working in harness.

Most of the rest of the coal is used as a fuel in industry and for home heating. It is difficult to burn coal without producing smoke. For home heating the coal is specially treated to make a smokeless fuel which can be burned in an open fire without causing air pollution.

When coal is heated strongly to a temperature of about 1000 °C, in the absence of air, it splits into three parts:

* coal gas, consisting of hydrogen, methane and carbon dioxide
* a mixture of tarry liquids, and
* a solid residue called coke.

Coke is used in blast furnaces to extract iron from its ores. It is also used as a source of carbon in the chemical industry and as a fuel. Coke is made from coal in special ovens heated by gas. Crushed coal is fed into the ovens and then left to heat up. One or two days later, when all the liquids and gases have been driven off, the doors at the front and back of the ovens are taken off and a large ram pushes out the glowing coke. The coke is immediately cooled by being sprayed with water.

Figure 13.18
Discharging hot coke from an oven.

The liquids condensed from the coke ovens are a useful source of chemicals. Many of the organic chemicals which are now made from oil (see Chapter **C2**) can also be made from coal. Coal reserves are much larger than oil supplies and so it is likely that coal will once again be an important source of chemicals as oil wells begin to run dry.

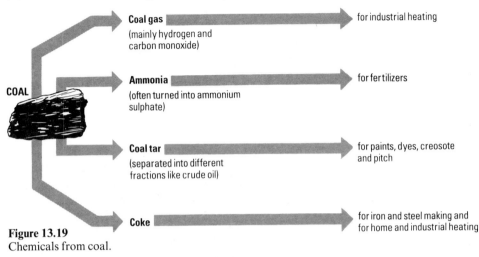

Figure 13.19
Chemicals from coal.

---

23 With the help of the fire triangle (figure 13.2) explain why:
a coal must be heated in the absence of air to make coke
b the coke must be quickly cooled when it is pushed out of the coke ovens.

24 In a blast furnace, the coke is the fuel which heats the furnace and also the reducing agent which extracts iron from its ore. Explain the chemistry of these reactions with the help of reference books.

25 Suggest possible uses for the mixture of gases driven off from coal during the manufacture of coke.

26 Why are organic chemicals now made mainly from oil rather than from coal?

Oil

It is difficult to find out how oil was formed. Oil is a liquid and it contains no fossils. The places where oil is found are not necessarily the places where it was formed because it can flow through porous rocks.

One theory is that oil was formed from the remains of tiny sea animals and plants. These sank to the bottom to form muddy sediments in places where the water was free from oxygen. Bacteria feeding on the remains removed the oxygen. Heat and pressure slowly turned the chemicals into oil.

The oil did not stay in the sediments in which it was formed. It flowed up through porous rocks until stopped by a cap rock. (See figure 13.20.)

The oil is not found in vast underground lakes as many people believe. It is trapped as tiny droplets in the pores between the grains of sedimentary rocks, like water in a sponge.

Crude oil as it comes out of the ground is of little use as a fuel. It is a thick, black liquid looking like dark treacle. As explained in Chapter **C2**, it is a mixture of many different hydrocarbons (see figure 2.5 on page 23). The number of carbon atoms in the molecules ranges from one to about fifty carbon atoms.

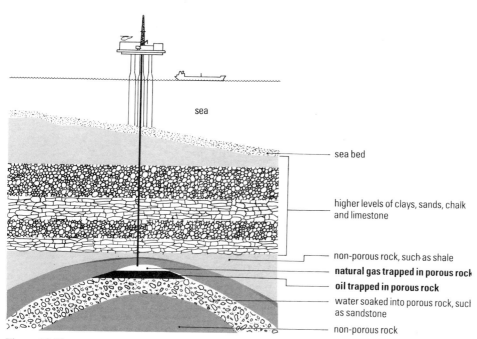

Figure 13.20
Gas and oil reservoirs under a cap rock. Gas is often found above the oil.

sea

sea bed

higher levels of clays, sands, chalk and limestone

non-porous rock, such as shale

natural gas trapped in porous rock

oil trapped in porous rock

water soaked into porous rock, such as sandstone

non-porous rock

Figure 13.21
Wellhead in the North Sea.

Figure 13.22
Drilling in progress on the Shell/Esso Brent B production platform.

Figure 13.23
Transporting the oil by tanker.

Figure 13.24
The trans-Alaska oil pipeline, with a pumping station in the background.

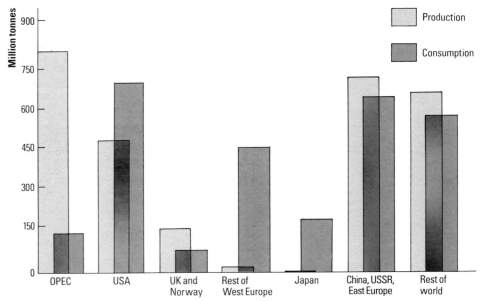

Figure 13.25
World production and consumption of oil in 1986.

27 Look at figure 13.25.
a Which are the three largest oil-producing regions?
b Which are the largest oil-consuming regions?
c In which regions are oil production and consumption nearly in balance?
d Why are large quantities of oil and oil products transported around the world?

28 Methane, ethane and propane are hydrocarbons found in oil (see section **C2.2**).
a What is the name of the family of hydrocarbons to which these compounds belong?
b Draw the graphical formulae of these three hydrocarbons.
c Why does decane have a higher boiling-point than hexane?
(See table 4 in the Data section and figure 8.29 on page 157.)

Oil has to be refined before it can be used. Figure 13.26 shows that a modern refinery is a complex place with its miles of pipes and shining columns. There are three main stages:

- separating
- converting
- purifying.

The oil is separated into fractions by distillation. The theory of this process is explained in section **C2.3**. About 90 per cent of the oil distilled is used to supply fuels and lubricants. Only about 10 per cent is used to make new chemicals, as described in Chapter **C2**.

The crude oil is heated in a furnace and pumped into the distillation column near the bottom. The vapours rise up the column and cool as they do so. The different hydrocarbons condense when they reach a level which is at a temperature below their boiling-points. The liquids collect on metal trays and run out of the column along pipes.

Designs of fractionating columns vary from one refinery to another. Often the oil is first split into four main fractions which are then processed and separated further to obtain the required output of the various products. The information in figure 13.28 is merely a summary of many complex processes.

Figure 13.26
The Esso refinery and chemical manufacturing complex at Fawley, Southampton.

Figure 13.27
Fractionating column at Sullom Voe, Shetland.

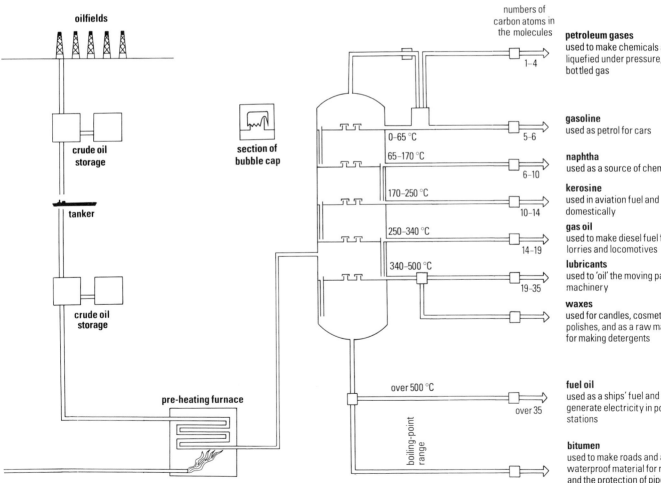

Figure 13.28
Products of distillation.

oilfields

crude oil
storage

tanker

crude oil
storage

pre-heating furnace

section of
bubble cap

numbers of
carbon atoms in
the molecules

0–65 °C

65–170 °C

170–250 °C

250–340 °C

340–500 °C

over 500 °C

boiling-point
range

1–4

5–6

6–10

10–14

14–19

19–35

over 35

petroleum gases
used to make chemicals and, when
liquefied under pressure, sold as
bottled gas

gasoline
used as petrol for cars

naphtha
used as a source of chemicals

kerosine
used in aviation fuel and
domestically

gas oil
used to make diesel fuel for buses,
lorries and locomotives

lubricants
used to 'oil' the moving parts of
machinery

waxes
used for candles, cosmetics and
polishes, and as a raw material
for making detergents

fuel oil
used as a ships' fuel and to
generate electricity in power
stations

bitumen
used to make roads and as a
waterproof material for roofing felt
and the protection of pipelines from
corrosion

additional
processes

The conversion stage of refining is needed to make sure that all the oil is turned into useful products. *Cracking* is used to split large molecules into smaller ones (see section **C2.4**). *Polymerization* is used to join up small molecules to make bigger ones (see section **C2.5**). *Reforming* is used to increase the yield of the fractions used to make petrol and also to improve the quality of the fuel.

The final stage of refining involves the removal of impurities such as sulphur compounds. If there is sulphur in a fuel it turns to the acidic gas sulphur dioxide when the fuel is burned.

---

29 Crude oil is sometimes called "black gold". Why is this an appropriate name?

30 Why are all the molecules about the same size in a fraction produced by distilling crude oil?

31 How many of the products of refining oil can you find in your home? How and where are they stored? What are they used for?

32 Crude oil and refinery products are transported by road tanker, rail tanker and by pipeline. What are the dangers involved in transporting these materials? What are the advantages and disadvantages of these different methods of transport?

33 Why is it necessary to remove sulphur compounds from fuels? (See section **C10.3**.)

Gas

Most of the natural gas found in rocks under the North Sea was formed at the same time as coal during the carboniferous period. The gas did not stay in the coal seams but escaped upwards into sandy sediments. These sediments were laid down when the area was a desert and the sea was shallow. At the same time salt was deposited as the sea evaporated and crystallized. The salt was important because gas could not get through it. The salt sealed the top of the porous sandstone, trapping the gas as it escaped from the newly formed coal.

Gas from the North Sea wells is brought ashore by pipeline to terminals on the east coast of Britain. The supply of methane from the Brent oil field is described in Chapter **C2**. The gas is blended with gas from other fields so that the mixture will produce the guaranteed amount of energy when burned. A trace of a smelly chemical is added so that gas leaks can be detected, because natural gas on its own has no smell.

The gas enters a national grid of high pressure, underground pipelines controlled from a centre in Hinckley, Leicestershire. Natural gas is mostly methane but it contains traces of other chemicals. Figure 13.30 gives a typical analysis.

Most natural gas is burned as a fuel. It also has an important chemical use in Britain: it is used to make ammonia. This is described in Chapter **C16**.

---

34 Since the world's reserves of coal, oil and gas are limited, we should perhaps consider controls so that each fuel is used efficiently. Which fossil fuel do you think should be used for the following?
Domestic heating, manufacturing chemicals, road, rail and air transport, generating electricity.

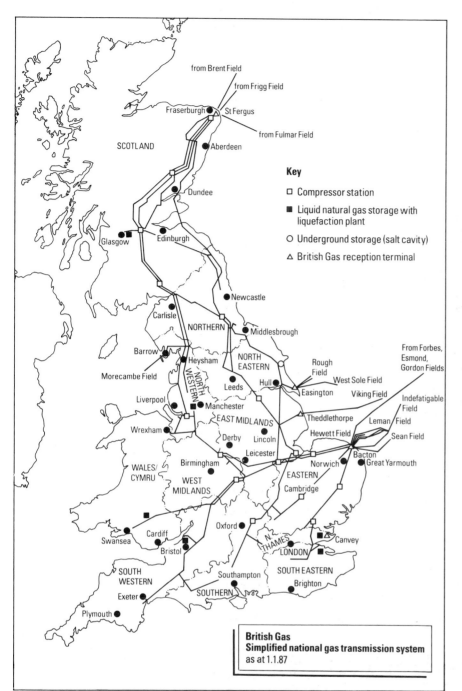

Figure 13.29

Gas	Percentage in natural gas
Methane	92.84
Ethane	3.27
Nitrogen	2.48
Propane	1.18
Carbon dioxide	0.23

Figure 13.30
The composition of natural gas.

Figure 13.31
Drill bits being taken out of a ceramic radiant tube furnace fired by gas.

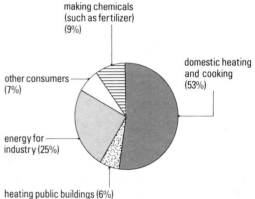

Figure 13.32
Gas use in Britain.

C13.4 Fossil fuels and the environment

*Biology Chapter **B**13 introduces you to ways of studying the natural environment. Biology Chapter **B**17 describes some of the ways in which industry can affect living things in the environment.*

*This section looks in more detail at the problems which can arise from the use of fossil fuels. Worksheet **C**13E will help you to investigate the topic of air pollution in more detail. If you have carried out the investigations on Worksheet **B**17B you will be in a better position to think about the effects of acid rain.*

Extracting fossil fuels from the earth, transporting them, and burning them may damage the environment in a number of ways.

When coal is mined, the waste rock is dumped on the surface in tips which may become a permanent and ugly part of the landscape. As far as possible the waste from tips is used for road construction and railway embankments.

Open cast mining can create ugly scars in the countryside. The land has to be restored for agriculture or forestry once the coal has been removed.

Figure 13.33
Open cast mine (left) in operation, (right) after restoration.

Another consequence of mining is subsidence of the land. This damages buildings and roads.

Figure 13.34
This house has been damaged by mining subsidence.

Drilling for oil under the sea is hazardous. An accident at a well head can lead to large amounts of oil gushing out into the water where it kills most living things.

Oil tankers may also pollute the sea. These ships use sea water to wash out their tanks, and some oil escapes into the water despite operating procedures designed to prevent this. Accidents may also happen when oil is being loaded on or off a tanker at a terminal.

Big tankers have been wrecked in the English Channel causing serious pollution along the coast. The first and most famous disaster involved the *Torrey Canyon*. This ran onto the rocks off Lands End in 1967, spilling 120 000 tonnes of oil. On that occasion it was not just the oil which caused problems. The detergents used to disperse the oil were toxic. Much effort has been spent since then on improving ways for controlling oil spills and for cleaning up or dispersing the oil.

Figure 13.35
This blowout in the Gulf of Mexico in 1979 destroyed part of the world's richest shrimp fishing ground and the jobs of 90 000 fishermen.

Figure 13.36
A floating boom designed to control spills of oil, seen here in practice at Sullom Voe, Shetland.

All fossil fuels contain carbon, so carbon dioxide is formed when they burn. It is estimated that about ten thousand million tonnes of carbon dioxide are added to the atmosphere each year in this way. Not all this carbon dioxide stays in the air. Some dissolves in the seas. Some is taken up by plants during photosynthesis. However, the concentration of carbon dioxide in the air is increasing. It has increased from about 310 parts per million in 1950 to 338 parts per million in 1980.

Carbon dioxide is not poisonous and it is needed for plant growth. You may be wondering why it matters that its concentration in the air is increasing. Have you noticed that cloudy nights are often warmer than clear ones? That is because water vapour in the air cuts down the amount of infra-red radiation which escapes into space carrying energy with it. Carbon dioxide has a similar effect. This means that an increasing level of carbon dioxide in the air leads to an increase in the average temperature at the surface of the earth. This is sometimes called the *greenhouse effect* because it is similar to the way in which the glass of a greenhouse stops infra-red radiation escaping.

This problem has been recognized for some time. At the beginning of this century the Swedish chemist, Svante Arrhenius, predicted that the mean temperature of the Earth would rise by about 4 °C if the concentration of carbon dioxide in the atmosphere doubled. Present estimates are similar.

Figure 13.37
A big power station on the River Trent – you can see the high stacks and the cooling towers by the river.

A rise of a few degrees may seem small, but it can be enough to cause significant changes in the climate. This could damage agriculture and food production. Another fear is that much of the ice at the poles might melt; this would cause a rise in the average sea level and widespread flooding.

Unfortunately it is very difficult to make reliable predictions. It is much more difficult to decide how burning fossil fuels will affect the climate in the next twenty years than it is to make a weather forecast in Britain for the next few days.

All the energy released when fossil fuels burn ends up spread out in the surroundings, warming up the earth, air and water. The total quantity of energy is small compared with the energy which comes from the Sun, so on a world scale the amount of warming is probably not important.

However, the use of fossil fuels is concentrated in the highly populated, industrial areas. Near power stations the waste energy can have an effect on the local environment. For example, the water taken from rivers for use as a coolant in power stations is returned to the river at a higher temperature, and this can change the balance of plant and animal life in the river.

---

35 How might the worldwide destruction of forests affect the level of carbon dioxide in the air? (See Biology sections **B**3.6 and **B**15.2.)

36 The carbon dioxide concentration in the air varies during the year. It is lowest in late summer and highest in late winter. Suggest a reason for this rise and fall.

37 The concentration of carbon dioxide is increasing at an average rate of about 0.8 parts per million (p.p.m.) per year. At this rate how long will it take for the level to rise from 338 p.p.m. to 500 p.p.m. (which is roughly double the estimated level before the Industrial Revolution)?

Figure 13.38 illustrates some of the other pollution problems which arise when fossil fuels are burned. Coal and oil contain sulphur compounds which form sulphur dioxide when they burn. In moist air the sulphur dioxide forms sulphuric acid which then results in acid rain.

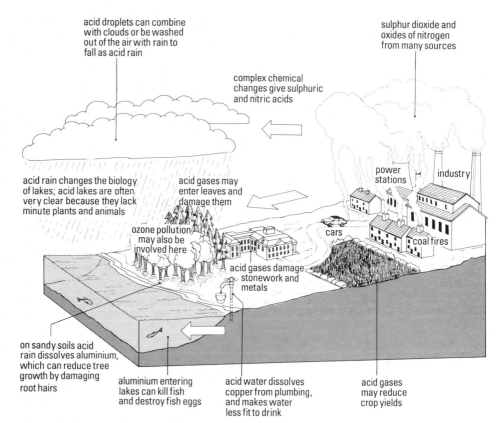

Figure 13.38
Acid gases and acid rain.

As you can see from figure 13.37, modern power stations now have tall stacks so that the waste gases are released high up. This means that the fumes are diluted with plenty of air before they reach the ground. It also means that much of the pollution is carried long distances by the wind.

In Sweden there are thousands of lakes which are almost lifeless because they have become too acid. Much of the acid comes from power stations in

Figure 13.39
This German machine sprays a mixture of magnesium and calcium carbonate powder onto the soil around trees affected by acid rain.

Gas	Approximate percentage in the exhaust from a car engine
Carbon dioxide	9
Carbon monoxide	5
Oxygen	4
Hydrogen	2
Hydrocarbons	0.2
Nitrogen oxides	0.2
Sulphur dioxide	trace

Figure 13.40
The composition of car exhaust gas.

other countries in Europe. British power stations have contributed to the problem because the prevailing westerly wind blows the polluted air to Sweden.

Other chemicals such as nitrogen oxides also contribute to acid rain. Nitrogen and oxygen do not react at low temperatures, but oxides of nitrogen are formed when fuels are burned at a high temperature in power stations and car engines.

The exhaust from cars contains other pollutants too. These include carbon monoxide and unburnt hydrocarbons. Nitrogen oxides react with hydrocarbons and oxygen in sunlight to form ozone. Scientists are still trying to find out whether some of the damage to European forests is caused by ozone produced in this way.

---

38 The table in figure 13.40 shows the percentage of various gases in the exhaust from a car engine.

a The gases listed in figure 13.40 add up to about 20 per cent of the exhaust gases. Which one gas would you expect to make up most of the remaining 80 per cent?

b Where does the carbon dioxide in the exhaust gas come from?

c Suggest reasons for the presence of carbon monoxide and unburnt hydrocarbons in the exhaust gases.

d How might hydrogen gas be formed from petrol in the cylinder of a car engine?

e The two main nitrogen oxides are NO and NO_2. Write two equations to show the formation of these gases from nitrogen and oxygen.

f Which is the most poisonous of the gases listed in figure 13.40?

g Why is it very important not to run a car engine in a closed garage?

We are now much more aware than we used to be of the many ways in which burning fuels may damage the environment. In Britain, the Clean Air Act of 1956 and the introduction of "smokeless zones" have had a dramatic effect on the amount of smoke in the air. People no longer suffer from the choking fogs which killed thousands in the 1950s. Buildings which were once blackened with soot have now been cleaned, and they stay clean. These measures show that it is possible to control the problems.

Methods for taking sulphur from fuels are now known, and sulphur dioxide can be removed from waste gases before they are discharged to the air. New methods for burning coal in power stations have been invented which run at a lower temperature and so reduce the formation of nitrogen oxides. Car engines can be designed so that they give out fewer pollutants and in some countries new cars have to be fitted with special exhaust systems which contain catalysts which help to get rid of harmful gases.

Concern about lead pollution has meant that there is now agreement in Europe to remove lead from petrol. It is intended that from 1991 all new cars in the EEC will run on lead-free petrol.

Chemists can contribute to the control of pollution. Sensitive methods of analysis are needed to identify pollutants and measure their amounts.

Figure 13.41
Monitoring air pollution.

---

39 Limestone can be used to remove sulphur dioxide from the waste gases of power stations. The limestone decomposes to give calcium oxide which then reacts with sulphur dioxide to make calcium sulphate. Write equations to show how this happens.

C13.5 How can fires be fought?

Preventing fires and putting out fires are major problems. In this section you can see the way in which a knowledge of chemistry can help to solve the problems. You may be asked to plan and carry out an investigation to study methods of fire-proofing fabrics.

We get most of our energy from the controlled burning of fuels. Sometimes there are accidents. Unwanted fires are a threat to life and property, and putting them out can be a serious problem.

Big fires are often in the news. The largest known forest fire burned from September 1982 to July 1983 in Borneo. This fire destroyed 36 000 square kilometres of forest and changed the local climate from drought to almost continuous monsoon.

Most of the methods used to control fires can be understood with the help of the fire triangle (figure 13.2). There are three ways of putting a fire out:

- removing the fuel
- removing, or limiting, the supply of oxygen
- cooling so that burning slows down and stops.

It is very important to use the correct type of extinguisher on a fire. Fires have been grouped into four classes to make it easier for people to decide how to tackle them in an emergency.

Class A: ordinary materials such as paper, wood, cloth and plastics
Class B: flammable liquids such as cooking fat, oil and petrol
Class C: flammable gases such as natural gas
Class D: metals.

Fire brigades have to be ready to cope with a wide range of fires. For instance, spilt chemicals following road accidents may catch fire. To help deal with this, lorries carrying chemicals have a coded information plate showing what they contain.

Figure 13.42
Water is sprayed onto a fire to cool it. The water evaporates and the steam cloud also helps to cut off the air supply.

Figure 13.43
Forests are planted with fire breaks. This helps to control fires because they cannot spread across an area with no fuel.

Figure 13.44
Dry powder extinguishers can be used for all types of fire depending on the powder. The powder forms a surface layer which cuts off the oxygen from the fuel. Ammonium phosphate is very effective on class A fires because it melts to form a crust over the burning material and so cuts off the air. Salts such as sodium chloride, potassium chloride and barium chloride are used to put out metal fires.

Figure 13.45
A fire blanket is made of material which will not burn. The blanket cuts off the oxygen from the fire.

Figure 13.46
The cylinder contains liquid carbon dioxide under pressure. The jet of dense gas surrounds the fire and cuts off the air supply. The gas is cold as it leaves the cylinder and so it also has a cooling effect.

Figure 13.47
Foam extinguishers form a blanket over the fire which prevents oxygen reaching the burning fuel. They also provide some cooling.

40 Why is it dangerous to use water to try to put out fires involving:
a oil (class B)
b metals such as magnesium (class D)
c electrical equipment?

41 What should you do to deal with the following situations?
a a small girl has stood too near a fire at bed-time and her nightdress has caught fire.
b you are cooking chips for the family meal and the oil catches alight
c water is dripping from a ceiling lamp and the electrical cable is beginning to smoke
d a paraffin heater has tipped over and flaming paraffin is spreading over the carpet.

42 The information in figure 13.48 shows the number of casualties in fires at home in Britain during a typical year.
a On a bar graph show the total, fatal casualties caused by room heaters, smoking materials, fires in grates, cookers, and children with matches.
b What steps could be taken in your home to reduce each of these risks?

Cause of accident	Casualties		
	Non-fatal	**Fatal**	**Total**
Electric, paraffin and gas room heaters:			
Electric	178	68	246
Paraffin	332	35	367
Gas	29	14	43
Smoking material	337	122	459
Fires in grates	375	66	441
Electric and gas cookers:			
Electric	278	6	284
Gas	244	16	260
Children with matches	138	26	164
Total	1911	353	2264

Figure 13.48
Casualties from fires in Britain in one year.

Clearly it is better to prevent fire than to have to try to put out a fire once it has started. It is now a legal requirement that children's night-clothes should be flameproofed. Fireproofing also has to be applied to the scenery on stage in theatres.

Fires in the home have become more dangerous now that furniture is filled with plastic foam instead of traditional materials. These foams give off poisonous fumes which suffocate people before they can escape from the fire. This is why it is worth spending more money to buy furniture with flame-proofed fabrics. These fabrics delay the start of a fire. This allows people more time to escape before they are overcome by toxic fumes.

BOX 3   Problem
How effective are flame-proofing agents?

P
PLAN

Some fabrics can catch fire quickly even if they just brush against a flame. Flame-proofing can increase the time it takes for materials to catch fire.

One flame-proofing agent is a solution of borax and boric acid in water. Plan an investigation to find out if this is effective. You can also compare fabrics which you have treated with others which have been flame-proofed commercially.

Does flame-proofing work on all the common types of fabric? Does washing affect flame-proof finishes? Does ironing have any effect?

C13.6   Fire on the farm

This section gives you a chance to use your knowledge of fires and fire fighting to think about a serious problem which can arise from the storage of large amounts of oil at refineries.

Figure 13.49 shows you a tank farm at a refinery. The lids of the storage tanks float on the surface of the oil in the tank so they move up and down with the level of liquid in the tank.

Tanks used to store oil and petrol are designed in this way for safety

Figure 13.49
The "tank farm" at Dalmeny, Scotland.

reasons. It would be dangerous to have an air gap between the lid of the tank and the top of the oil or petrol.

It is very difficult to make a lid that fits the inside of a tank exactly. So, instead, the lids are made slightly too small and then a flexible seal is used to fill the gap round the edge. Even with a seal some of the more volatile hydrocarbons do leak from the tank and are blown away in the wind.

On still days, a mixture of hydrocarbon vapour and air can build up above the lid in a tank which is not very full. This mixture could start a very serious fire if ignited by accident. So the oil companies operate very strict safety precautions to cut out all possible sources of ignition. They may even have sheep grazing round the tanks to keep the grass short rather than risk using motorized mowers.

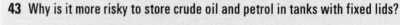

43 Why is it more risky to store crude oil and petrol in tanks with fixed lids?

44 What must be the properties of the material used to make the seals on the moving lids if they are to be effective and safe?

45 What suggestions can you make to reduce the risk of a flammable mixture of air and hydrocarbons collecting above the lid of a tank which is only partly full?

46 Thunderstorms can cause fires in tank farms. How do you think this happens? How could you reduce the risk of a thunderstorm starting a fire?

Figure 13.50
Fighting a fire in a refinery storage tank.

Despite all the precautions there have been serious fires in refinery storage tanks, and so there is now a good deal of experience of how to deal with them. Unfortunately mistakes have sometimes been made along the way. The fire fighters have not always chosen the best method for controlling the fires.

On some occasions fire fighters have even tried to put out the fire by spraying powerful jets of water into the tank. Not surprisingly, this has led to serious accidents. Fire fighters were killed when the burning tank they were trying to control suddenly boiled over and engulfed them in flaming oil. Why might this happen? How can it be prevented?

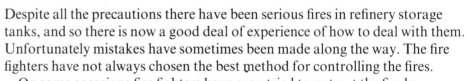

47 Why might some fire fighters have hoped that spraying water into the tank would put out the fire? (Hint: think in terms of the fire triangle.)

48 What will happen to any water which is sprayed into the tank but does not evaporate in the fire? Where will it go? (Hint: look at the data in figure 13.51.)

Hydrocarbon	Formula	Boiling-point in °C	Density in g/ml
Pentane	C_5H_{12}	36	0.63
Decane	$C_{10}H_{22}$	174	0.73
Eicosane	$C_{20}H_{42}$	344	0.79

Figure 13.51
Boiling-points and densities of selected hydrocarbons.

49 The hydrocarbons in oil start to burn when they evaporate and mix with air. This means that the hydrocarbons with low boiling-points burn first. How will this affect the density of the burning oil at the surface?

50 Put together your answers to questions **48** and **49** to suggest a reason for the fact that burning tanks which have been sprayed with water may suddenly boil over.

51 Suggest a method of changing the design of storage tanks so that they do not boil over if they catch fire and are sprayed with water.

52 Improved methods of controlling oil fires have been developed and are widely used in oil storage tank farms. What do you think they are?

Summary

1 Make a summary of the main ideas about fuels by answering these questions.

What is a fuel?

What is needed to make a fuel burn?

What are the products when hydrocarbons burn?

Which are the main fuels commonly used in Britain?

Which of the fuels you have listed are fossil fuels?

What are the properties of a good fuel for heating in homes?

Why is coal mainly used for generating electricity?

Why do most of the fuels used for transport come from oil?

What are the main ways by which your use of fuels affects the environment?

Why does the use of natural gas cause less pollution than coal or oil?

What precautions are taken to prevent accidental fires resulting from our use of fuels?

2 Burning, rusting and respiration all involve oxygen in the air. In what ways are these processes similar and in what ways are they different? (Refer to this chapter and also to sections **C**7.4 and Biology **B**9.2.) It may help you to make comparisons if you write word equations for the changes.

Chapter C14

Batteries

You may have wondered what batteries are made of and how they work. This chapter will tell you what they are made of. You can find out about the way in which batteries transfer energy electrically in Physics Chapters P16 and P17. You will find an explanation of how batteries work in Chapter C18.

C14.1 How were batteries invented?

One day in 1786 an Italian professor called Luigi Galvani was working in his laboratory with his wife and his assistant. One of them noticed that the legs of a dead frog gave a sudden kick when touched with the metal blade of a scalpel. Galvani was very excited by this observation.

Galvani decided to study the effect of lightning on muscles. He fixed some frog's legs with brass hooks to an iron grid outside his house and then connected the legs to the ground with long wires. When lightning flashed the leg muscles twitched and contracted as he had hoped they might.

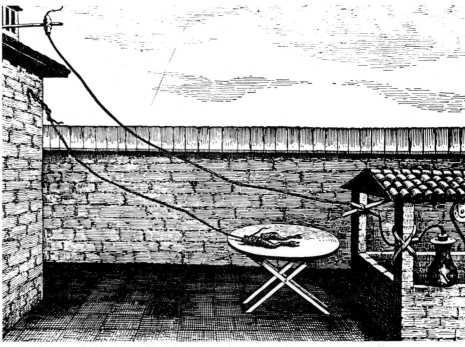

Figure 14.1
Galvani's experiments with lightning and frogs' legs.

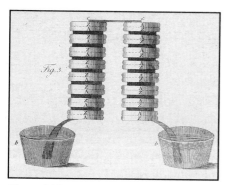

Figure 14.2
Volta's original pile.

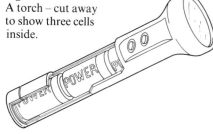

Figure 14.3
A torch – cut away to show three cells inside.

Sometimes Galvani had to wait a long time for lightning. On 20th September 1786 the sun shone brightly from a clear sky and Galvani was getting impatient. He looked at the frog's leg which his wife had freshly prepared and hung on the iron grid. The frog's legs were held with brass hooks and when he pressed these firmly against the iron of the grid he noticed that the muscles contracted just as they did when there was lightning.

Galvani took the frog indoors and pressed the hook against an iron plate. The result was the same. He did many more experiments and came to the conclusion that the bodies of dead animals must contain electricity. He called it "animal electricity".

A better explanation of Galvani's experiments was put forward by another Italian, Alessandro Volta who was Professor of Physics at the University of Pavia. Volta repeated Galvani's experiments and after two years of study he published a new theory.

He decided that the twitching of the muscles was caused by an electric current which was produced when two different metals touch something moist and are then connected together. Volta discovered that the size of the current depended on the kinds of metals used. He arranged the metals in a series, so that pairs of metals which were further apart in the series produced the greater electric currents. Volta's series was: zinc, lead, iron, copper, silver, gold.

In 1800 Volta described how he had invented a device which would produce a steady electric current. It came to be called a "voltaic pile" and it was the first battery.

Volta's pile consisted of a stack of pairs of discs of zinc and silver. Each pair of discs was separated by moist cloth or paper. It was this invention which made possible Humphry Davy's discoveries described in section **C4.3**. There is a picture of Davy's version of the pile in figure 4.26.

Each pair of metals with the solution between them in Volta's pile is called a *cell*. The whole collection of cells is called *a battery*. (The scientists borrowed the word battery from the army where a line of guns is still called a battery.)

In everyday conversation we commonly use the words cell and battery interchangeably. The bulb of the torch in figure 14.3 is lit by a *battery* of three *cells*.

You may be wondering about the use of the word "cell" here. The word is also used in biology to mean something which seems quite different. The use of the same word in prisons gives a clue to its meaning.

Notice that in all the uses of the word there is something held inside a small container. If you look at the picture of Davy's battery in figure 4.26 you can see that the parts of each cell are held in boxes. You know from biology that the parts of living cells are trapped inside a membrane. Prisoners are locked in behind bars.

C14.2   What types of cell do we use today?

Volta's experiment shows that an electric voltage is produced when two different metals are dipped into a conducting solution to form a cell.

The metal which is higher in the activity series becomes the negative electrode. The other metal forms the positive electrode.

Early cells were "wet" cells because they were filled with watery solutions. In modern "dry" cells the electrolyte is a paste and is not soaking wet.

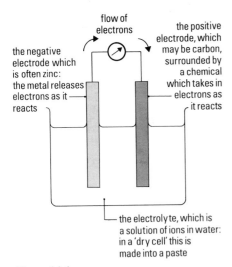

Figure 14.4
The essential features of a cell.

Figures 14.5 to 14.7 illustrate some of the variety of cells in use today.

In all cells, the flow of electric charge results from chemical reactions. This can be explained with the help of the same ionic theory used to describe what happens during electrolysis. You can study the theory in section **C**18.3.

The cells shown in figures 14.5 to 14.7 do not last forever. The chemicals which react to produce the flow of electrons eventually run out, then the battery has to be replaced. Batteries of this sort cannot be recharged and they are known as *primary cells*.

Primary cells are used because they are convenient but they are an expensive source of energy. Mains electricity costs about 5p for a kilowatt-hour. (See Physics section **P**12.1). It costs about £50 to transfer the same amount of energy using a zinc–carbon cell.

Energy from small button cells may cost as much as £7000 for a kilowatt-hour. Of course, these cells are not designed to supply large quantities of energy. They are used in low power instruments such as cameras, digital watches and calculators. They last for many months and are very convenient, so their cost is reasonable.

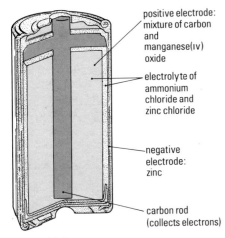

Figure 14.5
Zinc-carbon cell. This is the commonest and cheapest type of cell. The electrolyte is acidic. The negative electrode is zinc and the metal atoms turn into ions and dissolve when a current is taken from the cell. These dry cells have a relatively short life because the zinc slowly reacts even when no current is taken from the cell. The cells are liable to leak when old.

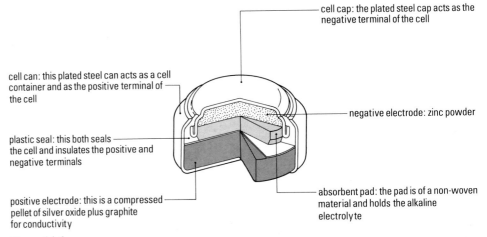

Figure 14.6
Silver oxide cell. Here too the electrolyte is potassium hydroxide and the negative electrode is zinc. These small cells have a long life. They maintain a steady voltage and are very suitable for use in watches, calculators and camera exposure systems.

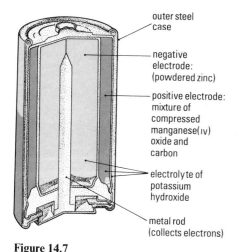

Figure 14.7
Alkaline manganese cell. The electrolyte is
potassium hydroxide. These cells last
longer than zinc-carbon cells and are less
likely to leak. Also higher currents can be
taken from them so they are suitable for
use in toys or cassette recorders which
have a motor taking a heavy or
continuous current.

C14.3   How can cells be used to store energy?

*The problem of storing the energy released when fuels are burned in power stations is discussed in
Physics Chapter P12. One possibility is hydroelectric pump storage schemes as described in
Physics section P12.4. Another possibility is to use storage cells which are described in this section.*
*Rechargeable cells are used in some electric vehicles. You can compare some of the advantages
and disadvantages of these vehicles compared with those that run on petrol with the help of
Physics Worksheet P18A.*

Primary cells such as those pictured in figures 14.5 to 14.7 are made of
expensive materials and so they are an expensive source of energy. Once they
have been used they cannot be recharged so they are thrown away.

It is often better to use *secondary cells* which can be recharged using mains
electricity. Rechargeable nickel-cadmium cells are now widely available. They
are more expensive to buy at first and a transformer is needed to recharge
them. However, they can be recharged over 500 times.

8a Find the current price of a typical zinc–carbon cell, and work out the cost of 500 of
these cells.
b Find the cost of a nickel–cadmium rechargeable cell and of a transformer to
recharge it. Estimate the total cost of buying the equipment and using the cell 500
times. (Ignore the low cost of the mains electricity used.)
c When is it better to use disposable cells? When might rechargeable cells be
preferred?

The most widely used type of rechargeable cell is the lead–acid cell used to
make car batteries. These cells store some of the energy from the burning
petrol ready to restart the engine and to light the lamps when the car is
stationary. (See figure 14.10.)

Lead–acid cells are also used for stand-by power. Stand-by power is an
emergency supply of energy which takes over if there is a power cut. This is
vital in hospitals – surgeons must be able to finish an operation if the mains
electricity supply fails. In industry and commerce, stand-by power is available
so that computers, telephones, and vital production processes can continue if

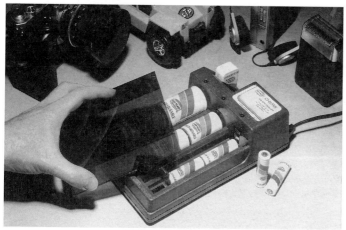

Figure 14.8
Domestic nickel-cadmium cells being recharged in a transformer.

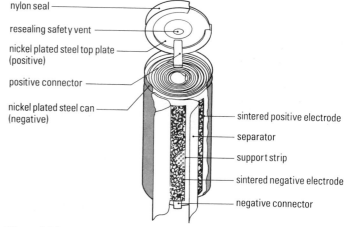

Figure 14.9
Nickel-cadmium cell.

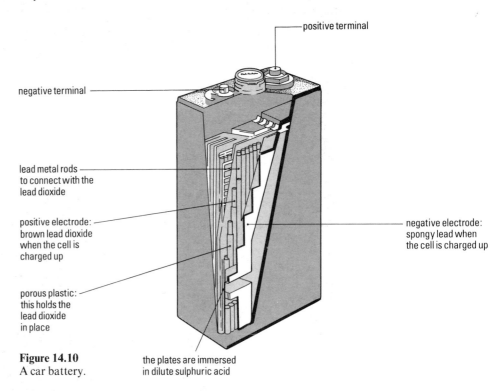

positive terminal

negative terminal

lead metal rods
to connect with the
lead dioxide

positive electrode:
brown lead dioxide
when the cell is
charged up

porous plastic:
this holds the
lead dioxide
in place

negative electrode:
spongy lead when
the cell is charged up

Figure 14.10
A car battery.

the plates are immersed
in dilute sulphuric acid

there is a break in the supply of mains electricity. It is also required in theatres and cinemas to guarantee emergency lighting at all times.

When a lead–acid cell is fully charged, the active material on the positive plates is lead oxide. The negative plates are metallic lead. The electrolyte is sulphuric acid. Chemical reactions take place as a current is drawn from the cell and they produce lead sulphate at both electrodes. If the cell is allowed to discharge completely, both plates are converted to lead sulphate. There is then no chemical difference between the two electrodes and so there is no cell voltage.

The reactions in the lead–acid cell use up some of the sulphuric acid in the electrolyte solution as it discharges. This means that during discharge the acid solution gets less dense. Measuring the density of the electrolyte solution with a hydrometer is the easiest way of checking the condition of a lead–acid cell.

Figure 14.11
Stand-by batteries.

Figure 14.12
Loading the battery used to power the Bedford electric van.

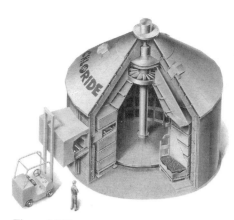

Figure 14.13
Sodium-sulphur battery designed to store off-peak electricity.

Lead–acid cells are also used for transport. They are commonly used in milk floats and fork-lift trucks. They are used at sea too, and some of the biggest storage batteries are in submarines. The biggest submarine batteries have 448 cells and weigh 250 tonnes.

The problem with lead–acid cells is that it takes a bulky and heavy battery to store a useful amount of energy. The Bedford electric van in figure 14.12 has a range of 130 km and can carry a load of 1 tonne. The unladen mass of the van is 2.5 tonnes and half of this is made up of the lead–acid battery.

New cells are being developed which can store more energy and are lighter. The sodium–sulphur cells now being tested should be able to double the range of electric vehicles with batteries which are only one third as massive. The chemistry of the new sodium–sulphur cells is explained in Chapter **C18**.

A large sodium–sulphur battery is being designed for use in the United States to store off-peak electricity from power stations. The problem with this type of battery is that it runs at temperatures between 270 and 410 °C. This means that they are not suitable for everyday use in the home. (See question 9.)

---

9 The sodium–sulphur cells in a battery are kept inside a double skin of stainless steel. The space between the steel layers is filled with glass fibre and is partially evacuated. The battery stays at its working temperature if 80 per cent of the stored energy is discharged every 24 hours and if it is then fully recharged.
a What is the purpose of the double skin of steel with its glass fibre and partial vacuum?
b Suggest an explanation of the fact that the batteries stay within the working temperature range of 270 to 410 °C if they are discharged and recharged every day.

Figure 14.14
The Apollo 15 command and service modules, seen from the lunar module.

C14.4   What is a fuel cell?

Secondary cells are a convenient way of storing energy transferred from power stations by an electric current. However, about 70 per cent of the energy released when fuels burn in power stations is wasted. Is there a more efficient way of making use of the energy from fossil fuels?

The challenge is to design a cell which is supplied with fuel and oxygen and which uses the energy from the reaction directly to create an electric voltage. This is called a *fuel cell*.

A fuel cell running on methane and oxygen would provide a much more efficient source of electric power than current power stations. So long as the fuel and oxygen are supplied, the cells will continue to run. Fuel cells and power stations are compared in figure 14.15 on the next page.

Fuel cells were an important feature of the Apollo programme which landed people on the Moon, because they were used to power the space craft. However more research and development is needed before fuel cells are in general use.

The problem is to invent a fuel cell which will run on cheap hydrocarbon fuels and air. The difficulty is to find suitable catalysts for the electrodes which will continue to work reliably.

In the Apollo fuel cells the fuel was hydrogen. The electrolyte was a solution

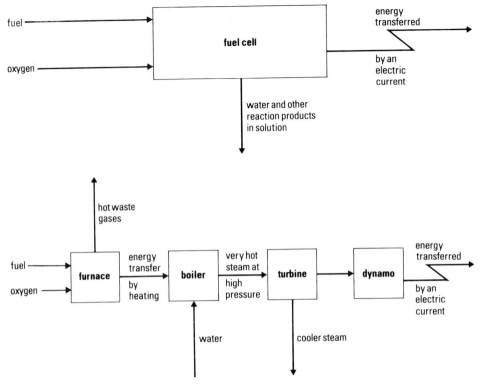

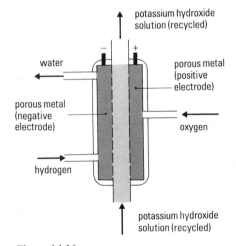

Figure 14.15
Comparing a fuel cell with a power station.

of potassium hydroxide. The reactions in the cells produced water which was used for drinking by the astronauts.

Figure 14.16 is highly simplified to illustrate the principle of a hydrogen–oxygen fuel cell. The secret of success lies in the design of the electrodes. They act as catalysts for the reaction in the cell as well as acting as the electrical conductors connecting the cell to outside circuits.

Figure 14.16
The essential features of a hydrogen/oxygen fuel cell.

Summary

1 Here is a list of devices which get their energy from cells and batteries. For each one, decide which type of primary or secondary cell you would recommend. Give the reasons for your choice in each case.

a toy car   **f** personal stereo
b battery razor   **g** space station
c invalid car   **h** portable computer
d hearing aid   **i** radio-controlled model aeroplane.
e system of fire alarm bells

Topic C5

Soil and agriculture

Chapter **C**15   **Soil**   256
Chapter **C**16   **Fertilizers**   268

Farmers today can grow much more food in this country than they could thirty years ago. Science and technology have helped to change the pattern of farming and chemists have played a big part in the increase of crop yields.

This topic starts by looking at the natural changes which take place as soil is formed. Soil is a precious resource which can easily be lost if carried off by floods or blown away by storms. We have to understand the soil if we are to be able to go on growing enough food.

There is now increasing concern about the large scale use of nitrate fertilizers. Why are fertilizers needed? What happens to fertilizers in the soil? Why does it matter if nitrates get into water supplies? To understand the issues you have to know something about the chemistry which you will learn by studying this topic.

Ploughing.

Chapter **C**15 **Soil**

*The elements needed for life, such as carbon and nitrogen are involved in natural cycles which you study in Biology (see Chapter **B**15). The energy which drives these natural cycles comes from the Sun. Above all we depend on plants which use the energy of the sun in photosynthesis (see Chapter **B**3) and so we rely on the soil in which plants grow and from which they get both water and minerals they need (see section **B**15.1).*

This chapter will help you to understand that the soil is one of the most important life-support systems on our planet because we depend on it for the production of most of our food.

Figure 15.1
The Earth from the Moon.

C15.1 Why study soil?

Figure 15.1 is one of the pictures of the Earth taken from the Moon which reminds us that we are isolated in space. As far as we know, the Earth is the only planet in the Solar System where there is life. The parts of the Earth and its atmosphere in which there are living things we call the *biosphere*.

Many chemical reactions take place in the oceans, in the atmosphere, and in the thin layer of soil and broken rock which together make up the biosphere. In the soil the three states of matter come together: solid mineral particles, liquid water, and gases from the air. The study of the chemistry of soil is part of the science called *environmental chemistry*.

Soil forms very slowly compared with your lifetime. It takes centuries to

Figure 15.2
The screes of Wastwater, Cumberland.

produce soil a few centimetres deep. Between 3000 and 12 000 years might be
needed to form a depth of soil equal to the length of this page.

Only about 10 per cent of the Earth's surface is suitable for agriculture and
much of the best land is already being farmed. Agriculture often speeds up
soil erosion. In many parts of the world, soil is being washed or blown away
much faster than it is being formed.

Figure 15.3
Severe soil erosion by water on a farm in the United States.

Figure 15.4
Soil erosion by wind: a "dust bowl" in the United States.

---

1 What is meant by the term *environment?* Which are the factors that affect the
conditions for living things in an environment? (See Biology section **B**13.1.)

2 Why do you think that agriculture can speed up soil erosion?

3 List some of the other activities in Britain which cause a loss of soil for growing
food. (Biology section **B**18.1 will give you some ideas.)

C15.2   How is soil formed?

*In Biology Chapter **B**3 you can read about the factors which affect the rate of photosynthesis, such
as temperature and concentration. In Chapter **B**5 you are able to investigate the way in which
these same factors control the rate of digestion. Here you will study the importance of similar
factors in determining the rate at which rocks weather and soil forms.*

*In the laboratory you will have the opportunity to plan and carry out investigations into the
factors affecting the rates of chemical changes helped by Worksheets **C**15 A, B, C, and D. This will
introduce you to some of the techniques used to measure the rates of reactions. You will also use
tables and graphs to display your results and so make use of some of the skills you have been
introduced to in Physics.*

*Before you read on, you may find it helpful to look again at section **C**4.1 so that you can recall
the meaning of terms such as: igneous rock, granite and basalt.*

Soil is formed by the weathering of rocks. In the end, the rocks are completely
broken down to soluble chemicals and colloidal particles in water. Soil is a
temporary stage in a journey of minerals from mountains to the sea. Soil
erosion speeds up the journey and wastes a precious resource.

The changes brought about by weathering are remarkable. Dead rock is changed into living soil. The changes are of two main kinds:

- physical changes which break the rock into smaller pieces
- chemical changes which decompose minerals.

These two types of change work together, and both are speeded up if there are plants and animals living and growing in the soil. When the rocks are broken into smaller pieces, the surface area increases and so more of the rock is exposed to air and water. Chemical reactions weaken the structure of rocks. Then it is easier for the rocks to be shattered by frost or broken open by the growth of plant roots.

The study of weathering is not easy. Few weathering processes can be seen in action because they are so slow. Scientists can analyse the rocks which are there at the start. They can also investigate the products formed after weathering. Then they have to try to work out how the products are formed from the rock, and this can be very difficult. It is also difficult to test the theories in the laboratory because the natural processes are slow and involve so many variables.

Physical weathering

Frost is a powerful rock breaker. This illustrates one of the remarkable properties of water. As water cools it contracts in volume like other liquids, until the temperature reaches 4 °C. Then something unusual happens. As the water cools further to 0 °C it expands.

Water trapped in a rock crevice can force open the crack as it freezes with a pressure of expansion well over a hundred times the pressure of the atmosphere.

Figure 15.5
Frost-shattered rock.

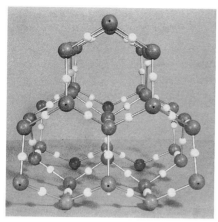

Figure 15.6
The structure of ice. Note the hexagonal
arrangement of molecules.

Figure 15.8
Frost-damaged stone.

Figure 15.9
The roots of this tree have grown into, and
widened, the crack in the rock.

Figure 15.7
Snow crystals.

If you look at the model of ice structure in figure 15.6 you can see that, as
water freezes, the molecules are arranged in an open network structure. The
structure has large empty spaces in it so that the molecules are more widely
spaced than they are in water. The hexagonal pattern you can see in
figure 15.6 is reflected in the shape of ice crystals in figure 15.7.

You do not have to climb mountains to see the way in which frost can
break up stone. Figure 15.8 shows an example of frost damage to building
stone, and you will probably be able to find examples near where you live.

The growing roots of shrubs and trees work their way into the crevices of
rock and build up big pressures as they do so. Cracks are widened as the roots
grow (figure 15.9).

4 What is the approximate size of colloidal particles? (See section **C**9.1.)

5 Why do householders have a problem with burst pipes during frosty weather in
winter?

6 Why does ice float on water? How does this help fish and other living things in
ponds during the winter?

Chemical weathering

Chemical weathering happens when rock is exposed to water, oxygen and acids. Have you ever broken open pebbles on a beach? You will have seen that the inside of the stones looks quite different from the outer surface. This is one of the reasons why geologists carry hammers. They have to break open pieces of rock so that they can see what they look like when they have not been weathered.

The speed of weathering depends on many factors, including the type of rock, the temperature, and the acidity of the water. Water in soil is acidic because of dissolved carbon dioxide. Some of the carbon dioxide comes from

---

BOX 1   Practical problem
Investigating the rate of reaction of an acid with a rock

The weathering of most rocks is a slow process. In the laboratory you can study the factors which affect the rate at which water and acids attack rocks in a shorter time – by investigating the reaction of hydrochloric acid with marble (a form of calcium carbonate). You will need to use hydrochloric acid which is more concentrated than the acids in soil.

You are asked to plan investigations to study how the following factors affect the rate of reaction:

- the temperature
- the concentration of the acid
- the particle size of the rock.

You can use your experiments to test some of the information given in this chapter. Is it true that a 10 °C rise in temperature will roughly double the rate of reaction? Does the reaction go faster if the acid concentration is increased? How much more concentrated must the acid be to double the rate of reaction? Is it true that breaking up a rock into small pieces increases the rate at which it reacts with acids?

This is the equation for the reaction of marble with hydrochloric acid:

$$CaCO_3(s) + 2HCl(aq) \longrightarrow CaCl_2(aq) + H_2O(l) + CO_2(g)$$

To follow the rate of this reaction you can either measure how quickly one of the reactants is used up, or measure the speed with which one of the products is formed.

The state symbols in the equation may help you to choose which reactant or product you are going to study and measure during the reaction.

You have to decide what measurements you are going to make. This will help you to choose a suitable arrangement of apparatus for your experiments. Table 7 in the Data section may give you some ideas about the apparatus for this investigation.

When planning your experiments you must think about all the factors which can vary, and decide how you are going to control them so as to make a fair test. If you allow too many properties to vary at once you will not be able to make sense of your results.

In each set of experiments you will want to vary just one factor while keeping all the others constant. Then you will make measurements to find out how the rate of the reaction changes when you alter the factor which you are varying.

Figure 15.10
Look at the difference between the inside and outside of a freshly broken rock sample.

the air, but most of it is formed by bacteria and fungi. These micro-organisms break down organic material. Organic acids and carbon dioxide are formed.

Chemical reactions go faster when the temperature rises. For many reactions the speed doubles for each 10 °C rise.

7 How much faster will rock weather in a tropical country with a mean annual temperature of 20 °C than in a country in a temperate zone with a mean annual temperature of 10 °C?

8 What is the name of the acid formed when carbon dioxide reacts with water? (See section **C10.3**.)

9 How is it that bacteria and fungi add to the level of carbon dioxide in soil water?

The weathering of igneous rocks is important because it results in the formation of clay. There are many types of igneous rock and the reactions involved in weathering are varied and complex. Figure 15.11 gives a simplified picture of the weathering of granite.

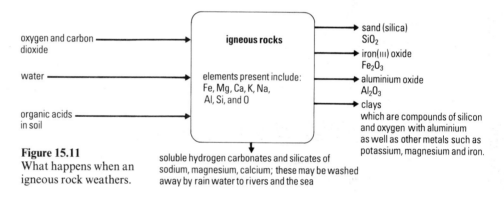

oxygen and carbon dioxide

water

organic acids in soil

igneous rocks

elements present include:
Fe, Mg, Ca, K, Na,
Al, Si, and O

sand (silica)
SiO_2

iron(III) oxide
Fe_2O_3

aluminium oxide
Al_2O_3

clays
which are compounds of silicon and oxygen with aluminium as well as other metals such as potassium, magnesium and iron.

soluble hydrogen carbonates and silicates of sodium, magnesium, calcium; these may be washed away by rain water to rivers and the sea

Figure 15.11
What happens when an igneous rock weathers.

Figure 15.12
Successful agriculture on the sides of a volcano (Tenerife).

In most climates, granites decompose slowly and release plant nutrients only gradually. Volcanic basalts weather faster and produce soils which are rich in nutrients. The sides of volcanoes have always been favoured for agriculture despite the risks.

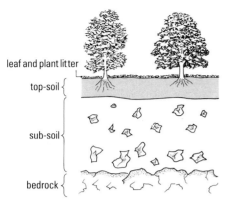

leaf and plant litter

top-soil

sub-soil

bedrock

Figure 15.13
A generalized soil profile.

Figure 15.14
A typical soil profile in Glen Torridon.

C15.3   What is in soil?

*You may find it helpful to re-read sections **B**2.4 and **B**15.2 in your Biology book before you study this section. This will remind you of some of the living organisms found in soil.*

Soil is a mixture of rock particles, air, water, living organisms, and the decaying remains of dead plants and animals.

● *Rock particles*
Rock sizes range from gravel (with particles greater than 2 mm in diameter) to clay (with particles less than 0.002 mm in diameter). The clay particles are so small that they are colloidal (see Chapter **C**9).

Chemical weathering continues at the surface of the rock particles in soil. This releases the minerals which plants need for growth. The larger the surface area the faster this process takes place. Figure 15.15 shows how the surface area increases as a piece of material is broken up into smaller and smaller pieces.

1cm
1cm
1cm

Figure 15.15
A piece of rock the size of a sugar lump (about 1 cm × 1 cm × 1 cm with a surface area of 6 cm²) ends up with the surface area of a large table (about 3 m × 1 m with a surface area of 30 000 cm²) when crushed to particles as fine as clay in the soil (about 0.002 mm across).

● *Air*
The roots of plants need oxygen to grow properly and so do most of the other organisms which live in the soil. In Chemistry we often ignore the nitrogen in the air because it is so unreactive. However, you will know if you have read Biology Chapter **B**15 that nitrogen gas in the soil is taken up by bacteria in the roots of some plants.

● *Water*
Water in the soil helps to release plant nutrients from rock particles. The water enters the soil mainly as rain and it may carry with it dissolved ions, as explained in the next chapter.

● *Living organisms*
One gram of fertile soil may contain tens of thousands of protists and algae, as well as huge numbers of bacteria and fungi. It is very much alive. More obviously it contains worms and other small animals which also contribute to its fertility.

The action of living things in soil produces carbon dioxide and organic acids which speed up the release of soluble salts from rock particles. Bacteria and fungi help to break down the dead remains of plants and animals so that the nutrients are recycled.

● *Roots of plants*

Roots bind the soil and help it to remain porous without being washed away. It is important for plants that their roots are held firmly in the soil in contact with water in which nutrients are dissolved.

● *Dead organic matter*

The breakdown of dead leaves and roots produces humus. This is a brown substance which with water can form a colloidal gel (see section **C**9.1). Humus binds together the rock particles, holds moisture, and can act as a reservoir of plant nutrients.

Figure 15.17
Aerating the soil.

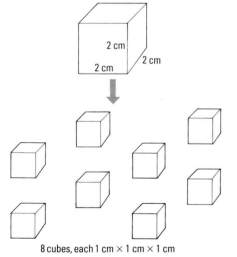

8 cubes, each 1 cm × 1 cm × 1 cm

Figure 15.16

10 Look back at figure 15.11. Which of the main plant nutrients become available in the soil as igneous rocks weather?

11 Figure 15.16 shows a 2-cm cube being divided into eight 1-cm cubes. What is the surface area of the one large cube? What is the **total** surface area of the eight small cubes?

12 Figure 15.17 shows a machine which treats games pitches to make the grass grow better. What is the machine doing, and why is it necessary?

13 How do earthworms contribute to the fertility of soil?

14 Most living things are inactive in soil below 6 °C. Activity increases to a maximum at about 30 °C but falls sharply at higher temperatures. Suggest explanations for the effect of temperature on the rates of natural changes in soil. (It may help if you remind yourself about some of the properties of enzymes described in Biology Chapter **B**5.)

15 Humus can form a colloidal gel. Explain why this helps the soil to hold water.

16 Why is soil such a favourable habitat for so many living things?

Clay and humus have a very important part to play in the fertility of soil. They hold plant nutrients in the soil so that they do not wash away. This is so

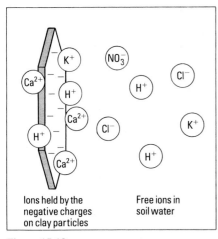

Figure 15.18
Cations held on clay.

because both clay and humus are colloids. The colloidal particles have a large surface area and are negatively charged.

Both clay and humus can behave like the ion exchange resins used in water softening. (See figures 10.30 and 10.31 in section **C**10.4.) The clay and humus particles are negatively charged and hold onto positive ions as shown in figure 15.18. So soil can be good at retaining positive ions but it is not so good at holding onto negative ions. This affects the use of nitrogen fertilizers, as explained in the next chapter.

C15.4   Why does soil pH matter?

*In Biology section **B**15.1 it is mentioned that the distribution of wild plants is very much affected by the nature of the soil. In this section you will examine some of the chemical reasons for this.*

The acidity or alkalinity of the soil is very important in determining how well different plants grow. Figure 15.19 shows the pH ranges which are best for the growth of a variety of flowers and vegetables.

Figure 15.19
The pH ranges preferred by a variety of garden plants.

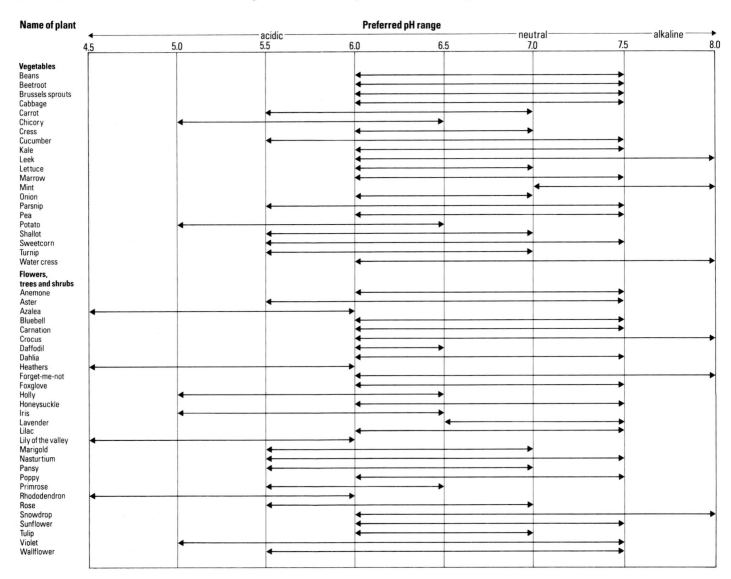

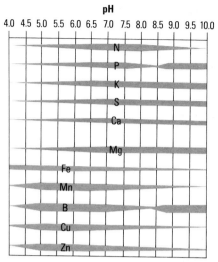

Figure 15.20
The availability of plant nutrients over a range of pH values.

17 Which pH range is best for most garden plants?

18 Which vegetables can be grown in: **a** an acid soil, and **b** an alkaline soil?

19 Which shrubs and flowers would you choose to grow in a garden with an acidic, peaty soil?

20 Which flowers and shrubs would you choose to grow in a chalky, alkaline soil?

One way of measuring the pH of the soil is to use a test kit which includes an indicator. If you are familiar with wild flowers you may not need a kit, because you will be able to tell whether the soil is acid or alkaline by looking to see which plants grow well.

Figure 15.20 shows that the pH affects the availability of mineral salts to plants. Plants take in the elements they need from the soil in the form of ions which are soluble in water. In figure 15.20, the thicker the horizontal band the more soluble the compound which contains the elements which plants need.

BOX 2   Practical problem
Measuring the pH of soil

Imagine that you have decided to manufacture and sell a kit for measuring the pH of soil which could be used by gardeners and growers of houseplants. The method is based on the use of a Universal Indicator.

You have done some research and discovered the following information.

● The soil or compost to be tested must be in small pieces.
● Soil samples must be taken from about 10 cm below the surface.
● The samples should be allowed to dry out before testing.
● No more than 10 parts of water should be added to 1 part of soil or compost.
● Soil acidity is fixed by the particles of clay and humus which must be present when the Indicator is added.
● The suspended particles hide the true colour of the Indicator and must be removed before using the colour chart to read off the pH.

First devise a safe and simple set of apparatus which could be used to measure the pH of soil. If possible, check your method in the laboratory.

Next draw up a set of instructions to be printed and supplied with the kit to explain how it should be used. Assume that the users of the kit will have little or no knowledge of chemistry.

C15.5   How can the pH of soil be controlled?

A common problem is that soil becomes too acid and so the pH has to be raised by adding lime. The lime used by gardeners and farmers may be powdered chalk or limestone; or it may be slaked (hydrated) lime which is calcium hydroxide. The chemical nature of these forms of lime is shown in figure 4.29.

Figure 15.21
Spreading lime.

GARDEN LIME

55% CaO

SAFE TO HANDLE

Improves fertility and cropping, encourages beneficial bacteria and soil organisms, sweetens acid soils, makes difficult clays easier to work and reduces club root (finger and toe) in brassicas. Assists decomposition, reduces fly and smell nuisance on the compost heap.

LIME IS LOST EACH YEAR by removal of growing crops, washing out by rain and acidification from smoke and industrial pollution.

CROPS THAT LIKE LIME include runner and French beans, all brassicas, (cabbage, cauliflower, sprouts etc.) beetroot, carrots, onions, leeks, lettuce, parsnips and spinach.

CROPS THAT DO NOT LIKE LIME include chalk or alkaline soils (ph over 7.0) high class lawns and turf, rhododendrons, azaleas, heathers, alpines, conifers, magnolias and other acid loving plants.

WHEN & HOW TO USE

GENERAL USE
Spread evenly over the area to be treated during winter or early spring. Fork in lightly after application. DO NOT MIX LIME AND FERTILIZERS, apply lime first and allow to weather, then spread fertilizers.

for very acid soil apply:
545-gms per square metre (16-oz. per square yard)
for new vegetable gardens apply:
272-gms per square metre (8-oz. per square yard)
for general maintenance apply:
136-gms per square metre (4-oz. per square yard)

Figure 15.22
Garden lime.

Another type of lime is quicklime (calcium oxide) which is unpleasant to handle because it is very reactive. It is good at neutralizing acids and is used to measure the neutralizing power of other forms of lime. Thus 100 g of garden lime with a *neutralizing value* of 55 will have the same effect as 55 g of pure calcium oxide.

Box 3 shows how quicklime and slaked lime neutralize acids. It also shows why the neutralizing value of slaked lime is less than that of calcium oxide.

21 Use the method explained in box 3 to calculate the neutralizing value of pure calcium carbonate.

22 The label on a packet of slaked lime sold for garden use shows that it has a neutralizing value of 66. Suggest reasons why the value is less than that calculated in the box.

BOX 3   Neutralizing values

Calcium oxide (quicklime) neutralizes hydrochloric acid as shown by this equation:

$$CaO(s) + 2HCl(aq) \longrightarrow CaCl_2(aq) + H_2O(l)$$

The equation for the reaction of calcium hydroxide (slaked lime) with hydrochloric acid is:

$$Ca(OH)_2(s) + 2HCl(aq) \longrightarrow CaCl_2(aq) + H_2O(l)$$

From these equations you can see that 1 mol of calcium oxide neutralizes the same amount of acid as 1 mol of calcium hydroxide.

The molar mass of calcium oxide $\quad = (40 + 16) \text{ g/mol}$
$$= 56 \text{ g/mol}$$

The molar mass of calcium hydroxide $= 40 + 2 \times (16 + 1) \text{ g/mol}$
$$= 74 \text{ g/mol}$$

This shows that 74 g of calcium hydroxide are needed to neutralize the same amount of acid as 56 g of calcium oxide.

So 100 g of calcium hydroxide will neutralize the same amount of acid as

$$\frac{100}{74} \times 56 \text{ g} = 75.7 \text{ g of calcium oxide.}$$

The neutralizing value of pure calcium hydroxide is about 76.

23 To raise the pH of a heavy clay soil from pH 4.5 to pH 6.5 requires about 1200 g/m² (grams per square metre) of powdered limestone (calcium carbonate). How much hydrated lime is needed to have the same effect? (See the box and your answer to question **21**.)

Summary

Produce an outline of all that you know about the factors which affect the rates of reactions in laboratories, in kitchens, in living things (biochemistry) and in the environment. Consider these factors:

- the concentration of the chemicals
- the temperature
- the light
- the size of any particles
- the effect of catalysts, including enzymes.

In addition to this chapter, these sections of your Chemistry and Biology textbooks will help you: **C**2.4, **C**2.6, **C**3.4, **C**13.5, **B**3.6, and **B**5.3.

Chapter **C**16 **Fertilizers**

Many farmers use large quantities of chemicals. They use fuels to drive farm machinery, and fertilizers to increase plant growth, as well as pesticides to control the weeds, insects and diseases which damage or destroy crops.

*In Biology Chapter **B**15 you can read about the mineral salts which plants need from the soil. Chapter **B**17 describes the use of pesticides and discusses some of the problems involved in their use. Chapter **B**17 also includes an account of the ways in which the use of fertilizers can affect the environment.*

*This chapter concentrates on the chemistry of fertilizers, their manufacture and uses. In the laboratory you will try out, on a small scale, some of the processes involved in making fertilizers. There is a large "energy cost" in making fertilizers, and you may be able to relate some of the ideas in this chapter to ideas about "energy cost" in Chapters **P**8 and **P**10.*

C16.1   Why do we need fertilizers?

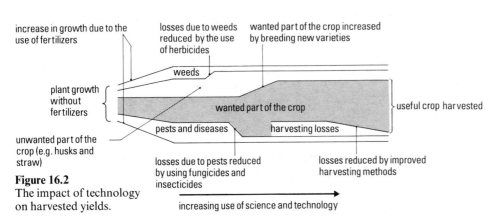

Figure 16.1
Spraying a crop with pesticide.

Figure 16.2
The impact of technology on harvested yields.

Figure 16.3
Heavy machinery is used in harvesting.

Crop	Extent to which the UK is self-sufficient in the crop (%)	
	1900	1983
Wheat	23	101
Barley	64	126
Oats	74	96
Potatoes	91	89
Sugar	0	54

Figure 16.4
Comparisons of British self-sufficiency in 1900 and 1983.

Major elements	Trace elements
Carbon	Iron
Hydrogen	Manganese
Oxygen	Copper
Nitrogen	Zinc
Phosphorus	Molybdenum
Sulphur	Boron
Potassium	Chlorine
Calcium	(Silicon)
Magnesium	(Cobalt)
(Sodium)	

Figure 16.5
Elements needed by plants. (The elements in brackets are not required by all plants.)

Crop	Mass of the element removed in kg/ha		
	N	P	K
Wheat grain	115	22	26
Wheat straw	7	2	31
Barley grain	72	14	13
Oat grain	72	13	18
Potato tubers	109	14	133
Sugar beet root	86	14	302
Sugar beet tops	117	15	165

Figure 16.6
The amount of nitrogen, phosphorus and potassium removed from soil by harvested crops.

Figure 16.2 shows how technology has improved crop yields.

It is the use of fertilizers which has done most to increase food production in Britain. We can now grow more food than we need to survive. Figure 16.4 shows the extent to which we have become self-sufficient in various foodstuffs during this century.

Some farm products are exported from Britain because we can now produce crops such as wheat as cheaply as anywhere in the world. We also import some food because we like to eat products which cannot be grown in our climate and because some foodstuffs can be imported more cheaply from other countries. Food production, imports and exports are all affected by the agricultural policies of the EEC.

1 Look at the information in figure 16.4. Which crop was grown in Britain in 1983 which was not grown in 1900? What alternative sources of supply were available in 1900?

2 What is the meaning of a figure of 126 per cent self-sufficiency in barley?

Fertilizers are made up of simple chemicals. They consist of various combinations of inorganic salts and they supply plants with the elements they need. The essential elements can be divided into two groups: the major elements which are needed on a large scale and the trace elements which are only needed in much smaller quantities. Most of these elements come from the soil, and a soil is fertile if it can supply crops with the required elements.

Plant roots take in nutrients from the soil as ions. The ions may be dissolved in soil water or held by particles of clay and humus. Figure B15.3 in Biology section B15.1 lists the main ions taken up by plants and shows why plants need them. Figure 16.5 on this page just lists the elements.

3 Three of the major elements (carbon, hydrogen and oxygen) do not come from minerals in the soil. Where do they come from? What is the name of the process involved?

4 Which elements does a plant need to make carbohydrates?

5 Why are soils in many areas of England unlikely to lack calcium?

6 What is the usual reason for spreading calcium compounds on farm land?

7 In which chemical forms is calcium usually added to soil?

8 Can you suggest a reason for the fact that it is seldom necessary to add sulphur compounds as fertilizers to the soil in the industrial regions of Britain?

9 Are the minor elements listed in figure 16.5 mainly metals or non-metals? Where are most of these elements to be found in the Periodic Table?

The purpose of the fertilizers used on farmland is to put back the nitrogen, phosphorus, potassium and other elements which are removed from the soil when crops are grown and harvested. The quantities involved are large, as figure 16.6 shows. One hectare is equal to ten thousand square metres, which is about the area of two large football pitches.

A long-running scientific experiment was started in 1842 by John Lawes who founded the Rothamsted Experimental Station for agricultural research in Hertfordshire. The experiment still continues and is designed to investigate the effect of fertilizers on crop yields. The experiment was set up in the Broadbalk field to study the growth of winter wheat.

The experiment has shown that nitrogen is the nutrient which does most to increase the yield of grain. This became clear soon after the first trials, and John Lawes published his findings even though it was not in his interest to do so – he owned a factory for producing phosphorus fertilizers.

Figure 16.7
Scientist at work in the Broadbalk field at the Rothamsted Experimental Station. He is applying radioactive-labelled fertilizers.

	Long-term treatment (since 1844) in kg/ha			Yield (1980–82) in tonnes/ha		Costs in £/ha	
	N	**P**	**K**	**Wheat**	**Potatoes**	**Wheat**	**Potatoes**
None	0	0	0	1.69	8.47	479	1468
N	96	0	0	3.68	8.30		
P, K	0	77	107	2.04	16.63		
N, P, K	96	77	107	6.60	38.57	575	1564

Figure 16.8
The results and costs from the Broadbalk field experiment.

10 What is the chemical connection between the statements **a** and **b** below?
a The growth of plants is controlled by enzymes.
b Nitrogen is the key nutrient for plant growth.
(Biology section **B15.4** may help you with this question.)

11 Look at the information in figure 16.8.
a Some plots of land have received no added fertilizer since 1844 but they still can produce crops. Where do the nutrients come from?
b What conclusions can you draw, from this information, about the effect of the three nutrients on plant growth?
c Calculate the cost per kilogram of growing potatoes without fertilizer, and the cost per kilogram of growing them with an NPK fertilizer. Does it seem to be worth spending money on fertilizers according to these figures?

C16.2 How has chemistry helped to solve the nitrogen problem?

*The nitrogen cycle and some of the natural processes which fix nitrogen are described in Biology section **B**15.4. This section tells you about a large-scale industrial method of fixing nitrogen.*

Only 0.03 per cent of the volume of the air is carbon dioxide, but this is enough for plants to use in photosynthesis. It seems surprising that plants can be short of nitrogen, since it makes up about 80 per cent of the air. The problem is that nitrogen is so inert that plants cannot use it.

Most plants take in nitrogen from the soil as nitrate ions. The nitrogen problem has been solved by finding ways of combining nitrogen gas with other elements to make compounds which plants can use. The process of making nitrogen compounds for this purpose is called *nitrogen fixation*.

The need to add to the natural reserves of nitrogen in the soil has been known for a long time. Just over a hundred years ago there was a flourishing trade based on a natural fertilizer – the sea-bird droppings which had built up over centuries on the rocky coasts and islands of South America. This manure was called *guano*. Eventually the supplies of guano ran out.

Figure 16.9
Only the potato plant in the middle has been fed with fertilizer.

Towards the end of the nineteenth century, farmers began to rely on the sodium nitrate deposits in Chile for their supplies of nitrogen fertilizer. Again the supplies were limited.

Meanwhile, the chemical industry was requiring more nitrogen to make dyes and explosives. So by the end of the nineteenth century there was a serious problem. A new method of fixing nitrogen was needed so that fertilizers could be made on a large enough scale to meet the growing demand.

Substance	Formula	Notes
Nitrogen gas	N_2	This makes up about 80 % of the air. It is chemically unreactive and cannot be used by most plants.
Ammonia	NH_3	Ammonia is the first compound to be formed when nitrogen is fixed either by the bacteria in the root nodules of legumes (peas, beans and clover) or by industrial methods.
Nitrate ions	NO_3^-	Most plants take in nitrogen in the form of nitrate ions from the soil. Ammonia is converted into nitrate ions by bacteria in the soil.
Proteins	one amino acid in a long chain	Most of the nitrogen taken up by plants is made into proteins. Proteins are chains of amino acids and each amino acid contains at least one nitrogen atom.

Figure 16.10
Forms of nitrogen.

12 What are the main ways by which nitrogen is *fixed* in the natural nitrogen cycle? (See Biology section **B15.4**.)

13 Before fertilizers were used, soil fertility was maintained by:

- crop rotation
- the use of farmyard manure
- resting the land from cropping (fallowing).

How do these methods help to maintain the fertility of the soil?

The work of Fritz Haber

The most important solution to the nitrogen problem was discovered by Fritz Haber (1868–1934), the son of a merchant of Breslau (now Wrocław), a town in what was then German Silesia and is now in Poland. After studying chemistry, the young Haber went into business but did not like it much. He gave it up and got a post as a lecturer at the technical college of Karlsruhe.

He knew about the efforts then being made to fix nitrogen and in 1904 he too began work at the problem. He was interested in the possibility of combining nitrogen and hydrogen to make ammonia.

nitrogen(g) + hydrogen(g) ⟶ ammonia(g)

At ordinary pressures, hardly any of the nitrogen and hydrogen combine, and at room temperature the reaction is very slow. Haber started by using theory to calculate the pressures and temperatures needed to produce as much ammonia as possible. He estimated that a pressure several hundred times atmospheric pressure and a temperature of several hundred degrees Celsius would be needed to produce even a small amount of ammonia.

Figure 16.11
Fritz Haber at work in his laboratory at Karlsruhe.

Other scientists disagreed with Haber, and experiments showed that a very small amount of ammonia could be made by reacting nitrogen and hydrogen at fairly low pressures. Haber revised his calculations. He also began to look for a catalyst to make the reaction faster: osmium seemed to give the best results, but he used platinum in his experimental apparatus.

In 1908, Haber found that he could get a yield of 8 per cent ammonia from nitrogen and hydrogen at 500 to 600 °C with a pressure of 175 atmospheres and with a catalyst. He was fortunate that the pressure needed to be no higher, because the laboratory compressor he was using had an upper limit of 200 atmospheres. He was also fortunate to have able and skilful assistants, because the design of the apparatus was complicated and Haber himself was clumsy with his hands.

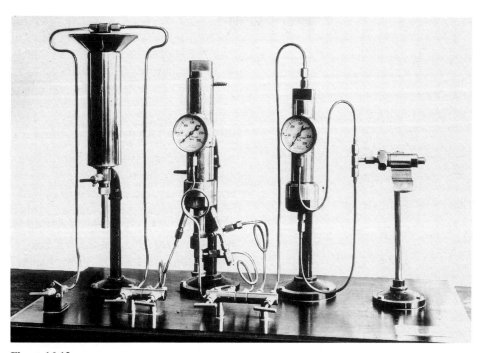

Figure 16.12
Haber's experimental apparatus.

Figure 16.13
Diagram of Haber's apparatus.

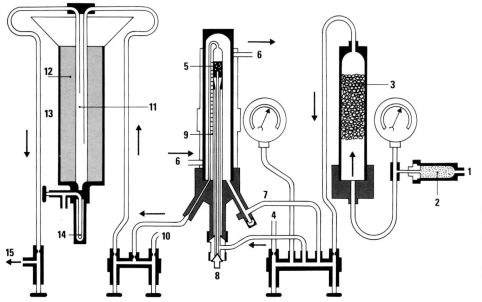

Key

1 entry of compressed gases from the circulating pump (nitrogen and hydrogen with traces of oxygen)

2 heated platinum asbestos to catalyse the reaction between hydrogen and traces of oxygen and so produce water

3 drier to remove the water produced in the reaction above

4 feed-in for fresh supply of nitrogen and hydrogen

5 converter

6 water cooler

7 pressurizer

8 heating elements

9 electrical heating

10 tube for the ammonia-holding gases

11 flow tube

12 cooling bath

13 heat exchanger

14 off-take for liquid ammonia

15 return of the high pressure gases to the circulating pump

When building his laboratory model, Haber made use of two ideas. The first of these was *continuous flow*. The reactants were fed in continuously at one part of the apparatus and the products were continuously drawn off at another, just like the conveyor belt system.

The second idea was to use a *heat exchanger*. An exchanger was built into the apparatus so that the hot ammonia leaving the reactor could heat up the incoming mixture of nitrogen and hydrogen. This saved valuable energy.

---

14 Compare a laboratory distillation apparatus with the apparatus used in industry to distil oil. Which process uses continuous flow?

15a Which part of a laboratory distillation apparatus acts as a heat exchanger?
b Where can you find heat exchangers in a car, in a kitchen, and in a hot water system?

16 Why do you suppose that Haber used platinum, rather than osmium, as the catalyst in his experimental reactor?

Scaling up the process

In 1909 the large dyestuffs manufacturer, Badische Anilin und Soda-Fabrik, bought the rights to the Haber process. The job of scaling up the laboratory model to a large manufacturing plant was given to Carl Bosch. Bosch's father owned a plumber's business, and this background was useful to his son, who had worked as a fitter for some time before going to university to study chemistry. In 1899 he entered industry as a chemical engineer.

Bosch and his team had to solve three main problems when creating a large scale version of Haber's apparatus:

● They had to design the reactor.
This had to stand up to high pressures and temperatures – much higher than had ever been met before in industry. The first trial reactor exploded when they tested it. They found that this happened because the steel was being weakened by reaction with the hot hydrogen under pressure.

Finally Bosch devised a double-walled reactor. The steel of the outer wall was strong enough to contain the high pressure gases. The steel used to make the inner wall did not react with hydrogen.

In 1913 Bosch was able to install a reactor 7 metres high and with a mass of 5 tonnes. Modern reactors may be about 20 metres high and have a mass of 170 tonnes.

● They had to find a suitable catalyst.
Haber's catalysts worked but were too expensive. Bosch had this problem thoroughly investigated and by the end of 1912 some 6500 experiments had been carried out. His assistants finally decided that the best results were achieved using an iron catalyst improved by adding potassium, aluminium and calcium. This type of catalyst is still used and has proved satisfactory at pressures up to 300 atmospheres.

● They had to devise a cheap way of getting nitrogen and hydrogen.
At first the hydrogen came from the electrolysis of water while nitrogen was obtained by distilling liquid air. These processes were too expensive on a large

Figure 16.14
The first synthetic ammonia plant at
Oppau.

scale. Bosch and his team succeeded in devising new methods of getting the
right mixture of nitrogen and hydrogen from air and water.

The decision to build a large plant was taken in 1912 and by the end of 1913
a complete works was operating at Oppau on the Rhine.

Haber was awarded the Nobel prize for Chemistry in 1918 for his
theoretical work on the Haber process. Bosch won the same prize in 1931 for
his achievements in developing new techniques for chemical engineering at
high pressures.

C16.3   The Haber process today

The Haber process is still the most important industrial method for fixing
nitrogen.

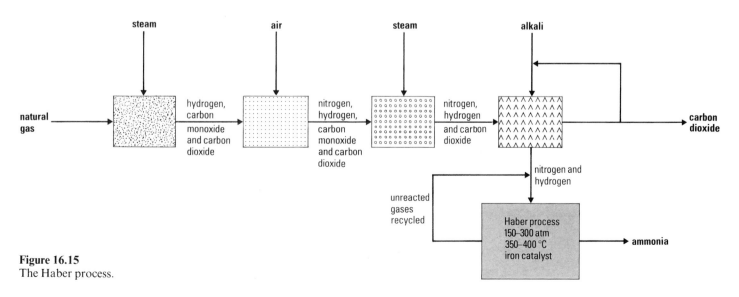

Figure 16.15
The Haber process.

Figure 16.16
A modern Haber process plant.

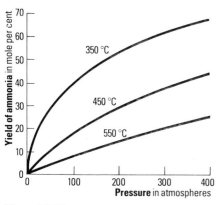

Figure 16.17
The yield of ammonia at different
temperatures and pressures.

The hydrogen required for the process is made by passing a mixture of
natural gas and steam over a nickel catalyst at a high temperature and at a
pressure of about 30 atmospheres. The main products are hydrogen and
carbon monoxide, but some carbon dioxide is formed at the same time.

methane(g) + steam(g) $\longrightarrow$ hydrogen(g) + carbon monoxide(g)

Next, air is added to the mixture of hydrogen and carbon monoxide. The
oxygen in the air reacts with some of the hydrogen, which is removed, leaving
the nitrogen. The volume of air added is controlled so that in the end the
amounts of nitrogen and hydrogen are in the correct proportions for the
Haber process.

Carbon monoxide and carbon dioxide must be removed because oxygen
and oxides "poison" the catalyst used in the Haber process. The carbon
monoxide is converted to carbon dioxide by reaction with more steam under
pressure in the presence of another catalyst.

carbon monoxide(g) + steam(g) $\longrightarrow$ carbon dioxide(g) + hydrogen(g)

The carbon dioxide is absorbed in a hot solution of alkali, leaving a mixture
of nitrogen and hydrogen for making ammonia.

The nitrogen and hydrogen are compressed to 150 to 300 atmospheres and
then passed over an iron catalyst at about 350 to 450 °C. The ammonia
formed is condensed to a liquid under pressure. Unreacted nitrogen and
hydrogen are recycled.

17 Write balanced symbol equations for the reactions involved in the manufacture of ammonia from natural gas, steam and air. (See table 5 in the Data section for the formulae.)

18 Carbon monoxide is a catalyst poison. Why is it also poisonous to people? (See Biology section **B8.5**.)

19 Which compound is oxidized and which is reduced
a when methane reacts with steam
b when carbon monoxide reacts with steam?

20 Why is an alkali chosen to absorb carbon dioxide? (See Chapter **C18**.)

21 What amount (in moles) of hydrogen is required to react with 1 mol of nitrogen?

22 Why do you suppose that hydrogen made from steam and natural gas is much cheaper than hydrogen made by the electrolysis of water?

23 What is the yield of ammonia if the Haber process operates at
a 150 atmospheres and 350 °C?
b 150 atmospheres and 450 °C?
c 300 atmospheres and 450 °C?
Comment on the effects of changing the temperature and pressure on the yield of ammonia. (See figure 16.17.)

24 What engineering and economic reasons can you think of for not running the Haber process at a pressure above 300 atmospheres?

25 About 40 000 000 kJ of energy are needed to make 1 tonne of ammonia. Which of the stages in the manufacture of ammonia seem to need a lot of energy?

C16.4   From ammonia to ammonium salts

Worksheet C16A asks you to investigate the problem of finding a suitable catalyst for oxidizing ammonia. This is an essential step in the manufacture of nitric acid. Nitric acid is needed to make fertilizers, dyes and explosives.

Here are some of the properties of ammonia:

● It is a gas at room temperature and pressure.
● It has a strong, eye-watering smell.
● It is very soluble in water.
● It is alkaline.

These properties make it quite difficult to deal with pure ammonia. Ammonia can be liquefied and injected into the soil as a fertilizer, but it is more usual to convert it into nitrogen compounds which are solid at room temperature. The box explains the chemical connection between ammon**ia gas** and ammon**ium salts**.

One of the commonest nitrogen fertilizers is ammon**ium** nitrate, which is made from ammon**ia** and nitric acid. The nitric acid is also made from ammonia, as outlined in figure 16.18.

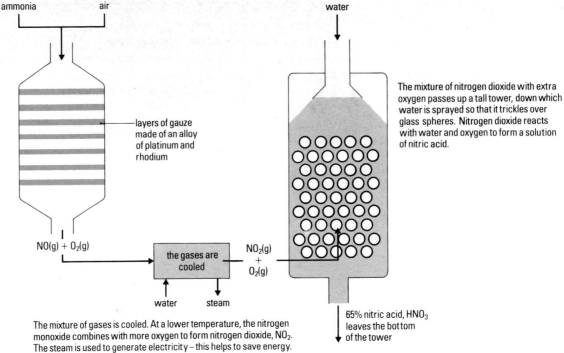

Ammonia and air are passed over layers of red hot platinum gauze. Energy given out by the reaction keeps the platinum hot.

The ammonia reacts with oxygen to form nitrogen monoxide, NO, which is a colourless gas.

There is more than enough air for this reaction, so that there is enough oxygen left to complete the later stages of the process.

layers of gauze made of an alloy of platinum and rhodium

$NO(g) + O_2(g)$

the gases are cooled

water   steam

$NO_2(g) + O_2(g)$

The mixture of gases is cooled. At a lower temperature, the nitrogen monoxide combines with more oxygen to form nitrogen dioxide, NO_2. The steam is used to generate electricity – this helps to save energy.

The mixture of nitrogen dioxide with extra oxygen passes up a tall tower, down which water is sprayed so that it trickles over glass spheres. Nitrogen dioxide reacts with water and oxygen to form a solution of nitric acid.

65% nitric acid, HNO_3 leaves the bottom of the tower

Figure 16.18
The manufacture of nitric acid.

BOX 1   Ammonia and ammonium salts

Ammonia is a compound of nitrogen and hydrogen. These are two non-metal elements and as usual we find that a compound of non-metals consists of molecules. The molecules are small. The forces between them are weak. So ammonia is a gas at room temperature and pressure.

An ammon**ia** molecule, NH_3

Ammonia reacts with acids to form ammon**ium** salts. With hydrogen chloride ammonia forms ammonium chloride, NH_4Cl.

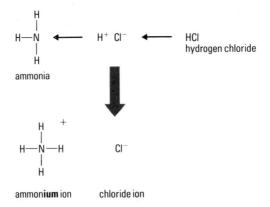

ammonia   HCl hydrogen chloride

ammon**ium** ion   chloride ion

You can see that the hydrogen chloride molecule splits into a hydrogen ion and a chloride ion. The hydrogen ion sticks onto an ammonia molecule to form an ammonium ion.

26 Suggest an explanation for the fact that ammonia and hydrogen chloride, which are both gases, react together to form a solid product. (Think in terms of the forces between molecules and the forces between ions. The box opposite will help you.)

27 Write symbol equations to show what happens when ammonia reacts with:
a nitric acid to form ammonium nitrate, and
b sulphuric acid to form ammonium sulphate.
(You will find the formulae you need in table 5 of the Data section, and the box on the opposite page will help you.)

28 Nitric acid is needed to make explosives. The word "explosive" may make you think of war. However, explosives are needed in peacetime too. Describe some of the peacetime applications of explosives.

C16.5   How are fertilizers manufactured?

*Worksheet **C16B** shows you how to make a sample of a fertilizer. You may be asked to work out the cost of making your product which you can then compare with the prices of commercial fertilizers.*

Figure 16.19 is an aerial view of the ICI works at Billingham, where fertilizers are manufactured on a large scale. Figure 16.20 gives an outline of some of the manufacturing processes. Figure 16.21 is a map to show where Billingham is situated in the North East of England. Note that the potassium chloride is mined at Boulby, which is one of the places where alum used to be manufactured (see Chapter **C4**).

Figure 16.19
The fertilizer complex at Billingham.

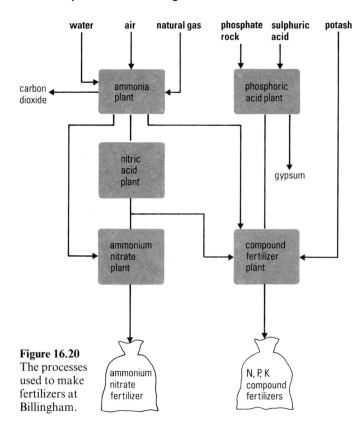

Figure 16.20
The processes used to make fertilizers at Billingham.

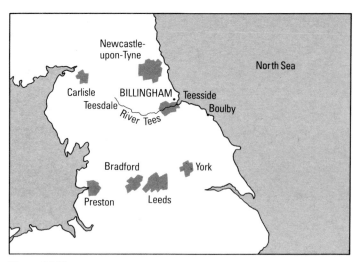

Figure 16.21
The position of Billingham.

Figure 16.22
Ammonium nitrate prills.

● *Nitrogen fertilizers*

The nitrogen fertilizer made at Billingham is ammonium nitrate, which is sold under the trade name Nitram. The chemistry involved is described in section **C16.4**. Ammonium nitrate is very soluble in water, and a little water makes the particles stick together to produce hard lumps which are difficult to break up. Farmers want their fertilizers to be free-flowing so that they can be spread easily by machinery. For this reason the ammonium nitrate is converted into small hard spheres called *prills*. The prills are easy to handle and they contain a drying agent to stop caking.

The prilling tower at Billingham is shown in figure 16.19. Molten ammonium nitrate is sprayed into the top of the tower like water from a watering can. The droplets cool and solidify as they fall. They are collected from the bottom and carried away to be packed into bags.

● *Phosphorus fertilizers*

The minerals in phosphate rock are very insoluble in water. To produce a fertilizer the phosphate rock is treated with sulphuric acid to make phosphoric acid. The phosphoric acid is then neutralized with ammonia to make ammonium phosphate.

● *Potassium fertilizers*

Potassium fertilizers are often referred to as "potash" which is a vague term for potassium compounds. Potassium chloride is widely used but farmers still give it the very old-fashioned name "muriate of potash".

Potassium chloride is mined from a seam of rock called sylvinite which is 5 to 10 metres thick and lies 1100 metres below the surface at Boulby. Sylvinite consists of about 40 per cent potassium chloride. The manufacturing process involves purification to remove the unwanted impurities, which are mainly sodium chloride and clay.

● *Compound fertilizers*

The three main nutrients required by plants are nitrogen, phosphorus and potassium. Compound fertilizers contain these nutrients in varying proportions. They are made from ammonia, nitric acid, phosphoric acid and potassium chloride.

Notice that there are two uses of the word compound. Ammonia, nitric acid and phosphoric acid are *chemical compounds* because they consist of elements joined. *Compound fertilizers* are so called because they supply all three of the main elements needed by plants together.

Figure 16.23
Compound fertilizers.

Figure 16.24
The storage silo at Billingham.

Fertilizers are manufactured at Billingham day and night throughout the year, but farmers apply fertilizers to the soil over a short period of the year. This means that very large quantities of fertilizers have to be stored until they are needed.

29 Look at the position of Billingham on the map in figure 16.21. With the help of figure 16.20, suggest reasons for the fact that Billingham is a major centre for the manufacture of fertilizers.

30 The name ''muriate of potash'' would have made more sense to Humphry Davy (see figure 4.26) than it does to a modern chemist. Can you find out, from reference books, why this name is still sometimes used for potassium chloride?

31 Look at the formula of ammonium nitrate. Is it a chemical compound? Is it a compound fertilizer?

32 The ranges of values for the percentages of the elements in fertilizers in figure 16.25 are due to the presence of impurities. Use the method explained in section **C**4.6 (box 7) to calculate what the percentages for the fertilizers in figure 16.25 would be if the compounds were pure.

33 What is the main advantage of fertilizers according to figure 16.26 on the next page?

34 What are the advantages of using manures rather than fertilizers?

	Percentage by mass		
	Nitrogen	**Phosphorus**	**Potassium**
Fertilizer			
Ammonium nitrate	33 to 34.5	–	–
Ammonium phosphate	11 to 12.5	21 to 24.5	–
Potassium chloride	–	–	41.5 to 52.5
Manure			
Farmyard manure	0.6	0.13	0.58
Digested sewage sludge	1.4	0.44	0.08
Bone meal	3.6	10.3	–

Figure 16.25
A comparison of fertilizers with manures.

Figure 16.26 shows a typical label on a bag of fertilizer. You can see that the nutrients present are listed as nitrogen, phosphorus oxide and potassium oxide. This way of showing the amounts of elements was used in early laws about fertilizers, and it has remained in use even though fertilizers are not made of oxides.

16.8.24

EEC FERTILIZER

NPK FERTILIZER	**16.8.24**
TOTAL Nitrogen (N)	16.0%
Ammoniacal Nitrogen	11.2%
Nitric Nitrogen	8.8%
Phosphorus Pentoxide (P₂O₅)	
Soluble in Neutral Ammonium Citrate + Water	8.0%(3.5%P)
Soluble in Water	7.2%(3.1%P)
Potassium Oxide (K₂O)	
Soluble in Water	24.0%(19.9%K)

Weight 750 kg   1650 lb net

Imperial Chemical Industries PLC, Agricultural Division,
P.O. Box 1, Billingham, Cleveland TS23 1LB, England

Figure 16.26
Label from a fertilizer bag.

---

35 Check that the label in figure 16.26 is correct by showing from the formula that if the fertilizer contains the equivalent of 24 % potassium oxide it will contain 19.9 % potassium.

36 What percentage by mass of potassium chloride, KCl, is equivalent to 24 % potassium oxide as on the label in figure 16.26?

37 What do you understand by the terms "ammoniacal nitrogen" and "nitric nitrogen" on the label shown in figure 16.26? If the percentages of these two forms of nitrogen are 11.2 % and 8.8 %, why is there a total of only 16 % nitrogen in the fertilizer?

C16.6   What happens to fertilizers in the soil?

*Biology section **B**17.5 describes some of the problems which arise when fertilizers are used on a large scale. The main problem is caused by nitrates. This section will help you to understand why it is mainly the nitrogen fertilizers which cause problems, and not the potassium or phosphate fertilizers.*

When farmers apply fertilizers to their land they hope that the nutrients will be taken up by the crops to improve growth. They do not want to pay for nutrients to be washed away by rain water.

Phosphate forms insoluble salts with the ions of iron and aluminium at low pH values and with calcium above pH 6.5. Long-term experiments suggest that very little phosphate is wasted. About a quarter of the phosphate applied is taken up by crops in the first season. The remainder stays combined in the soil and is used by crops in later years. Little or no phosphate is washed from the soil by rain water.

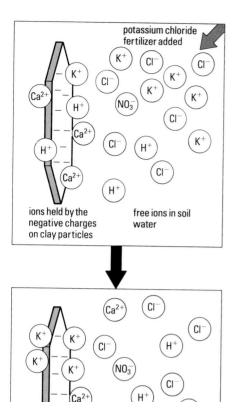

Figure 16.27
How potassium ions are held on cation-exchange sites on clay.

ions held by the
negative charges
on clay particles

free ions in soil
water

potassium ions exchange places with
other positive ions on the clay. So most
of the added potassium is held in the soil.

The same is normally true of potassium. The potassium ions exchange with calcium and hydrogen ions held on clay and humus in the soil. This creates a reserve of potassium ions which is available to plant roots.

Most nitrogen fertilizers used in Britain contain ammonium ions or nitrate ions (or both). Ammonium ions are positively charged and so they can be held by clay or humus in a similar way to potassium ions (see figure 16.27). However, ammonium ions are converted to nitrate ions by bacteria in the soil. So most of the nitrogen added to soil is changed to nitrate. This is the form of nitrogen which is most easily taken up by plants, but unfortunately it is also the form which is most easily lost from the soil by leaching and denitrification. (See figure 16.28.)

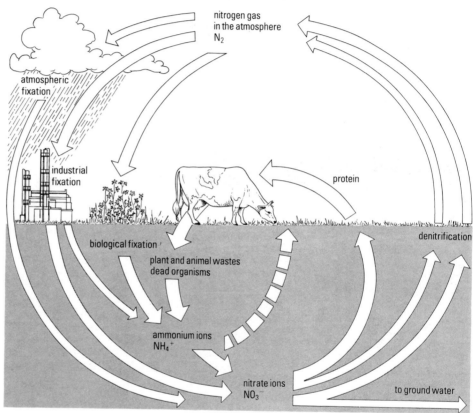

Figure 16.28
What happens to nitrogen in the soil.

Figure 16.29 sums up the results of recent research to find out what happens to the nitrogen added to the soil by fertilizers. There is considerable interest in the subject at the moment because the levels of nitrate in drinking water are rising. Some people are worried that this may have a long-term effect on our health, although there is no definite evidence to support this at present.

The equipment used to trace the movement of water and nutrients in the soil is called a lysimeter. A lysimeter is a small section of a field which is separated from the rest of the soil by a glass fibre cylinder. The amounts of fertilizer added to the soil surface of the lysimeter can be recorded. Drainage from the soil is collected so that the amount of nutrient leached from the soil can be measured.

Typically the top layer of soil contains about 10 000 kg of nitrogen per hectare mainly as organic compounds. Fertilizers are applied at the rate of about 200 kg per hectare. For research purposes the fertilizer nitrogen is labelled with the heavy isotope nitrogen-15 so that its movement through the

10% is lost by leaching

10% is lost by denitrification

25% becomes part of the soil organic matter

55% taken up by the crop

Figure 16.29
The fate of nitrogen fertilizers when applied to cereal crops.

soil can be followed. This makes it possible to discover how much of the fertilizer nitrogen is taken up by the crop, how much is temporarily stored in roots, how much is immobilized in humus, and how much is lost by leaching or denitrification.

*The term isotope is explained in Chapter **C18**. Other examples of the use of isotopes as tracers in research are given in Biology section **B8.2** and Worksheets **B9A** and **B9B**, and in Physics Chapter **P3**. The nitrogen-15 isotope is not radioactive, but the atoms can be detected because they are slightly heavier than ordinary nitrogen atoms.*

Figure 16.30
Installing a lysimeter.

Figure 16.31
Lysimeter in use: studying the effect of water-logging on crop growth.

Studies of this sort help to show the movement of fixed nitrogen in the soil. Worldwide studies can be made to make rough estimates of the extent to which the Haber process for fixing nitrogen is modifying the natural nitrogen cycle. To some extent the figures are guesses based on many small scale studies. The amount of nitrogen fixed by the Haber process is having a big effect on the natural nitrogen cycle.

C16.7   Are fertilizers a good thing?

You may have heard about EEC food mountains. You may also have seen reports of health fears about the rising level of nitrates in water supplies. The use of fertilizers has become controversial. This section gives you some of the arguments used for and against the use of fertilizers in farming.

Thanks to the use of fertilizers and other farm chemicals we can now grow plenty of many crops. However, there are people who think that an agriculture based on intensive farming with a high input of fertilizers and other farm chemicals is bad for the environment, wasteful of energy and harmful to our health. Here are some of the arguments put forward by those who are for and against modern agriculture.

Energy use in 1981	PJ
Fertilizer manufacture	85
All other agricultural use	380
Food processing and distribution	1400
Home energy use	590
Total UK energy use (including agriculture, industry, transport, and domestic uses)	8369

1 PJ = 1 000 000 000 000 000 J

Figure 16.32
The pattern of energy consumption in Britain.

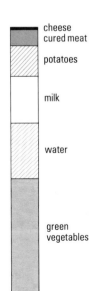

cheese
cured meat

potatoes

milk

water

green vegetables

Figure 16.33
Sources of nitrate in the diet.

For:

● Fertilizers used correctly increase crop yields dramatically. They allow many more people to be fed from the same area of land.

● The use of fertilizers improves the nutritional value of crops, especially the protein content.

● The use of fertilizers reduces the cost of food.

● Fertilizers add nothing harmful to the soil. The nutrients released by fertilizers are the same as those found naturally in soil.

● It is important to distinguish clearly between fertilizers and other chemicals such as pesticides. The purpose of fertilizers is to support life while the purpose of pesticides is to destroy it selectively.

● At the moment there is no medical evidence to support the EEC limit on the nitrate content of drinking water. Nitrate is not a toxic substance and there is no risk to human health. There is no evidence to connect nitrate intake with stomach cancer.

● Most of the nitrate in our diet comes from eating green vegetables. Of the total nitrate consumed in the UK, on average about 80 % comes from foodstuffs and only 20 % from drinking water.

● Fertilizer manufacture uses only about 1 % of all the fossil fuel energy used in Britain. Far more energy is used in processing, packaging and distributing food than in food production. (See figure 16.32.)

Against:

● Fertilizers are so easy to use that we are now wasting the nutrients in animal manure and sewage. Organic farming may give crop yields which are about 20 % lower, but this does not matter now that we have large surpluses of many crops in Europe.

● There is growing demand for organically grown food and many people are willing to pay the extra cost.

● Lettuces grown using organic nitrogen fertilizer maintain their weight, yield and protein content but contain less nitrate than those fertilized with mineral nitrogen.

● Nitrates travel slowly through the soil. The nitrate levels in underground aquifers now reflect farming practice of about 20 years ago. Already the recommended nitrate levels in drinking water are being exceeded in some areas. It will cost the water authorities millions of pounds to keep nitrate levels below the maximum level allowed.

● Use of fertilizers is part of an energy intensive system of food production. Estimates suggest that the energy cost of providing food using the methods of British farmers, processors and distributors is the equivalent of burning one tonne of coal per head per year.

38 Burning 1 tonne of coal produces about 30 000 000 000 J of energy. The population of Britain is about 60 000 000. Use this information and the estimates in figure 16.32 to show that it takes the energy equivalent of 1 tonne of coal to feed one person in Britain.

39 Are you for or against the use of inorganic fertilizers? Write a letter to the editor of your local newspaper putting forward your views.

Summary

1 Ammonia is a very important chemical. Summarize its main properties under these headings:

- colour
- smell
- solubility in water
- pH of its solution in water
- reactions with acids
- oxidation to oxides of nitrogen.

Include word and symbol equations for the reactions you mention.

2 Copy and complete figure 16.34, which shows the main processes used to convert raw materials to fertilizers.

Raw material	Where the raw material comes from	What the raw material is used to make
Nitrogen	_____	Used to make ammonia. Some of the ammonia is converted to _____ acid. Then ammonia and _____ acid are mixed to make _____ which is used in _____ and compound fertilizers
_____	From local reservoirs	
Methane	_____	
Sulphur	Imported	Used to make sulphuric acid
_____	Imported	Treated with sulphuric acid to make phosphoric acid. The phosphoric acid is neutralized with _____ to make ammonium phosphate
_____	Mined at Boulby	Purified to obtain the potassium chloride needed to make NPK compound fertilizers

Figure 16.34
The main processes used to convert raw materials to fertilizers.

3 Summarize what you think about the use of fertilizers and other farm chemicals. Use this chapter to help you, but also add information from Chapter **B**17 as well as what you have read in the newspapers or seen on television.

Topic C6

The Periodic Table, atoms and bonding

Chapter **C17**   **The Periodic Table**   288
Chapter **C18**   **Atoms and bonding**   298

The first five topics in this book tell you about the importance of chemistry in our homes, in industry, in living things and in the environment. This final topic concentrates on the theory of chemistry. You are unlikely to study all of this topic at one time, and your teacher may tell you to leave out some sections. You may well be asked to look at some sections in these two chapters early in your chemistry course.

The topic brings together ideas which have been introduced gradually in the earlier chapters. You will find many references to other parts of this book and to your Biology and Physics books. These references, and the questions, are included to show how the ideas of science link together.

A wood carving of Dmitri Ivanovich Mendeléev by Y. N. Rukavishnikov of the Ukraine.

The Periodic Table

In this first chapter of Topic C6 you can find out how chemists discovered a useful way of classifying the hundred and six elements. Notice how long it took for the idea of the Periodic Table to be generally accepted. We give the credit for the discovery of the Periodic Table to Mendeléev, but he was building on the work of others.

You can compare the classification of the elements with the classification of living things described in Biology Chapters B1 and B2.

C17.1 How was the Periodic Table discovered?

If you look in a dictionary to find out how to spell "biennial" for example, you automatically open the book near the beginning. If you next want to find how to spell "parallel", you quickly skip a great chunk of the book because you have a pattern in your head which helps you to know where to look among all the thousands of words.

Just as twenty-six letters of the alphabet can form the thousands of words in a dictionary, so the hundred and six chemical elements can be arranged to form millions of compounds, each element and compound with its own set of properties. The problem for chemists, when many new elements were discovered, was to find a pattern to help them make sense of so much new information.

In section **C1.3** and in figure 4.26 you can read about the discovery of new elements in the first half of the nineteenth century. At the time a few chemists began to feel that a pattern might exist to help make sense of these exciting discoveries. They noticed that some of the elements had similar properties. They knew about Dalton's work on atomic theory and began to wonder if there might be a connection between the chemical properties of elements and their atomic masses.

One of the first attempts to investigate this problem was made by a German chemist called Johann Döbereiner (1780–1849). He studied groups of three elements which were similar chemically such as chlorine, bromine and iodine. He noted that bromine seemed just half-way in its properties between chlorine and iodine. He also noticed that the atomic mass of bromine was the mathematical mean of the other two elements. He wondered if this was a coincidence and went on to look for other examples. He called these groups of three elements "triads". Two of his triads are shown in figure 17.1 using modern values for atomic masses.

Döbereiner's triads were not taken very seriously because most of the known elements could not be fitted into the scheme. Many thought that his

Element	Atomic mass in u (approximate)
Ca	40
Sr	88
Ba	137
Cl	35.5
Br	80
I	127

Figure 17.1
Examples of Döbereiner's triads.

1 H	8 F	15 Cl
2 Li	9 Na	16 K
3 Be	10 Mg	17 Ca
4 B	11 Al	and so on
5 C	12 Si	
6 N	13 P	
7 O	14	

Figure 17.2
A modern version of part of
Newlands' table.

findings were just coincidence. They could not see why the properties of elements should be connected to their atomic masses.

Döbereiner's ideas were published in 1829. It was not until 1864 that the English chemist, John Newlands (1838–98) came up with another attempt to find a pattern among the elements. He arranged all the fifty or so known elements in order of atomic mass. Newlands wrote: "The eighth element starting from a given one, is a kind of repetition of the first, like the eighth note in an octave of music." He called his rule the "Law of octaves".

Look at figure 17.2. If you count sodium as one, and then continue counting along the rows in the table, you get to the similar element potassium as number eight. You can also see that one of Döbereiner's triads appears in the first part of Newlands' table, and that the others could be found in the full table.

Newlands' ideas were also not taken very seriously by many chemists. His law of octaves stopped working after the first seventeen elements. Most people felt that it was all just chance and were put off by the idea of a link with musical scales.

---

1 Use your knowledge of their chemistry and table 3 in the Data section to show that lithium, sodium and potassium support Döbereiner's idea of triads.

2 With the help of the dates in table 2 of the Data section, draw up a table of the elements using only those discovered before 1864. To what extent does your table support Newlands' "Law of octaves"?

It was in 1869 that a Russian chemist, Dmitri Mendeléev, published the table on which all later versions of the Periodic Table have been based. He succeeded where others had failed because he realized that Newlands had made two big mistakes. One mistake was to assume that all the elements had already been discovered. The other was to think that the "Law of octaves" would work throughout the Table with seven elements in every row. Mendeléev realized that the later rows in the Table have to be longer than those at the beginning.

Mendeléev stated his *periodic law* as follows: "When the elements are arranged in order of atomic mass, similar properties recur at intervals."

To understand the name *Periodic Table*, picture a pendulum swinging backwards and forwards. The movement of the pendulum is *periodic* – it traces out a repeating pattern with time. Now look at the Periodic Table on page 316 and run your eyes from left to right along the horizontal rows which we now call *periods*. The properties of the elements and their compounds show repeating patterns. The most obvious regular change is from metals at the left of each period to non-metals at the right, then back to metals again.

Waves are also periodic, as you know from Chapter **P14**. A German chemist, Lothar Meyer (1860–95), was interested in the same problems as Mendeléev and he published a graph showing wave-like, repeating patterns a year later. He also discovered that the periods differ in length. A modern version of his graph is shown in figure 17.4.

Meyer arrived at his graph by working out the volume of equal numbers of atoms of each element when solid. So the vertical (y) axis on the graph shows the volumes of one mole of each element. Along the horizontal (x) axis is plotted the atomic mass of the elements.

Figure 17.3
Dmitri Mendeléev.

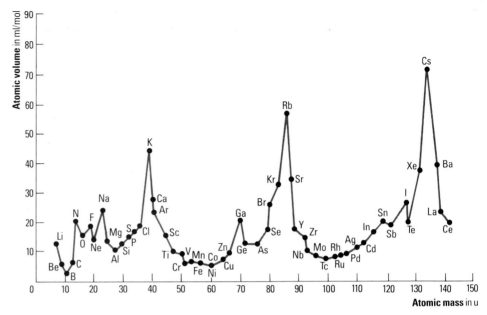

Figure 17.4
Modern version of Lothar Meyer's atomic volume curve.

---

3 Label the elements on a shortened version of the Periodic Table (Worksheet **C17A**) to show whether they are metals, non-metals, or ''in-between'' elements.

4 Write down the formulae of the chlorides of as many elements as possible in to a shortened version of the Periodic Table. You will find the formulae in table 5 of the Data section.

5 Predict the formulae of:
beryllium chloride, boron chloride, gallium chloride, and germanium chloride.

6 Plot a bar chart to find out whether the melting-points of the elements show a periodic pattern. Plot the melting-points on the vertical axis and the atomic masses (for the elements hydrogen to calcium) on the horizontal axis.

C17.2   Patterns and problems

You may find it helpful to re-read section C1.3 before you go further in this chapter. This will remind you of the work of Dalton, the units used to measure atomic masses, and the chemical meaning of the terms group *and* period.

Missing elements

Mendeléev made the bold move of leaving gaps in his table for undiscovered elements. He even used his table to predict the properties of these unknown substances. For example, there was a gap between silicon and tin in group 4 of the table. Mendeléev gave the name ekasilicon to the element he expected to fill the gap. Figure 17.5 shows how his predictions compare with the properties of germanium, which was found to fill the gap in 1886.

Mendeléev was equally successful in predicting the properties of other missing elements including gallium (discovered 1875), and scandium

Mendeléev's predictions about ekasilicon	Observed properties of germanium
Will be a light grey metal	It is a dark grey metal
Will form an oxide (EsO_2) with a high melting point	Forms a white oxide (GaO_2) which melts at 1116 °C
The chloride will boil at a temperature below 100 °C and have a density of about 1.9 g/ml	The chloride boils at 86.5 °C and its density is 1.887 g/ml

Figure 17.5
A comparison of Mendeléev's predictions with the properties of germanium.

(discovered 1879). So within a few years it was impossible to doubt the usefulness of Mendeléev's table, even though there were still problems which could not be solved.

Elements which seem to break Mendeléev's law

Mendeléev organized his table so that elements with similar chemical properties were put in the same *group*. He found that to do this he sometimes had to break the rule of putting the elements in order of atomic mass. If you look at the Periodic Table in the Data section you will see that iodine (atomic mass 126.9 u) comes after tellurium (atomic mass 127.6 u).

At the time, Mendeléev could not explain why there was this break in the pattern. It was not until much more was known about atomic structure that it was possible to understand why these, and a few other elements, seemed to break the rules. (See section **C18.2**.)

A new group

We now know that, when Mendeléev published his Table, there was a whole group of elements missing. At that time no one had any idea that they existed. Then argon was identified by William Ramsay (1852–1916). The atomic mass of argon is 39.9 u. Keeping the elements in order of atomic mass suggests that argon should be placed in the Periodic Table between potassium (atomic mass 39.1 u) and calcium (atomic mass 40.1 u).

Chemically this does not make sense. Argon is chemically unreactive. It forms no compounds. Looking at the patterns of properties of the elements, and the formulae of compounds, it seems much more sensible to place argon between chlorine and potassium. So argon is another element which breaks the rules.

Atomic mass in u	32.1	35.5	39.9	39.1	40.1
Symbol	S	Cl	Ar	K	Ca
Charge on ion	2−	1−	0	1+	2+
Formula of chloride	SCl_2	ClCl	No chloride	KCl	$CaCl_2$

Figure 17.6
The pattern of properties which puts argon between chlorine and potassium.

But, if Mendeléev's ideas were correct, argon could not exist on its own in the Periodic Table. Instead of there just being one of these new elements, there had to be a whole family of them. Ramsay set out to look for the other members of the group. He confirmed the existence of helium in 1895, and in 1898 he separated neon, krypton and xenon from liquid air. All these unreactive gases belong to the group which we now call the *noble gases*. In this way another prediction based on Mendeléev's theories turned out to be correct.

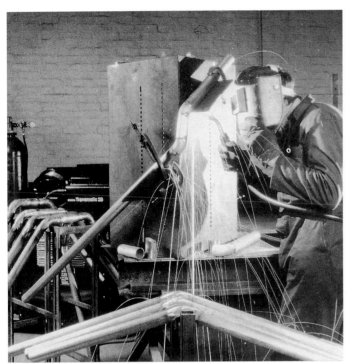

Figure 17.7
Welding stairway handrails using a helium-based gas to exclude air from the arc and the weld.

Figure 17.8
Tungsten-filament bulb production. They are filled with argon, a noble gas.

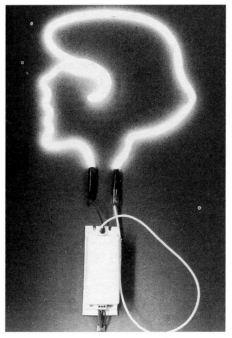

Figure 17.9
Neon strip lighting.

Figure 17.10
Testing a small xenon-filled flash bulb.

7 Two of the pairs of "law-breaking elements" are tellurium and iodine, potassium and argon. What is the other pair of "law-breakers" in the modern Periodic Table?

8 Suggest reasons for the use of helium in welding as shown in figure 17.7.

9 The family name for the noble gases used to be the *inert gases*. Why were they called "inert" in the first place? Why did the name have to be changed? (You will have to ask your teacher or consult reference books to find the answers.)

10 Ramsay confirmed the existence of helium on Earth, but it was first discovered by astronomers in 1868. With the help of reference books find out how it was that astronomers, rather than chemists, discovered a new element.

C17.3 Groups and periods

This section reviews the chemistry of groups 2 and 7. When you answer the questions in this section you will find that you already know a good deal about these elements. You should find that the Periodic Table helps to make sense of their chemistry.

The main use of the Periodic Table today is to bring order to the mass of information which has been discovered about all the elements and their compounds. The elements in the table are related to each other in two ways: vertically in columns called *groups* and horizontally in rows called *periods*.

Groups

The groups consist of elements which have similar properties. One group has already been mentioned in this chapter: the group of noble gases. They are all colourless gases and they are all extremely unreactive.

Magnesium and calcium are two metals which form many important compounds which are mentioned frequently in the other topics in this book. Both metals belong to group 2 in the Periodic Table. Figure 17.11 includes some of the chemistry of these metals which you have met previously. You can see that they are similar, but not identical.

Element	Appearance	Reaction with oxygen	Carbonate	Hydroxide	Ion
Magnesium	grey metal	burns brightly on heating with a white flame to form white oxide, MgO	white solid $MgCO_3$ which is insoluble in water	white solid, $Mg(OH)_2$, which is slightly soluble in water giving an alkaline solution; used as an antacid	Mg^{2+} causes hardness in water
Calcium	grey metal	burns brightly on heating with a red flame to form a white oxide, CaO	white solid $CaCO_3$ which is insoluble in water	white solid, $Ca(OH)_2$, which is slightly soluble in water giving an alkaline solution; used to control soil acidity	Ca^{2+} causes hardness in water

Figure 17.11
The similarities between magnesium and calcium.

11 When were the group 2 metals discovered? Who discovered them and how? (See table 1 in the Data section and figure 4.26.)

12 Which minerals consist of the carbonates of calcium or magnesium? (See figure 4.7.)

13 Write a symbol equation for the reaction of magnesium oxide with water to form magnesium hydroxide.

14 With the help of an equation, explain the use of magnesium hydroxide as an antacid. (See section **C**12.4.)

15 In what ways is calcium hydroxide useful as a soil conditioner? (See section **C**15.5.)

16 Why do calcium and magnesium ions cause hardness in water? (See section **C**10.4.)

17 Copy and complete the table in figure 17.12. Fill in the first row of the table by making predictions about strontium from what you know of the chemistry of magnesium and calcium. Then fill in the second row with the help of reference books and your teacher. How accurate are your predictions?

	Appearance of the element	**Reaction with oxygen**	**Colour, solubility, and formula of the carbonate**	**Formula and properties of the hydroxide**	**Symbol for the ion**
Predicted properties of strontium					
Known properties of strontium from reference books					

Figure 17.12

Element	**State at room temperature**	**Structure**	**Effect on vegetable dyes**	**Symbol for ion**	**Appearance, solubility and formula of the compound with sodium**	**Appearance, solubility, formula and effect of light on the compound with silver**
Chlorine	yellow–green gas	molecular Cl_2	chlorine in water bleaches rapidly	Cl^-	white, crystalline solid, NaCl, soluble in water	white, insoluble solid, AgCl; turns grey in the light
Bromine	dark red liquid turning to an orange vapour	molecular Br_2	bromine in water bleaches slowly	Br^-	white crystalline solid, NaBr; soluble in water	pale yellow, insoluble solid, AgBr; slowly darkens in the light
Iodine	grey solid turning to a violet vapour on warming	molecular I_2	iodine is hardly soluble but the solution will bleach on warming	I^-	white crystalline solid, NaI; soluble in water	yellow, insoluble solid, AgI; light-sensitive

Figure 17.13
The similarities between chlorine, bromine, and iodine.

Chlorine, bromine and iodine are three non-metals which belong to the group of elements called the halogens. Figure 17.13 is a table which includes aspects of the chemistry of these elements. Like the members of most families, the halogens are in many ways similar, but they are not identical.

18 How do you explain the observations that chlorine is a gas and that bromine and iodine are easily vaporized?

19 Predict the products at the electrodes when **a** molten sodium bromide and **b** aqueous sodium bromide are electrolysed. (See section **C**4.3.)

20 What use is made of the fact that the silver compounds of the halogens are sensitive to light?

21 Copy and complete the table in figure 17.14: fill in table by making predictions from what you know of the chemistry of chlorine, bromine, and iodine.

Figure 17.14

	State at room temperature	Structure	Effect on vegetable dyes	Symbol for ion	Appearance, solubility, and formula of the compound with sodium	Appearance solubility, formula, and effect of light on the compound with silver
Predicted properties of fluorine						

Periods

Periods are the horizontal rows in the Table. The periods start with metals on the left and end with non-metals on the right. Each complete period is finished off with a noble gas.

There are two elements in the first period, eight elements in the second and third periods, but eighteen in the fourth. There is a close connection between the periods in the Table and the structure of atoms, as explained in the next chapter.

Figure 17.15
There is a change from metallic to non-metallic properties across a period. Here sticks of sodium and phosphorus are being tested for electrical conductivity.

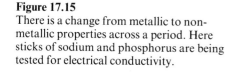

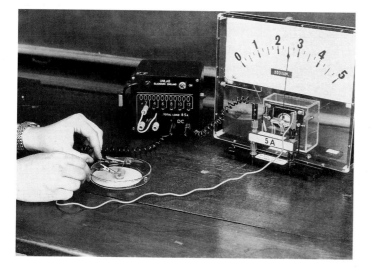
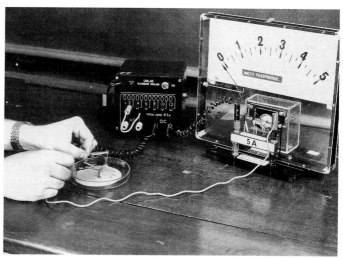

Figure 17.16
The change from alkaline hydroxides on the left to acids on the right when the oxides of elements react with water.

							2 He
3 LiOH lithium hydroxide	4 Be	5 B	6 H_2CO_3 carbonic acid	7 HNO_3 nitric acid	8 O	9 F	10 Ne
11 NaOH sodium hydroxide	12 $Mg(OH)_2$ magnesium hydroxide	13 Al	14 Si	15 H_3PO_4 phosphoric acid	16 H_2SO_4 sulphuric acid	17 $HClO_4$ chloric acid	18 Ar
19 KOH potassium hydroxide	20 $Ca(OH)_2$ calcium hydroxide						

Earlier chapters in this book have mentioned the chemistry of some of the oxides of the elements. There is a pattern to this chemistry which becomes clearer if looked at with the help of the Periodic Table.

Look at figure 17.16. In groups 1 and 2 on the left of the table, the reactive metals form oxides which combine with water to make alkaline hydroxides. The manufacture of sodium hydroxide from brine by electrolysis is described in Chapter **C4** and so is the manufacture of calcium hydroxide from limestone. Chapters **C10** and **C15** mention the use of calcium hydroxide as an alkali to control the pH of water and soil. The use of magnesium hydroxide as an antacid is investigated in Chapter **C12**.

On the right of the table, many of the oxides of non-metals react with water to form acids. Carbon dioxide forms carbonic acid in water and this is responsible for the hardness of water in limestone regions as explained in Chapter **C10**. Chapter **C10** also explains that sulphur dioxide turns into sulphuric acid in the presence of air and water. This is the origin of acid rain. The effect of acid rain on the environment is discussed in Biology section **B**17.7.

22 In Biology section **B3.2** you are asked to design an experiment to see what happens if plants grow in the light but without carbon dioxide. On Worksheet **B3A** you are told that you can use soda lime to remove carbon dioxide from the air. Soda lime is a mixture of sodium hydroxide and calcium hydroxide. Explain why soda lime can be expected to react with carbon dioxide.

23a How does nitrogen dioxide get into the atmosphere? (See section **C13.4** and Biology sections **B15.4** and **B17.7**.)
b Write a balanced equation for the reaction of nitrogen dioxide with water and oxygen to form nitric acid.

Summary

1 Draw up a time chart to show some of the main events leading to the discovery of the Periodic Table. (Figure 4.13 shows you one way of setting this out.)

2 Copy and complete the table in figure 17.17 to contrast a typical group 2 metal (magnesium) with a typical group 7 non-metal (chlorine).

Property	Magnesium (a typical group 2 element)	Chlorine (a typical group 7 element)
Appearance and state at room temperature		
Melting point in °C		
Electrical conductivity		
Structure		
Symbol for the ion		
Formula and acid–base properties of the compound formed when its oxide dissolves in water		

Figure 17.17

Chapter **C**18

Atoms and bonding

*As you know from Physics Chapter **P**3, the discovery of radioactivity helped to revolutionize our understanding of the structure of atoms. It is no longer possible to think of atoms as solid lumps which cannot be broken apart. This chapter starts with the picture of atoms described in Physics section **P**3.7. It then goes further and shows that a knowledge of how the electrons are arranged in atoms can begin to explain much of the chemistry introduced in the other chapters of this book.*

If you go on to study chemistry at a more advanced level, you will realize that one of the major themes in the subject is the attempt to explain the properties of elements and compounds in terms of atomic structure.

This is a very theoretical chapter. You will probably not study it all at once, and your teacher will tell you which parts you are expected to understand. You will find many references to other parts of this book and to sections in your Physics and Biology books. These references, and the questions, are included to help you to revise topics which you have studied earlier in your science course.

C18.1 **Atomic structure**

We now imagine that atoms have a minute positive nucleus surrounded by a cloud of negatively charged electrons.

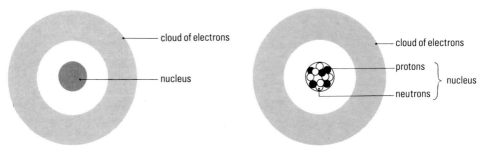

Figure 18.1
The structure of an atom.

Figure 18.2
The particles which make up atoms.

The nucleus consists of protons which are positively charged and neutrons which are uncharged. Protons and neutrons have the same mass of 1 u. The mass of an atom is concentrated in its nucleus because the mass of the electrons is very, very small.

In an atom the number of electrons is equal to the number of protons. Overall the positive and negative charges cancel out and the atom is electrically neutral.

An extra idea is needed to explain many of the important properties of elements and the arrangement of the elements in the Periodic Table. This part of the theory suggests that the electrons are arranged in a series of *shells* around the nucleus.

Think of a table tennis ball – the white plastic is like an egg shell only spherical. Now imagine a series of larger and larger plastic balls one outside the other with space between them. Finally (and this is the hard bit) get rid of the plastic and shrink the whole thing down to the size of an atom. Just think of the shells as regions in space where electrons can be. The shells only exist if there are electrons in them!

Each shell can only contain a limited number of electrons. When one shell is full the electrons go into the next shell. Evidence for this theory can be obtained by measuring the amount of energy needed to remove electrons from atoms.

There are eleven electrons in a sodium atom. The quantities of energy needed to remove these electrons one by one have been measured and are represented by the areas of the squares in figure 18.3. You can see that one electron is removed quite easily. The next eight are more difficult to remove. The last two electrons are obviously held very powerfully by the nucleus.

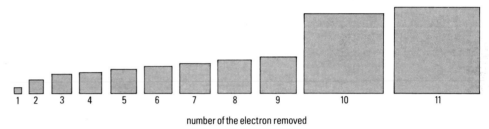

number of the electron removed

Figure 18.3
The amounts of energy needed to remove the electrons one by one from a sodium atom.

This supports the idea that the electrons in a sodium atom are arranged in three shells, as shown in figure 18.4.

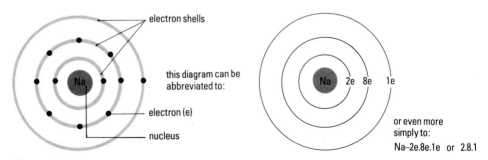

Figure 18.4
Three ways of representing the electrons in shells in a sodium atom.

The one electron furthest from the nucleus is the easiest to remove. Then there are eight in the middle and another two nearest to the nucleus. These two are very strongly held by the positive charge at the centre.

The first shell which is closest to the nucleus can hold up to two electrons. The second shell can hold eight. If there are more electrons they occupy further shells, although after the first twenty elements the arrangements become increasingly complicated. The number and arrangement of the electrons in the atoms of the first twenty elements are shown in figure 18.5.

Element	Number of protons in the nucleus	Number of electrons in each shell			
		Shell 1	Shell 2	Shell 3	Shell 4
Hydrogen	1	1			
Helium	2	2			
Lithium	3	2	1		
Beryllium	4	2	2		
Boron	5	2	3		
Carbon	6	2	4		
Nitrogen	7	2	5		
Oxygen	8	2	6		
Fluorine	9	2	7		
Neon	10	2	8		
Sodium	11	2	8	1	
Magnesium	12	2	8	2	
Aluminium	13	2	8	3	
Silicon	14	2	8	4	
Phosphorus	15	2	8	5	
Sulphur	16	2	8	6	
Chlorine	17	2	8	7	
Argon	18	2	8	8	
Potassium	19	2	8	8	1
Calcium	20	2	8	8	2

Figure 18.5
The atomic structures of the first 20 elements.

You can work out the structure of an atom given two numbers:

- the atomic number, and
- the mass number.

The *atomic number* is the number of protons in the nucleus. In an atom, the number of electrons is equal to the number of protons, so the atomic number also tells you how many electrons there are in an atom.

The *mass number* is the **total** number of protons and neutrons in the nucleus.

Box 1 shows how these numbers can be added to the symbols for the elements.

BOX 1   Symbols for atoms

The atomic number and mass number of an element can be added to its chemical symbol. The correct arrangement is as shown here:

mass number ⟶ A
X ⟵ symbol
atomic number ⟶ Z

The atomic number of sodium is 11 and its mass number is 23. It is represented as: $^{23}_{11}\text{Na}$

Box 2 gives examples showing how to work out the numbers of protons, neutrons and electrons in an atom or ion from the atomic number and mass number.

BOX 2 Using atomic numbers and mass numbers

Example 1
How many protons, neutrons and electrons are there in the atom: $^{39}_{19}K$?

Answer
The number of protons $=$ the atomic number
$= 19$

The number of neutrons $=$ the mass number $-$ the atomic number
$= 39 - 19$
$= 20$

The number of electrons $=$ the number of protons
$= 19$

So in an atom of $^{39}_{19}K$ there are 19 protons, 20 neutrons and 19 electrons.

Example 2
How many protons, neutrons and electrons are there in the ion: $^{9}_{4}Be^{2+}$?

Answer
The number of protons $= 4$

The number of neutrons $= 9 - 4$
$= 5$

The ion has a double positive charge because it has lost two electrons, so there are not enough electrons to balance the charge on the nucleus.

The number of electrons $= 4 - 2$
$= 2$

So in the ion $^{9}_{4}Be^{2+}$ there are 4 protons, 5 neutrons and 2 electrons.

---

1 Answer this question on a copy of the shortened version of the Periodic Table provided on Worksheet **C17A**. Below the symbol for each element write the arrangement of electrons in shells. (For example, write "2.8.1" for sodium.)

2 How many protons, neutrons and electrons are there in these atoms and ions: $^{14}_{6}C$, $^{20}_{10}Ne$, $^{234}_{91}Pa$, $^{127}_{53}I$, $^{1}_{1}H^{+}$, $^{35}_{17}Cl^{-}$?

3 Draw diagrams to show the arrangement of electrons in shells in these atoms: $^{9}_{4}Be$, $^{19}_{9}F$, $^{24}_{12}Mg$.

C18.2 Atomic structure and the Periodic Table

*If you read section C17.1 you will find that it took a long time before the idea of the Periodic Table was accepted by scientists. They could not understand why there should be a connection between the chemical properties of an element and the mass of its atoms. The theory of atomic structure described in this chapter **can** explain the arrangement of the elements in the Periodic Table. This is* *one of the successes of the theory. It is part of the evidence which makes the theory seem reasonable.*

The exceptions to Mendeléev's law

Mendeléev knew that he had to put iodine in group 7 and tellurium in group 6, but he did not know why (see section **C**17.2). Now we know the answer. As you can see if you look at the modern version of the Periodic Table in the Data section, the elements are arranged strictly in order of atomic number. There are no exceptions to this rule.

The first element in the table is hydrogen with one proton in its nucleus, next comes helium with two, and so on. Tellurium comes before iodine because there are 52 protons in a tellurium atom and 53 protons in an iodine atom. Tellurium atoms have more neutrons than iodine atoms. This increases their atomic mass but does not affect chemical properties. The modern version of Mendeléev's law states: "When the elements are arranged in order of **atomic number** similar properties recur at intervals."

Groups

If you study figure 18.5 and answer question **1**, you can see why elements in the same group have similar, but not identical, properties. Look, for example, at the elements in group 2 (Be, Mg, Ca). You can see that they are similar because they all have two electrons in the outer shell. They differ because as you go down the group the number of inner, full shells increases.

When atoms react it is the electrons in the outer shell which are involved. So elements with the same number of outer electrons have similar properties.

If you examine the Periodic Table in the Data section you will note that for each main column the group number is the same as the number of electrons in the outer shells of the atoms. Tellurium has similar chemical properties to the other elements in group 6 because there are six electrons in the outer shell of a tellurium atom. Iodine belongs in group 7 because it has seven outer electrons, like the other halogens.

Periods

You can also see that there is a connection between the horizontal rows in the table and the structure of the atoms. The first row fills the first shell. Then the second shell is filled across the second row from lithium (2.1) to neon (2.8).

At this point the situation becomes more complicated for reasons which you will understand if you go on to more advanced chemistry courses. The third shell starts to fill from sodium to argon, but then two electrons go into the fourth shell with potassium and calcium, before ten more electrons go into the third shell from scandium to zinc. This accounts for the place of the transition metal elements in the table.

There is no simple explanation of why the transition elements come immediately after calcium. A detailed study of the energies of the electrons in atoms is needed to account for the positions of the elements in the Periodic Table beyond calcium.

Isotopes

*You will find it helpful to read Physics sections **P**3.5 and **P**3.6 before you study the next part of this section. The information on Worksheet **P**3B will also help you.*

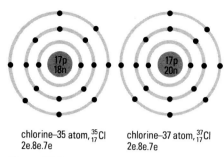

chlorine–35 atom, $^{35}_{17}$Cl
2e.8e.7e

chlorine–37 atom, $^{37}_{17}$Cl
2e.8e.7e

Figure 18.6
The structure of the two main isotopes of chlorine.

The term *isotope* is based on two Greek words: *isos* meaning "same" and *topos* meaning "place". The term was invented because the isotopes of an element belong in the "same place" in the Periodic Table.

Chemically all the atoms of an element **are** the same, but it turns out that they can differ in mass. Chlorine has two main isotopes with masses 35 u and 37 u. The atomic number of chlorine is 17, so both isotopes have 17 protons in the nucleus. The different masses are due to the different numbers of neutrons in the nuclei, as shown in figure 18.6.

Chlorine-35 atoms and chlorine-37 atoms both have the same chemical properties because they both have 17 electrons arranged in a 2.8.7 pattern. Both atoms have seven electrons in the outer shell and it is the outer electrons which decide the chemistry of an element. They are isotopes in the "same place" in the Periodic Table because they have the same atomic number.

There are usually three chlorine-35 atoms to every one chlorine-37 atom in the gas, and so the **average** atomic mass is 35.5 u, which is the value quoted in table 3 of the Data section.

BOX 3   Atomic masses

In natural chlorine there are three chlorine-35 atoms for every chlorine-37 atom.

So the average atomic mass of natural chlorine

$$= \frac{(3 \times 35\,\text{u}) + 37\,\text{u}}{4}$$

$$= 35.5\,\text{u}$$

Neither of the isotopes of chlorine is radioactive. Some other elements which are not usually radioactive may have isotopes which are. An example is carbon. Most natural carbon is carbon-12 which is not radioactive. However, there is a small percentage of radioactive carbon-14 present. This "tiny proportion" of carbon is mentioned in Physics section **P**3.6 to explain how measurements of radioactivity can be used to work out the age of materials.

Carbon-14 is also used in *tracer* experiments. In Biology section **B**8.2 you can find out how radioactive carbon is used to "trace" the movements of sugars in plants. Figure 8.3 in your Biology book shows how photography helps to find out where the radioactive carbon travels to in a plant.

Experiment B in Biology Worksheet **B**9A uses a similar method to find out what happens to the carbon atoms in sugar molecules during respiration. In this case the presence of the radioactive isotope is detected using a Geiger–Müller tube.

Radioactive sulphur-35 can be used to study the uptake of sulphate ions by barley seedlings as explained in Biology Worksheet **B**9B. There are other examples of the use of tracers on Physics Worksheet **P**3B.

The movement of isotopes can be followed even if they are not radioactive. This is possible because a sensitive instrument called a mass spectrometer can separate and detect atoms with different masses.

The isotopes of the elements with high atomic numbers, such as uranium and protactinium, are all radioactive. Figure 3.23 in Physics section **P3.6** illustrates an experiment to measure the half-life of protactinium-234. If you have seen this experiment you may have wondered why there is protactinium in a solution of a uranium compound.

Uranium-238 decays by giving off alpha particles. An alpha particle consists of two protons and two neutrons. So when a uranium-238 atom decays, it loses a total of four particles from the nucleus. Its mass number goes down by 4 units. At the same time, its atomic number decreases by 2 units because it has lost two protons. The result is an atom with mass number 234 and atomic number 90. This is thorium-234 as shown in box 4.

BOX 4   Alpha and beta decay

Example 1
What happens when an atom of uranium-238 decays by giving off an alpha particle?

Answer
The Periodic Table on page 316 shows that the atomic number of uranium is 92. The symbol for an atom of uranium-238 is: $^{238}_{92}U$.

An alpha particle consists of two protons and two neutrons. The loss of two protons means that the atomic number of the new atom is two units less than for a uranium atom. The Periodic Table shows that the atom formed must be an isotope of thorium.

The loss of four particles from the nucleus means that the mass number goes down by four units. So the new atom is thorium-234.

$$^{238}_{92}U \longrightarrow {}^{234}_{90}Th + \text{an alpha particle}$$

Example 2
What happens when an atom of thorium-234 gives off a beta particle?

Answer
A neutron turns into a proton when an atom gives off a beta particle. The mass number is unchanged because the total number of neutrons and protons does not alter.

The atomic number increases by one unit because there is now one more proton. In this example the new element is protactinium, as can be seen from the Periodic Table.

$$^{234}_{90}Th \longrightarrow {}^{234}_{91}Pa + \text{a beta particle}$$

Thorium-234 is also radioactive, and box 4 shows what happens when it decays by giving off a beta particle. A neutron turns into a proton when an atom emits a beta particle (see Physics section **P3.7**). This means that the mass number is unchanged because the total number of protons and neutrons does not alter. However the atomic number (the number of protons) increases by 1 unit. So an atom of thorium-234 turns into an atom of protactinium-234.

Notice that when the atoms of a radioactive element lose alpha particles the new element formed is two places to the left in the Periodic Table. The atomic number goes down by two units. When an element decays by beta decay the new element is one place to the right because the atomic number increases by one unit.

---

4 Draw diagrams in the style of figure 18.6 to represent the two main isotopes of: $^{20}_{10}Ne$, $^{22}_{10}Ne$.

5 Copper consists of 69 % copper-63 and 31 % copper-65. What is the average atomic mass of copper?

6 Work out what is formed when the following atoms decay: in each case, show what happens in the style of the examples in box 4. (Look up the atomic numbers in table 2 of the Data section.)
a Thorium-232 gives off alpha particles.
b Protactinium-234 gives off beta particles.
c Lead-212 gives off beta particles.
d Thallium-208 gives off alpha particles.

C18.3   Atoms into ions

Some theories in science are very powerful because we can use them to explain many things. The idea that atoms can turn into ions is one of these powerful theories, and is used in Biology and Physics as well as in Chemistry.

This section gives you a chance to revise some of the examples of ionic theory which you have met during your science course. It explains how atoms can turn into ions and then goes on to look at the two examples in more detail.

What can ionic theory explain?

Ionic theory was first thought of by Michael Faraday to explain electrolysis. Figure 5.34 in section C5.4 tells you something of Faraday's work. In that section there is a series of questions which you may have answered. The questions should help you see how ionic theory can account for some of the properties of sodium chloride, and other ionic compounds. The theory of electrolysis is explained in more detail later in this section.

Boxes 2 and 3 in section C5.4 show you how to work out the formulae of ionic compounds knowing the charges on the ions. A list of common ions is included in table 6 of the Data section.

Topic **C3** in this book uses ionic theory to explain the reactions of acids and alkalis. We think that all acid solutions contain hydrogen ions, H^+, (box 1 on page 185 in section C10.3). We also think that alkaline solutions contain hydroxide ions (box 1 on page 205 in section C11.2). The theory is that, in a neutralization reaction, the hydrogen ions from the acid react with hydroxide ions from the alkali to form neutral water (see box 2 on page 219 in section C12.4).

Ionic theory also helps us to understand how water becomes hard and how it can be softened by ion exchange (see section C10.4). Similar ideas explain how the weathering of rocks can add to the fertility of soil. Chapter **B15** includes a list of the ions needed by plants. Section C16.5 describes how fertilizers are manufactured to provide these ions, which plants require for growth.

The alpha, beta and gamma rays given off by radioactive materials are called ionizing radiation (see Physics section **P3.2**). This is because they have enough energy to break up molecules into ions. The instruments which detect and measure radioactivity respond to the ions produced by the radiations.

---

7 Work out the formulae of these ionic compounds:
a aluminium bromide
b iron(III) sulphate
c magnesium phosphate
d sodium sulphide.

8 Write an equation to show what happens when hydrogen ions from an acid are neutralized by hydroxide ions from an alkali, using the reaction of nitric acid with sodium hydroxide as the example.

9 Which are the three main elements supplied to plants by fertilizers? In which ionic form are they taken up by plants from the soil?

10 Compare, with the help of diagrams, the natural ion exchange processes in the soil (see sections **C15.3** and **C16.6**) with the use of ion exchange to soften water (see section **C10.4**).

11 How can ionic theory explain the formation of precipitates? Use some of the following examples to illustrate your answer: the appearance of rust stains when clothes are washed in water which contains iron compounds (see section **C10.3**); the formation of scum when washing with soap in hard water (see section **C10.4**); the use of mordants in dyeing (see section **C11.2**); and the formation of insoluble phosphates when fertilizers are added to soil (see section **C16.6** and Worksheet **C15D**).

12 How could you demonstrate by experiment that there are ions in a candle flame? (See Physics section **P3.1**.)

13 Use ionic theory to explain how a Geiger–Müller tube works (see Physics section **P3.2**).

How do atoms turn into ions?

You may have wondered how it is possible for an atom to turn into an ion. Why do metals form positive ions while non-metals form negative ions? Is there a reason for the fact that all the metals in group 1 of the Periodic Table form $1+$ ions while the metals in group 2 form $2+$ ions?

Knowledge of the Periodic Table and atomic structure helps to explain what happens. Each period finishes with a noble gas: He (2), Ne (2.8), Ar (2.8.8) and so on. The noble gases are very unreactive. So we can expect that atoms or ions with the same arrangement of electrons as the noble gas atoms will also be unreactive.

As you already know, ionic compounds consist of a metal combined with one, or more, non-metals. Take sodium chloride as an example. Figure 18.7 shows how sodium and chlorine atoms turn into ions when they react.

You will notice that the sodium atom loses its outer electron, so in a sodium ion the number of protons is one more than the number of electrons. The ion therefore carries a $1+$ charge. You will also notice that the sodium ion has the same number, and arrangement, of electrons as the nearest noble gas which is neon.

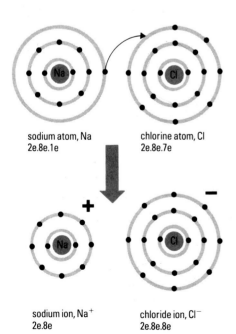

sodium atom, Na
2e.8e.1e

chlorine atom, Cl
2e.8e.7e

sodium ion, Na$^+$
2e.8e

chloride ion, Cl$^-$
2e.8e.8e

Figure 18.7
How ions are formed when sodium reacts with chlorine.

The chlorine atom gains one electron, so that the ion formed carries a 1 − charge. This gives the chloride ion the same number and arrangement of electrons as the noble gas argon.

It appears that when atoms react to form ions they gain, or lose, electrons in such a way that, the ions formed have the same electron arrangement as a noble gas. Reactive atoms turn into ions. This is the answer to the question at the head of section **C**5.4. Sodium chloride is so different from sodium and chlorine because the ions in the compound are much less reactive than the atoms in the elements.

---

14 Draw diagrams to show how atoms turn into ions when these compounds are formed from their elements: lithium fluoride, magnesium oxide, sodium oxide and magnesium chloride.

15 Look up the charges formed by simple ions in table 6 of the Data section. Write the charges onto a copy of the shortened version of the Periodic Table (Worksheet **C**17A). (Simple ions are ions which are formed from just one atom.)

16 Look at the full Periodic Table in the Data section and predict the charges on the simple ions of these elements: caesium (atomic number 55), strontium (atomic number 38), gallium (atomic number 31), selenium (atomic number 34) and astatine (atomic number 85).

17 Why do metals form positive ions while non-metals form negative ions?

18 Why do all the metals in group 1 form 1 + ions while all the metals in group 2 form 2 + ions?

19 Write down the number of protons, neutrons and electrons in **a** a sodium ion and **b** a neon atom. In what ways are a sodium ion and a neon atom the same, and how do they differ?

20 Write down the number of protons, neutrons and electrons in **a** a chloride ion and **b** an argon atom. In what ways are a chloride ion and an argon atom the same, and how do they differ?

21 Why are ionic compounds crystalline solids with high melting-points? (See section **C**5.4.)

22 Show that the symbol $^4_2\text{He}^{2+}$ represents an alpha particle.

How can electrolysis be explained?

Electrolysis turns ions back into atoms. Metal ions are positively charged and so they are attracted to the cathode which is negative. Non-metal ions are negatively charged so they are attracted to the anode. This is explained in figure 18.8 using molten sodium chloride as the example.

The cathode is negatively charged because electrons are flowing into it from the battery or power pack. Sodium ions arriving at the cathode are given electrons, and this turns them back into neutral sodium atoms.

The anode becomes positively charged as electrons flow out of it to the battery or power pack. When chloride ions arrive at the anode they lose their extra electron and turn back into neutral atoms. But chlorine atoms do not exist singly – they join up in pairs to form chlorine molecules.

These changes can be described using equations.

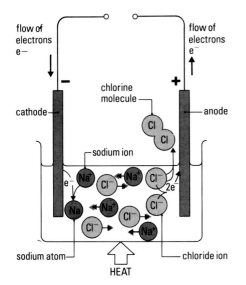

Sodium ions are attracted to the cathode where they gain electrons and turn back into sodium atoms.

Chloride ions are attracted to the anode where they lose electrons and turn back into atoms. The atoms join in pairs to form chlorine gas.

Figure 18.8
How ions turn back into atoms during the electrolysis of molten sodium chloride.

Understanding electrolysis

Two students are looking at the electrolysis of copper(II) chloride solution.

Figure 18.9

Figure 18.9 (continued)

At the cathode: $Na^+ +$ an electron $\longrightarrow$ Na
           ion   (supplied by   atom
                           the cathode)

At the anode: $Cl^- \longrightarrow$ Cl + an electron
          ion   atom   (taken away by
                              the anode)

then: $Cl + Cl \longrightarrow \quad Cl_2$
                             molecule

23 Why do ionic compounds conduct when molten, or when dissolved in water, but not when solid?

24 Why do the changes seen during electrolysis take place at the surface of the electrodes and not in between them?

25 Study figure 18.9 and then write equations using symbols and words to explain what happens at the electrodes during the electrolysis of copper(II) chloride. (Box 2 on page 188 in section **C**10.3 tells you the meaning of the (II) in the name copper(II) chloride.)

26 Write equations using symbols and words, to show what happens during the electrolysis of molten aluminium oxide (see section **C**4.5).

How do cells and batteries work?

You may find it helpful to re-read Chapter C14 before you study this section. This will remind you of the design of cells and batteries.

At the moment the most widely used rechargeable cell is the lead–acid cell. In the future it may be that new sodium–sulphur cells will become much more common. Here we shall use the sodium–sulphur example to explain how cells work. The chemistry is simpler than in more familiar cells.

An electric current is a flow of electrons. This is discussed in Physics section **P**19.3. How can a chemical reaction produce a current?

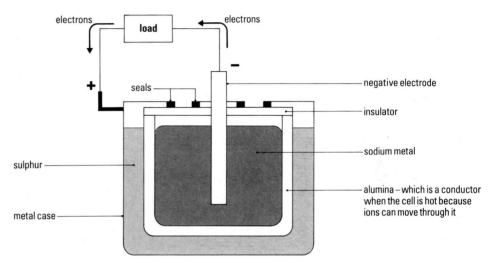

Figure 18.10
The design of a sodium-sulphur cell.

The sodium–sulphur cell is based on the reaction of sodium metal with sulphur. Atoms turn into ions when these two elements react. This is similar to the reaction of sodium with chlorine described in section **C**5.4.

$$2Na + S \longrightarrow (2Na^+ + S^{2-})$$
sodium   sulphur   sodium sulphide

To understand how this reaction can be used to produce a flow of electrons, it helps to split the equation into two parts. One part is to show what happens to the sodium atoms. The other part is to show what happens to the sulphur atoms.

Sodium: $2Na \longrightarrow 2Na^+$ + two electrons given to the sulphur atom
          two sodium   two sodium
          atoms   ions

Sulphur: S + two electrons from the sodium atoms $\longrightarrow$ S^{2-}
           sulphur atom   sulphide ion

When sodium reacts directly with sulphur, the electrons jump from the sodium atoms to the sulphur atoms as sodium sulphide is formed. The reaction is exothermic. The reaction releases energy which heats up the surroundings.

In a sodium–sulphur cell, the sodium and sulphur are separated by alumina so that they cannot react directly. The electrons are forced to flow out of the electrode dipping into the sodium, through a wire, to the electrode dipping into the sulphur.

Now the energy from the reaction can be transferred more usefully. The cell runs at 2 volts. This is a measure of the amount of energy available from the reaction.

More and more of the atoms turn into ions as the cell discharges. When the cell is "flat"; all the sodium and sulphur has reacted and the cell has to be recharged.

Recharging the cell is an example of electrolysis. Electrolysis turns ions back into atoms, as explained in the previous section. So electrolysis can be used to turn the sodium ions back to sodium and the sulphide ions back into sulphur. Then the cell is ready for use again.

Chemically the idea of the sodium–sulphur cell is quite simple. Making a reliable cell for commercial use is much more complicated. The cells run at 270 to 410 °C (see section **C14.3**). Early versions of the cell were unreliable because the seals leaked. Also the alumina which divides the two parts of the cell was liable to crack.

---

27 Why do you think that sodium–sulphur cells have to be kept at 270 to 410 °C when in use?

28 Write equations to show what happens at the electrodes during the recharging of a sodium–sulphur cell.

29 In all three cells illustrated in figures 14.5 to 14.7 the negative electrode is zinc. What do you think happens to the zinc atoms when a current is being taken from these cells? Write an equation for the change, which is the same in all three cells.

C18.4   Bonding in molecules

Another important idea is the theory that the bonding which holds together the atoms in molecules is much stronger than the weak attractive forces between the molecules. You may find it helpful to re-read section C5.3 before you study this section.

What can molecular theory explain?

Look at the picture of water boiling in figure 18.11. This may help to remind you of your studies of Physics and cooking in Chapter **P2**.

The energy being transferred to the water goes to separating the molecules, $H_2O(l)$, from each other so that they become steam, $H_2O(g)$. The molecules do not change – they are just freed from each other as they evaporate.

Much more energy is needed to break up water molecules into hydrogen atoms and oxygen atoms. This does **not** happen as water evaporates. The molecules of steam are the same as the molecules of water.

This idea is very important. The bonds **within** molecules are strong so that molecules do not break up easily. The forces **between** molecules are weak so that it is easy to separate them. Molecular substances are often liquids or gases at room temperature.

These ideas are used in Physics Chapter **P2**. Pressure cookers, refrigerators and aerosol sprays all depend on molecular liquids and gases. The theory also helps to explain processes such as diffusion, osmosis and dialysis.

Figure 18.11
Water boiling in a glass beaker.

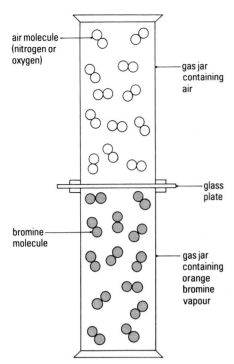

Figure 18.12
An experiment to demonstrate diffusion.

Diffusion is a spreading out and mixing process. We imagine one set of molecules jostling its way through a second set of molecules. In the lungs oxygen molecules diffuse from the air we breathe through the nitrogen molecules to the alveoli where they are taken up by the blood. Carbon dioxide diffuses in the opposite direction. This is described in Biology Chapter **B7**.

Not all molecules are small. Polymer molecules, for example, are very long. There are natural polymers such as starch and cellulose. There are synthetic polymers such as polythene and nylon. The properties of polymers are described in Chapter **C8**.

The difference between the strong bonds in the molecules and the weak forces between the molecules is true for big molecules as well as small ones. You can remind yourself of how the theory explains the difference between thermoplastics and thermosets by reading section **C8.3**.

Many of the older dyes used to colour cloth are held to the fibres by weak forces between the molecules. New bright and fast dyes were discovered by a research group trying to develop colours which would react with the fibres of cloth. These reactive dyes are held by the strong chemical bonds which link atoms in molecules. This discovery is described in outline in section **C11.4**.

---

30 How does molecular theory explain these facts?
a Water evaporates faster when it is hotter.
b The pressure inside a pressure cooker is higher than atmospheric pressure while food is being cooked in it.
c An aerosol paint can feel cold after it has been sprayed for some time.

31 Figure 18.12 represents a diffusion experiment. The diagram shows where the molecules are at the start. Draw two more diagrams: one diagram to show how you picture the arrangement of the molecules a few minutes after the glass plate has been removed; and a second diagram to show the arrangement after several hours.

32 Use diagrams to explain what happens to the molecules during dialysis and osmosis. Why do these processes depend on the fact that the forces between molecules are weak? (See section **C9.2** and Biology sections **B12.5** and **B10.1**.)

33 A candle snaps if you try to bend it, and it is easy to scrape bits of wax off a candle with a finger nail. A rod of polythene bends without breaking and is hard to scratch. Wax and polythene are both made of hydrocarbon molecules. Why are they different? (See section **C8.3** and especially figures 8.29 and 8.30.)

What are the bonds in molecules?

A knowledge of atomic structure can help to make sense of the bonding in molecules. Here we will show that in simple cases it is possible to explain the number of bonds formed by different atoms. If you look back to figure 5.14 you will see how many bonds are formed by a number of common atoms.

When non-metal atoms combine to form molecules, they do so by sharing electrons. Take hydrogen as an example: each atom has one electron in the first shell. Two hydrogen atoms combine to form a hydrogen molecule by electron sharing.

By sharing electrons each atom can be considered to have the electron arrangement of helium with two electrons in the first shell. This type of strong bonding in molecules is called *covalent* bonding. "*Co*" means "together or

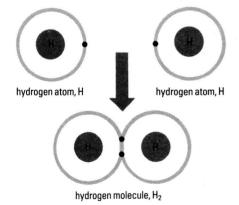

Figure 18.13
Covalent bonding in hydrogen.

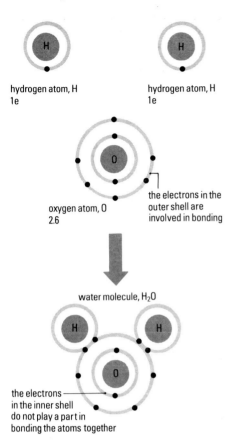

oxygen atom, O
2.6

the electrons in the
outer shell are
involved in bonding

water molecule, H₂O

the electrons
in the inner shell
do not play a part in
bonding the atoms together

Figure 18.14
Covalent bonding in water.

joint" while the Latin word *valentia* means strength. So we have strength by sharing.

Other examples of covalent bonding are shown in figures 18.14 and 18.15. If you look at the diagrams you will see that in each case all the atoms in the molecules have in their shells the same number of electrons as a noble gas atom. This happens in many, but by no means all molecules.

All molecules are held together by covalent bonds, but there are also important materials with covalently bonded giant structures, such as diamond and silica (see section **C5.3**).

---

34 Draw diagrams to show the covalent bonding in these molecules: fluorine, F_2; hydrogen chloride, HCl; ammonia, NH_3; ethane, C_2H_6.

35 Try to extend this theory of bonding to explain how double bonds form in these molecules: oxygen, O_2; carbon dioxide, CO_2; and ethene, C_2H_4.

The theory of bonding given in this section explains the number of bonds formed by a limited number of non-metal atoms. As described here the theory does not explain why electron sharing holds atoms together. If you go on to study chemistry at a more advanced level, you will find out more about covalent bonding.

Summary

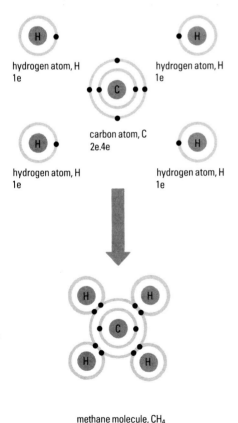

carbon atom, C
2e.4e

methane molecule, CH₄

Figure 18.15
Covalent bonding in methane.

1 Copy and complete the diagram in figure 18.16.

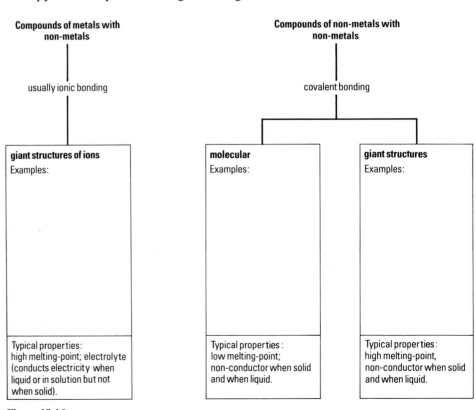

Figure 18.16
Summary of bonding in compounds.

Data section

Table 1   **The Periodic Table**

Table 2   **The elements**
Atomic number, name, symbol used in equations, structure, date of discovery

Table 3   **The elements**
Atomic number, name, atomic mass, molar mass, melting-point, boiling-point, density, symbol

Table 4   **Properties of some organic compounds**
Name, formula, structure, melting-point, boiling-point, solubility in water

Table 5   **Properties of some inorganic compounds**
Name, formula, structure, melting-point, boiling-point, solubility in water

Table 6   **Charges on ions**

Table 7   **Chemical apparatus**

Table 1 — **The Periodic Table**

Period	Group 1 ALKALI METALS	Group 2 ALKALINE EARTH METALS						
1								1 H Hydrogen 1.0 u
2	3 Li Lithium 6.9 u	4 Be Beryllium 9.0 u						
3	11 Na Sodium 23.0 u	12 Mg Magnesium 24.3 u				TRANSITION ELEMENTS SERIES		
4	19 K Potassium 39.1 u	20 Ca Calcium 40.1 u	21 Sc Scandium 45.0 u	22 Ti Titanium 47.9 u	23 V Vanadium 50.9 u	24 Cr Chromium 52.0 u	25 Mn Manganese 54.9 u	26 Fe Iron 55.9 u
5	37 Rb Rubidium 85.5 u	38 Sr Strontium 87.6 u	39 Y Yttrium 88.9 u	40 Zr Zirconium 91.2 u	41 Nb Niobium 92.9 u	42 Mo Molybdenum 95.9 u	43 Tc Technetium (99 u)	44 Ru Ruthenium 101.1 u
6	55 Cs Caesium 132.9 u	56 Ba Barium 137.3 u	57–71 see below Lanthanum	72 Hf Hafnium 178.5 u	73 Ta Tantalum 181.0 u	74 W Tungsten 183.9 u	75 Re Rhenium 186.2 u	76 Os Osmium 190.2 u
7	87 Fr Francium (223 u)	88 Ra Radium (226 u)	89–103 see below Actinium	104 Unq Unnilquedium				

LANTHANIDE SERIES	57 La Lanthanum 138.9 u	58 Ce Cerium 140.1 u	59 Pr Praseodymium 140.9 u	60 Nd Neodymium 144.2 u	61 Pm Promethium (147 u)	62 Sm Samarium 150.4 u	63 Eu Europium 152.4 u
ACTINIDE SERIES	89 Ac Actinium (227 u)	90 Th Thorium 232.0 u	91 Pa Protactinium (231 u)	92 U Uranium 238.1 u	93 Np Neptunium (237 u)	94 Pu Plutonium (242 u)	95 Am Americium (243 u)

Key

57	Atomic number
La	Symbol
Lanthanum	Name
138.9 u	Atomic mass

Group 3	Group 4	Group 5	Group 6	Group 7 HALOGENS	Group 8 NOBLE GASES
					2 He Helium 4.0 u
5 B Boron 10.8 u	6 C Carbon 12.0 u	7 N Nitrogen 14.0 u	8 O Oxygen 16.0 u	9 F Fluorine 19.0 u	10 Ne Neon 20.2 u
13 Al Aluminium 27.0 u	14 Si Silicon 28.1 u	15 P Phosphorus 31.0 u	16 S Sulphur 32.1 u	17 Cl Chlorine 35.5 u	18 Ar Argon 39.9 u

27 Co Cobalt 58.9 u	28 Ni Nickel 58.7 u	29 Cu Copper 63.5 u	30 Zn Zinc 65.4 u	31 Ga Gallium 69.7 u	32 Ge Germanium 72.6 u	33 As Arsenic 74.9 u	34 Se Selenium 79.0 u	35 Br Bromine 79.9 u	36 Kr Krypton 83.8 u
45 Rh Rhodium 102.9 u	46 Pd Palladium 106.4 u	47 Ag Silver 107.9 u	48 Cd Cadmium 112.4 u	49 In Indium 114.8 u	50 Sn Tin 118.7 u	51 Sb Antimony 121.8 u	52 Te Tellurium 127.6 u	53 I Iodine 126.9 u	54 Xe Xenon 131.3 u
77 Ir Iridium 192.2 u	78 Pt Platinum 195.1 u	79 Au Gold 197.0 u	80 Hg Mercury 200.6 u	81 Tl Thallium 204.4 u	82 Pb Lead 207.2 u	83 Bi Bismuth 209.0 u	84 Po Polonium (210 u)	85 At Astatine (210 u)	86 Rn Radon (222 u)

64 Gd Gadolinium 157.3 u	65 Tb Terbium 158.9 u	66 Dy Dysprosium 162.5 u	67 Ho Holmium 164.9 u	68 Er Erbium 167.3 u	69 Tm Thulium 168.9 u	70 Yb Ytterbium 173.0 u	71 Lu Lutetium 175.0 u
96 Cm Curium (247 u)	97 Bk Berkelium (245 u)	98 Cf Californium (251 u)	99 Es Einsteinium (254 u)	100 Fm Fermium (253 u)	101 Md Mendelevium (256 u)	102 No Nobelium (254 u)	103 Lr Lawrencium (257 u)

Table 2

The elements

In this table the elements are listed in alphabetical order.

You can use this table to help you find an element in table 3 where they are arranged in atomic number order.

The structure is only shown for the commoner elements which you may meet during your chemistry course.

Atomic number	Element	Symbol used in equations	Structure	Date of discovery
89	actinium	Ac		1899
13	aluminium	Al	giant (atoms)	1827
95	americium	Am		1944
51	antimony	Sb		ancient
18	argon	Ar	molecular (single atoms)	1894
33	arsenic	As		1250
85	astatine	At		1940
56	barium	Ba	giant (atoms)	1808
97	berkelium	Bk		1949
4	beryllium	Be	giant (atoms)	1798
83	bismuth	Bi		1753
5	boron	B		1808
35	bromine	Br_2	molecular	1826
48	cadmium	Cd	giant (atoms)	1817
55	caesium	Cs	giant (atoms)	1860
20	calcium	Ca	giant (atoms)	1808
98	californium	Cf		1950
6	carbon	C	giant (atoms)	ancient
58	cerium	Ce		1803
17	chlorine	Cl_2	molecules	1774
24	chromium	Cr	giant (atoms)	1797
27	cobalt	Co	giant (atoms)	1735
29	copper	Cu	giant (atoms)	ancient
96	curium	Cm		1944
66	dysprosium	Dy		1886
99	einsteinium	Es		1952

Atomic number	Element	Symbol used in equations	Structure	Date of discovery
68	erbium	Er		1843
63	europium	Eu		1896
100	fermium	Fm		1953
9	fluorine	F$_2$	molecular	1887
87	francium	Fr		1939
64	gadolinium	Gd		1880
31	gallium	Ga		1875
32	germanium	Ge		1886
79	gold	Au	giant (atoms)	ancient
72	hafnium	Hf		
2	helium	He	molecular (single atoms)	1868/1895 (see page 293)
67	holmium	Ho		1879
1	hydrogen	H$_2$	molecular	1766
49	indium	In		1863
53	iodine	I$_2$	molecular	1811
77	iridium	Ir		1803
26	iron	Fe	giant (atoms)	ancient
36	krypton	Kr	molecular (single atoms)	1898
57	lanthanum	La		1839
103	lawrencium	Lw		1961
82	lead	Pb	giant (atoms)	ancient
3	lithium	Li	giant (atoms)	1818
71	lutetium	Lu		1907
12	magnesium	Mg	giant (atoms)	1808
25	manganese	Mn	giant (atoms)	1774
101	mendeleevium	Md		1955
80	mercury	Hg		ancient
42	molybdenum	Mo		1778
60	neodymium	Nd		1885
10	neon	Ne	molecular (single atoms)	1898
93	neptunium	Np		1940
28	nickel	Ni	giant (atoms)	1751
41	niobium	Nb		1802
7	nitrogen	N$_2$	molecular	1772
102	nobelium	No		1957
76	osmium	Os		1803
8	oxygen	O$_2$	molecular	1774
46	palladium	Pd		1803
15	phosphorus			1669
	white	P$_4$	molecular	
	red	P	giant (atoms)	

Atomic number	Element	Symbol used in equations	Structure	Date of discovery
78	platinum	Pt	giant (atoms)	1735
94	plutonium	Pu		1940
84	polonium	Po		1898
19	potassium	K	giant (atoms)	1807
59	praseodymium	Pr		1885
61	promethium	Pm		1947
91	protactinium	Pa		1917
88	radium	Ra		1898
86	radon	Rn		1900
75	rhenium	Re		1925
45	rhodium	Rh		1803
37	rubidium	Rb		1861
44	ruthenium	Ru		1844
62	samarium	Sm		1879
21	scandium	Sc		1879
34	selenium	Se		1817
14	silicon	Si	giant (atoms)	1824
47	silver	Ag	giant (atoms)	ancient
11	sodium	Na	giant (atoms)	1807
38	strontium	Sr		1808
16	sulphur	S_8	molecular	ancient
73	tantalum	Ta		1802
43	technetium	Tc		1937
52	tellurium	Te		1783
65	terbium	Tb		1843
81	thallium	Tl		1861
90	thorium	Th		1828
69	thulium	Tm		1879
50	tin	Sn	giant (atoms)	ancient
22	titanium	Ti		1825
74	tungsten	W		1783
92	uranium	U		1841
23	vanadium	V		1830
54	xenon	Xe	molecular (single atoms)	1898
70	ytterbium	Yb		1878
39	yttrium	Y		1843
30	zinc	Zn	giant (atoms)	1746
40	zirconium	Zr		1824

Table 3

The elements

In this table the elements are arranged in order of atomic number. All the elements with atomic numbers from 1–36 are included, together with selected elements with higher atomic numbers.

The values for the atomic masses and molar masses are accurate enough for most calculations. More accurate values are given in table 1.

The abbreviation "sub" shows that the solid sublimes instead of melting at the temperature listed.

The densities of gases are given at atmospheric pressure and 25 °C.

Atomic number	Element	Atomic mass in u	Molar mass in g/mol	Melting-point in °C	Boiling-point in °C	Density in g/ml	Symbol
1	hydrogen	1	1	−259	−253	0.000 08	H
2	helium	4	4	−270	−269	0.000 17	He
3	lithium	7	7	181	1331	0.5	Li
4	beryllium	9	9	1283	2487	1.9	Be
5	boron	11	11	2027	3927	2.5	B
6	carbon	12	12				C
	-diamond			3550	4827	3.5	
	-graphite			3720^{sub}		2.3	
7	nitrogen	14	14	−210	−196	0.001 17	N
8	oxygen	16	16	−219	−183	0.001 33	O
9	fluorine	19	19	−220	−188	0.001 58	F
10	neon	20	20	−248	−246	0.000 84	Ne
11	sodium	23	23	98	890	0.97	Na
12	magnesium	24	24	650	1117	1.7	Mg
13	aluminium	27	27	659	2447	2.7	Al
14	silicon	28	28	1410	2677	2.3	Si
15	phosphorus (white)	31	31	44	281	1.8	P
16	sulphur	32	32				S
	-monoclinic			119	445	1.96	
	-rhombic			113		2.07	
17	chlorine	35.5	35.5	−101	−34	0.002 99	Cl
18	argon	40	40	−189	−186	0.001 66	Ar
19	potassium	39	39	63	766	0.86	K
20	calcium	40	40	850	1492	1.6	Ca

Atomic number	Element	Atomic mass in u	Molar mass in g/mol	Melting-point in °C	Boiling-point in °C	Density in g/ml	Symbol
21	scandium	45	45	1400	2477	3.0	Sc
22	titanium	48	48	1677	3277	4.5	Ti
23	vanadium	51	51	1917	3377	6.1	V
24	chromium	52	52	1903	2642	7.2	Cr
25	manganese	55	55	1244	2041	7.4	Mn
26	iron	56	56	1539	2887	7.9	Fe
27	cobalt	59	59	1495	2877	8.9	Co
28	nickel	59	59	1455	2837	8.9	Ni
29	copper	64	64	1083	2582	8.9	Cu
30	zinc	65	65	419	908	7.1	Zn
31	gallium	70	70	30	2237	5.9	Ga
32	germanium	73	73	937	2827	5.4	Ge
33	arsenic	75	75	613^{sub}		5.7	As
34	selenium	79	79	217	685	4.8	Se
35	bromine	80	80	−7	58	3.1	Br
36	krypton	84	84	−157	−153	0.003 46	Kr
47	silver	108	108	961	2127	10.5	Ag
50	tin	119	119	232	2687	7.3	Sn
53	iodine	127	127	114	184	4.9	I
55	caesium	133	133	29	685	1.9	Cs
56	barium	137	137	710	1637	3.5	Ba
74	tungsten	184	184	3377	5527	19.3	W
78	platinum	195	195	1770	3827	21.5	Pt
79	gold	197	197	1063	2707	19.3	Au
80	mercury	201	201	−39	357	13.5	Hg
82	lead	207	207	328	1751	11.3	Pb
92	uranium	238	238	1135	4000	19.1	U

Table 4

Properties of some organic compounds

Organic compounds are carbon compounds. In this table the substances listed are arranged according to type as far as possible.

The abbreviation "dec" means that the compound decomposes before it melts or boils.

Compound	Formula	Structure	Melting-point in °C	Boiling-point in °C	Solubility in water
Alkanes					
methane	CH_4	molecular	-182	-161	insoluble
ethane	C_2H_6	molecular	-183	-88	insoluble
propane	C_3H_8	molecular	-188	-42	insoluble
butane	C_4H_{10}	molecular	-138	-0.5	insoluble
pentane	C_5H_{12}	molecular	-130	36	insoluble
hexane	C_6H_{14}	molecular	-95	69	insoluble
decane	$C_{10}H_{22}$	molecular	-30	174	insoluble
eicosane	$C_{20}H_{42}$	molecular	37	344	insoluble
Unsaturated hydrocarbons					
ethene	C_2H_4	molecular	-169	-104	insoluble
propene	C_3H_6	molecular	-185	-48	insoluble
phenylethene	C_8H_8	molecular	-30	145	insoluble
Alcohols					
methanol	CH_3OH	molecular	-98	65	soluble
ethanol	C_2H_5OH	molecular	-114	78	soluble
Acids					
methanoic acid (formic acid)	HCO_2H	molecular	9	101	soluble
ethanoic acid (acetic acid)	CH_3CO_2H	molecular	17	118	soluble
Carbohydrates					
glucose	$C_6H_{12}O_6$	molecular	146	dec	soluble
fructose	$C_6H_{12}O_6$	molecular	103	dec	soluble
sucrose	$C_{12}H_{22}O_{11}$	molecular	185	dec	soluble
starch	$(C_6H_{10}O_5)_n$	polymer	dec		insoluble
Useful solvents					
propanone (acetone)	CH_3COCH_3	molecular	-95	56	soluble
trichloroethane	CH_3CCl_3	molecular	-30	74	insoluble

Table 5

Properties of some inorganic compounds

The compounds are in alphabetical order.

The following abbreviations are used in this table:
dec – which means that the compound decomposes before it melts or boils
h – which means that the crystals are often hydrated so that they give off water on heating before they melt

The solubility is indicated approximately using these abbreviations:
i – insoluble
sl.s – slightly soluble
s – soluble
vs – very soluble
r – means that the compound reacts with water

All the compounds are white or colourless unless otherwise stated.

Compound	Formula	Structure	Melting-point in °C	Boiling-point in °C	Solubility	Notes
aluminium chloride	$AlCl_3$	giant (ions)	sub		r	
aluminium hydroxide	$Al(OH)_3$	giant (ions)	300	dec	i	
aluminium oxide	Al_2O_3	giant (ions)	2015	2980	i	
ammonia	NH_3	molecular	−78	−34	vs	
ammonium chloride	NH_4Cl	giant (ions)	sub		s	
ammonium nitrate	NH_4NO_3	giant (ions)	170	dec	vs	
ammonium sulphate	$(NH_4)_2SO_4$	giant (ions)	dec		s	
barium chloride	$BaCl_2$	giant (ions)	963	1560	s	h
barium sulphate	$BaSO_4$	giant (ions)	1580		i	
calcium carbonate	$CaCO_3$	giant (ions)	dec		i	
calcium chloride	$CaCl_2$	giant (ions)	782	2000	s	h
calcium hydroxide	$Ca(OH)_2$	giant (ions)	dec		sl.s	
calcium nitrate	$Ca(NO_3)_2$	giant (ions)	561	dec	vs	h
calcium oxide	CaO	giant (ions)	2600	3000	r	
carbon monoxide	CO	molecular	−205	−191	i	
carbon dioxide	CO_2	molecular	sub		sl.s	
copper(II) chloride	$CuCl_2$	giant (ions)	620	dec	s	h, green
copper(II) nitrate	$Cu(NO_3)_2$	giant (ions)	114	dec	vs	h, blue

Compound	Formula	Structure	Melting-point in °C	Boiling-point in °C	Solubility	Notes
copper(I) oxide	Cu_2O	giant (ions)	1235		i	red
copper(II) oxide	CuO	giant (ions)	1326		i	black
copper(II) sulphate	$CuSO_4$	giant (ions)	dec		s	h, blue
hydrogen bromide	HBr	molecular	-87	-67	vs, r	
hydrogen chloride	HCl	molecular	-114	-85	vs, r	forms hydro-chloric acid in water
hydrogen iodide	HI	molecular	-51	-35	vs, r	
hydrogen oxide (see water)						
hydrogen peroxide	H_2O_2	molecular	0	150	vs	
hydrogen sulphide	H_2S	molecular	-85	-60	sl.s	
iron(II) chloride	$FeCl_2$	giant (ions)	667	sub	s	yellow–green
iron(III) chloride	$FeCl_3$	giant (ions)	307	dec	s	h, orange
iron(III) oxide	Fe_2O_3	giant (ions)	1565		i	red
iron(II) sulphate	$FeSO_4$	giant (ions)	dec		s	pale green
lead(II) chloride	$PbCl_2$	giant (ions)	501	950	sl. s	
lead(II) nitrate	$Pb(NO_3)_2$	giant (ions)	dec		s	
lead(II) oxide	PbO	giant (ions)	886	1472	i	yellow
lead(IV) oxide	PbO_2	giant (ions)	dec		i	brown
lead(II) sulphate	$PbSO_4$	giant (ions)	1170		i	
magnesium carbonate	$MgCO_3$	giant (ions)	dec		i	
magnesium chloride	$MgCl_2$	giant (ions)	714	1418	s	h
magnesium nitrate	$Mg(NO_3)_2$	giant (ions)			vs	h
manganese(IV) oxide	MnO_2	giant (ions)	dec		i	black
manganese(II) sulphate	$MnSO_4$	giant (ions)	700	dec	s	h, pink
concentrated nitric acid	HNO_3	molecular	-42	83	vs, r	
nitrogen hydride (see ammonia)						
nitrogen oxide	NO	molecular	-163	-151	sl.s	
nitrogen dioxide	NO_2	molecular	-11	21	s	brown
potassium bromide	KBr	giant (ions)	730	1435	s	
potassium chloride	KCl	giant (ions)	776	1500	s	
potassium hydroxide	KOH	giant (ions)	360	1322	vs	
potassium iodide	KI	giant (ions)	686	1330	vs	
potassium manganate(VII)	$KMnO_4$	giant (ions)	dec		s	purple
potassium nitrate	KNO_3	giant (ions)	334	dec	vs	
silicon dioxide	SiO_2	giant (atoms)	1610	2230	i	
silver bromide	AgBr	giant (ions)	432	dec	i	pale yellow
silver chloride	AgCl	giant (ions)	455	1550	i	
silver iodide	AgI	giant (ions)	558	1506	i	yellow
silver nitrate	$AgNO_3$	giant (ions)	212	dec	vs	

Compound	Formula	Structure	Melting-point in °C	Boiling-point in °C	Solubility	Notes
sodium bromide	$NaBr$	giant (ions)	755	1390	s	
sodium carbonate	Na_2CO_3	giant (ions)	851	dec	s	h
sodium chloride	$NaCl$	giant (ions)	808	1465	s	
sodium hydrogencarbonate	$NaHCO_3$	giant (ions)	dec		s	
sodium hydroxide	$NaOH$	giant (ions)	318	1390	s	
sodium nitrate	$NaNO_3$	giant (ions)	307	dec	vs	
sodium sulphate	$NaSO_4$	giant (ions)	890		s	
sulphur dioxide	SO_2	molecular	−75	−10	vs, r	
sulphur trioxide	SO_3	molecular	17	43	r	
concentrated sulphuric acid	H_2SO_4	molecular	10	330	vs, r	
titanium(IV) oxide	TiO_2	giant (ions)	1830		i	
zinc oxide	ZnO	giant (ions)	1975		i	yellow when hot
zinc sulphate	$ZnSO_4$	giant (ions)	740	dec	s	
water	H_2O	molecular	0	100		

Table 6

Charges on ions

Positive ions (cations)			Negative ions (anions)		
Charge	Cation	Symbol	Charge	Anion	Symbol
1+	copper(I)	Cu^+	1−	bromide	Br^-
	hydrogen	H^+		chloride	Cl^-
	potassium	K^+		hydroxide	OH^-
	silver	Ag^+		iodide	I^-
	sodium	Na^+		nitrate	NO_3^-
2+	calcium	Ca^{2+}	2−	carbonate	CO_3^{2-}
	copper(II)	Cu^{2+}		oxide	O^{2-}
	iron(II)	Fe^{2+}		sulphate	SO_4^{2-}
	magnesium	Mg^{2+}		sulphide	S^{2-}
	zinc	Zn^{2+}		sulphite	SO_3^{2-}
3+	aluminium	Al^{3+}	3−	nitride	N^{3-}
	iron(III)	Fe^{3+}		phosphate	PO_4^{3-}

Table 7   Chemical apparatus

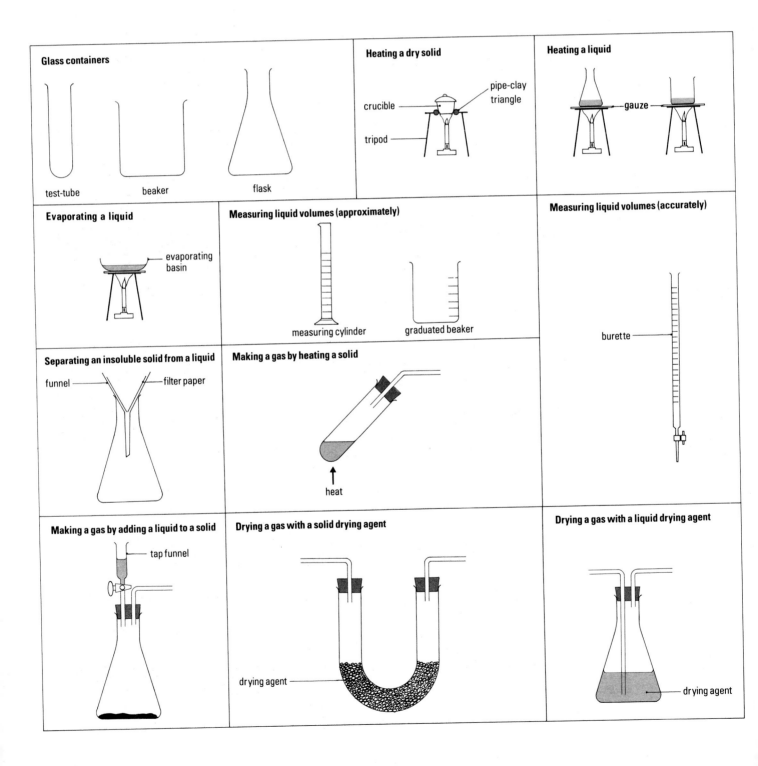

Glass containers

test-tube   beaker   flask

Heating a dry solid

crucible   pipe-clay triangle

tripod

Heating a liquid

gauze

Evaporating a liquid

evaporating basin

Measuring liquid volumes (approximately)

measuring cylinder   graduated beaker

Measuring liquid volumes (accurately)

burette

Separating an insoluble solid from a liquid

funnel   filter paper

Making a gas by heating a solid

heat

Making a gas by adding a liquid to a solid

tap funnel

Drying a gas with a solid drying agent

drying agent

Drying a gas with a liquid drying agent

drying agent

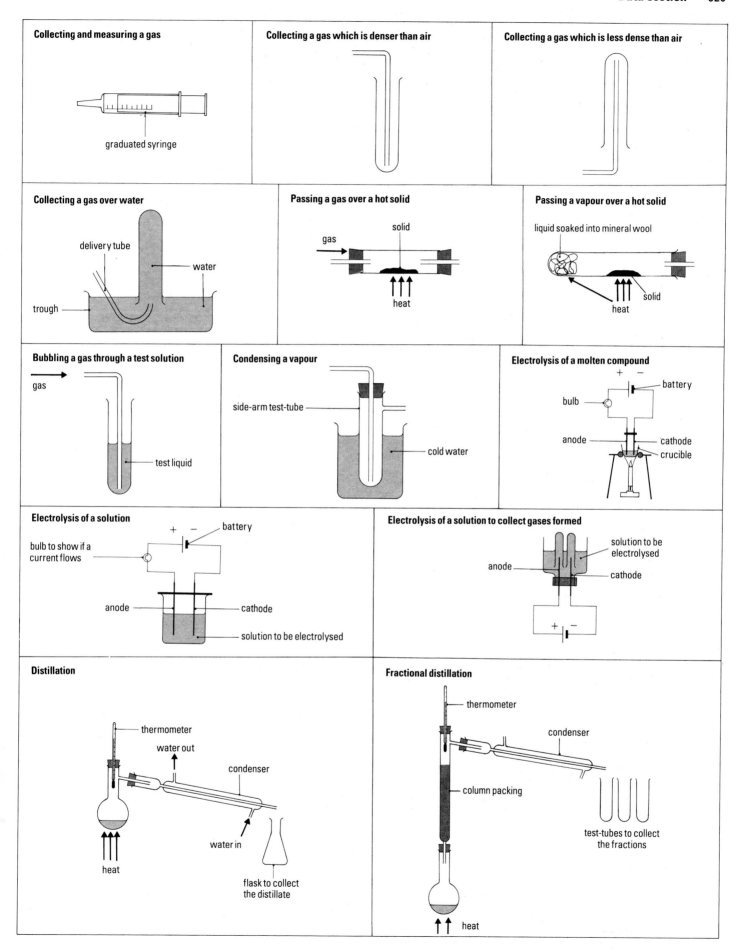

Collecting and measuring a gas

graduated syringe

Collecting a gas which is denser than air

Collecting a gas which is less dense than air

Collecting a gas over water

delivery tube

water

trough

Passing a gas over a hot solid

solid

gas

heat

Passing a vapour over a hot solid

liquid soaked into mineral wool

solid

heat

Bubbling a gas through a test solution

gas

test liquid

Condensing a vapour

side-arm test-tube

cold water

Electrolysis of a molten compound

+ −

battery

bulb

anode

cathode

crucible

Electrolysis of a solution

+ −

battery

bulb to show if a
current flows

anode

cathode

solution to be electrolysed

Electrolysis of a solution to collect gases formed

solution to be
electrolysed

anode

cathode

+ −

Distillation

thermometer

water out

condenser

water in

heat

flask to collect
the distillate

Fractional distillation

thermometer

condenser

column packing

test-tubes to collect
the fractions

heat

Index

absorption spectrometer, 81
acid rain, 185, 240–42
acids, 296
  as catalysts, 55
  neutralization, 219–20
  properties, 185
addition polymerization, 147
aerosols, 165, 168
air
  in soil, 262, 263
  noble gases from, 292
  prevention of pollution, 242
alanine, 54
alcohol, see ethanol
alcohols, data, 323
alizarin, 47, 201, 202, 203
  synthetic, 207
alkali metals, 15
alkaline manganese cell, 251
alkalis, 205, 219, 296
  production, 66, 67, 68–9
alkanes, 24–35, 323
alloy steels, 139
alloys, 129
  properties, 138–40
  structure, 141
  uses, 129–30, 139–40
alpha decay, 304
alum, 82
  as a mordant, 61, 203, 205
  industry, 61–6
aluminium, 83
  extraction, 78–81
  recycling, 85
  uses, 78
aluminium nitrate, 100
aluminium oxide, 78–9, 83
amino acids, 53–4, 148
ammonia, 277–9
  alkaline properties of solution, 205
  manufacture, 232, 236, 272–7
  properties, 277
ammonite, 63
ammonium nitrate, 277
ammonium salts, 278
amount of substance, 107–9
  calculations using concept, 116, 191, 221, 267
analgesics, 215–18
analysis, 7, 81–3
  mine air, 231
  water, 200
annealing (glass), 124
antacids, 218–21
antibiotics, 222
aquifers, 184
argon, 291
Arrhenius, Svante, 239
"artificial silk", 154
aspirin, 215–18
atmosphere, 10
  see also air
atomic mass unit, 15
atomic masses, 14, 15, 288–9, 291
  data, 323–4
  of isotopes, 303
atomic nucleus, 298
atomic number, 300–301, 302, 303, 304
atomic volume, 289–90
atoms, 11, 14
  structure of, 298–305
Avogadro constant, 108

balancing equations 16, 17, 18–19
ball-and-spring models, 16, 24, 25, 39, 40, 41, 89
basalt, 59, 261

batteries, 248–54
  Davy's, 71
bauxite, 77, 78–9
Benedict's solution, 52
benzene, ring structure, 4
beta decay, 304–5
Billingham works, 279–80, 281
biochemical oxygen demand (BOD), 199
biosphere, 256
biotechnology, 56
blast furnace, 75
blow moulding, 160
blown glass, 114, 122
boiling, 312
boiling-points, 25–6, 90, 323–6
bonding, 90
  metallic, 134
  within molecules, 24–5, 90–92, 157, 312–14
borosilicate glass, 114
Bosch, Carl, 274–5
bottle bank, 127
bottles, 122, 128
Braefoot Bay terminal, 31, 35, 36, 38
Bragg, Sir Lawrence, 102–4, 138
Bragg, Sir William, 102–4
brass, etched, 131
bricks, 125
  refractory, 121
bridges, 87, 134, 141
brightening agents, 182–3
brine, 67–8
Broadbalk field experiment, 270
bromine
  properties, 294–5
  reaction with ethene, 40
  structure, 91
bronzes, 140
Brownian motion, 170
bubble raft, 138
burning, 224–6
butane, 24, 26–7, 29–30

caesium chloride, structure, 104
caffeine, 214, 215
calcium, 293
calcium carbonate, 74–7
  in hard water, 192–4, 195–6
  reaction with acid, 260
  uses, 68, 74–5, 241
  see also chalk; limestone
calcium hydroxide, 74
  neutralizing value, 266–7
  uses, 74, 187, 265
calcium oxide (quicklime), 74
  neutralizing value, 266–7
  uses, 74, 75
calcium sulphate, 194, 195
  see also gypsum
Calgon, 197
camping gas, 27
cans, 135, 142
car battery, see lead–acid cell
car exhaust gas, 242
carbohydrates, 48–53, 323
  composition, 48
  structure, 49–51
carbon
  absorbent, 186–7
  diamond, 92–3, 132, 138
  graphite, 92, 94, 95
  data, 323
  naming, 147
carbon compounds, 22
carbon dioxide, 239, 260–61, 244, 276
carbonates, 220

catalysts, 55, 130
  for ethanol synthesis, 41
  for Haber process, 273, 274, 276
  for polymer cross-linking, 157
  in car exhausts, 242
  see also enzymes
cells (electrical), 249–50
  early history, 248–9
  fuel, 253–4
  ionic theory, 311–12
  primary, 249–50
  secondary, 251–3
cells (plant), walls, 51, 53
celluloid, 152–3
cellulose, 51–3, 210
cellulose acetates, 152–4, 167
ceramics, 87
  composition, 118
  fabrication of articles from, 124–6
  properties, 119, 121–2
  uses, 120–2
chalk, 57, 58, 265
charcoal, 186–7
chemical apparatus, diagrams, 328–9
chemical industry, 4–6
  economic importance, 4–5, 62, 68–9
  location, 31, 36–8, 61, 75–6, 80, 279–80
  people in, 6–7, 43–4, 158–62
  safety in, 36, 44, 246
  use of energy by, 80, 126–8, 210, 274
chemists, 1, 4, 6, 7, 214
chemotherapy, 222
chlorine, 294–5
  isotopes, 303
  production, 68
  reaction with sodium, 96–7, 306–7
  uses, 69
  as disinfectant, 187, 190–91
chloroethene, 42
cholera, 215
cinchona bark, 216
clay
  ceramics from, 87, 118, 124–6
  formation, 261
  in soil, 263–4, 283
cloud seeding, 175
coal
  as a fuel, 63, 230–31
  as a source of chemicals, 206, 232
  mining, 230–1, 238
  origins, 76, 230
coal gas, 232
coal tar, 232
coal tar dyes, 206
cochineal, 203, 204
codeine, 215
coffee, 214
coke, 231, 232
collagen, 145
colloids
  characteristic properties, 169–171
  coagulation of, 176–7
  formation of, 171–4
  in soil, 263, 264
combustion (burning), 224–6
compound fertilizers, 281
compounds, 10–13
  composition by mass, 83
compression moulding, 161–2
concentrations, 191
condensation polymerization, 148, 150–51
continuous phase, 168

copper
  crystals, 132–3
  in malachite, 82
  wire, 135
copper(II) chloride, electrolysis, 308–10
copper–zinc alloy, crystals, 132
corrosion, 141–4
cosmetics, 165, 174
cotton, 149
  dyes for, 208, 209–11
cotton–polyester blend, 212
covalent bonding, 313–14
cracking (hydrocarbons), 28, 32–6, 39, 236
Crick, Francis, 106
cryolite, 79
crystals, 97
  determination of structures, 102–6
  metals, 130–35
cupro-nickel, 140
cysteine, 54

Dalton, John, 14
Davy, Humphry, 70–71, 73, 230, 249
decane, cracking, 39
density (data), 323–4
descaler (kettle), 96
detergents, 180–82
  soap, 180, 181–2, 194–5
  soapless, 180–81, 195
dialysis, 170–71
diaminohexane, 148
diamond, 138
  structure, 92–3, 132
dibromoethane, 40
diffusion, 313
direct dyes, 208
disease prevention, 214–15
disperse phase, 168
distillation, 26, 28–9, 35, 234–5
DNA (deoxyribonucleic acid), 106
Döbereiner, Johann, 288
double bond, 39, 40
drinking vessels, 87
drugs, 47, 213–22
dry cleaning, 179
dyes, 61, 313
  fastness of, 201, 204–5, 206–7, 209–10
  natural, 47, 201–204
  synthetic, 66, 206–212

E numbers, 173
Earth
  age of, 57–8
  element abundance in, 11
  element distribution in, 10
egg white, 169, 170
electrical insulators, 121–2
electrolysis, 70–73
  industrial applications, 68, 78–80
  ionic theory, 98, 99, 307–10
  of molten compounds, 72, 78–80, 97, 99
  of solutions, 68, 73
electrons
  in covalent bond, 313–14
  shells, 299–300, 302
electrostatic precipitator, 177
elements, 10–13
  data, 318–20, 323–4
  discovery, 14–15, 70–71
  structures, 90–94, 95, 96
  symbols, 11, 300
  see also Periodic Table

emulsifiers, 173, 182
emulsions, 164, 165–6, 168, 172–4
energy consumption, 285
energy saving, 127–8, 197
energy transfer, in burning fuel, 228
energy use by industry, 80, 126–8, 210, 274
energy values, fuels, 229
enzymes, 54, 55–6, 145, 169, 182–3
equations, 15–19
　balancing, 16–19, 106–107
　calculations from, 110–12, 221
　model, 16, 33, 40–41
　symbol, 16, 17, 106–7
　word, 12–13
erosion, 58, 257
etching, 131
ethane, 24, 25, 26–7, 28
　cracking, 32–6
ethanol (alcohol), 94, 214
　manufacture, 40–41
　molecular models, 89
　products of burning, 225
ethene (ethylene)
　laboratory production, 39
　manufacture, 32–5
　molecular structure, 39
　polymerization, 41–4
　reactions, 38–42
　uses, 39, 40–1
eutrophication, 199
evaporation, 312
exothermic reactions, 44
extrusion
　aluminium, 81
　cellulose acetate, 153–4
　clay, 125
　plastics, 159–60
　polythene film, 44

Faraday, Michael, 97, 98, 305
fats, 40, 166
fertilizers, 199, 268–86
　composition, 281, 282
　effect on plant growth, 270, 271
　fate in soil, 282–4
　issues related to the use of, 284–5
　manufacture, 279–81
fibres
　natural, 47, 149
　synthetic, 153–4
Fife Ethylene Plant, 32–5
fire extinguishers, 244
fire triangle, 224, 243
fireclay, 121
fires, 243–7
flame calorimeter, 228
flame-proofing, 245
flaring (gas), 37
Float Process, 123–4
fluoridation of water, 191–2
fluorite, 60
foams, 168, 169, 183
　solid, 165, 168
food, 47, 229
food additives, 173
food packaging film, 146, 154
formulae
　calculations based on, 83
　graphical, 24–5
　molecular, 24–5
　of ionic compounds, 99–100
Forth bridge, 141
fossil fuels, 229–37
　and environment, 237–42
fossils, 58, 63, 75, 230
fractional distillation, 26, 28–9, 35, 234–5
Franklin, Rosalind, 106
frost, weathering by, 258–9
fructose, 55
fuel cells, 253–4
fuels, 225, 226–37
　criteria for selection, 226–8
　fossil, see fossil fuels

galena, 60
gallium, 91
Galvani, Luigi, 248–9
galvanized steel, 130, 142
gasoline, natural, 30
gastric juice, 220–21
gelatine, 167
gels, 167
geological timescale, 57–8
germanium, 290–91
giant structures, 90, 314
　carbon, 92–4, 95
　ceramics, 118
　glass, 115, 117
　metals, 92, 132–3, 134, 136–8
　silicon dioxide and silicates, 94, 95
　sodium chloride, 97
glasses, 118
　fabrication of articles from, 122–4
　manufacture, 114
　properties, 113–14, 119, 121–2
　recycling, 126–8
　structure, 115
　uses, 113–14, 119–22, 123
glazes, 125–6
glucose, 48–52, 55, 171
　polymerization of, 51
　structure, 49, 50
glycine, 54
glycogen, 51, 145
golden syrup, 55
grains (metal crystals), 131–2
granite, 59, 94, 95, 261
graphical formulae, 24, 25
graphite
　properties, 92, 94
　structure, 94–5
　uses, 92
greenhouse effect, 239–40
groups in the Periodic Table, 15, 302
　group 1 (alkali metals), 15
　group 2, 293–4
　group 7 (halogens), 15, 294–5
　group 8 (noble gases), 291–2
guano, 271
gypsum, 60, 194

Haber, Fritz, 272–4, 275
Haber process, 272–7
haematite, 60
haemoglobin, 145
halogens, 15, 294–5
henna, 204
hexanedioyl dichloride, 148
hip joint, artificial, 129
Hodgkin, Dorothy, 105–6
Hofmann, August, 206, 207
Hofmann, Felix, 216
honey, 55
Humber bridge, 134
humus, 263–4, 283
Hutton, James, 58
Hyatt, John, 152
hydrated (slaked) lime, 74
　neutralizing value, 266–7
　uses, 74, 187, 265
hydrazine, 226
hydrocarbons
　as fuels, 29–30, 229, 234–7
　bonding in, 22, 24–5
　cracking, 28, 32–6, 39, 236
　data, 323
　fractional distillation, 26, 28–9, 35, 234–5
　molecules of, 23–5, 33, 39, 40, 41–2, 157
　storage, safety of, 30, 245–6
　see also alkanes; benzene; ethene
hydrogen
　fuel, 253–4
　ions, 185, 219, 296
　molecule, 313
　reactions: with oxygen, 12, 13, 16

　with nitrogen, 272–7
hydrophilic groups, 181
hydrophobic groups, 181
hydrosphere, 10
hydroxide ions, 205, 219, 296

ice, 259
igneous rocks, 59, 261
indigo, 47, 201–2, 203
industry
　impact on the environment, 38, 75–6, 198, 200, 238–42
　see also chemical industry; mining
injection moulding, 160–61
inorganic compounds
　data, 324–6
　naming, 61, 188
insulators, electrical, 121–2
insulin, 105–6
inversion of sugars, 55
iodine, 294–5
　position in Periodic Table, 291, 302
　structure, 92, 136–7
ion exchange
　in soil, 264, 283
　water softening by, 197–8
ionic compounds
　bonding in, 96–7, 306–7
　formulae, 99–100
　molar masses, 109–10
　names, 188
　structure of, 96–9, 103–104
ionic theory, 96–7, 99–100, 305–7
　and electrical cells, 311–12
　and electrolysis, 307–10
ionizing radiations, 306
ions
　charges on, 99–100, 188, 327
　colloid coagulation by, 176–7
　data, 327
　formation, 96–7, 306–7
　in glass, 115, 117
　in the soil, 265, 269, 271
iron
　rusting, 141, 142, 143–4
　structure, 136–7
　see also steel
iron compounds, in water, 186
iron(III) hydroxide, precipitation, 189
isotopes, 302–5

jam, 55

kaolinite, 118
Kekulé, F. A., 4
keratin, 145
kettle fur, 195
　removal, 196–7
kidney failure, 171
kinetic theory, 170

lactose, 48
Laue, Max von, 103
Lavoisier, Antoine, 14
Lawes, John, 270
lead, 129
lead–acid cell, 251–3
lead(II) bromide, electrolysis of molten, 72
lead glass, 114
lead pollution, 242
light scattering, 169–70
lime, 187, 265–7
limestone, 57, 58, 59
　and water hardness, 192, 193–4, 195–6
　as raw material, 68, 74–7, 115
　for soil pH control, 265
　issues relating to its extraction, 75–6
logwood, dye from, 204
lysimeter, 283

madder, 47, 202, 203
magma, 59
magnesium, 293
　ions, in hard water, 195
　sacrificial anode, 142
magnesium bromide, 99
magnesium carbonate, 219, 220, 241
magnesium hydroxide, 219–20, 221
malachite, 82
manure, 271, 282
maps, as models, 89
mass number, 300–301, 304
materials and materials science, 86–8
　see also ceramics, glasses, metals, and polymers
mauveine, 206–7, 208
medicines, 213–14
melting-points, 90
　of elements, 321–2
　of inorganic compounds 324–6
　of organic compounds, 323
membrane cell, 68
membranes, 52, 53, 171
Mendeléev, D. I., 287, 289
metallic bonds, 134
metals, 129–44
　corrosion, 141–4
　extraction, 78–80
　fabrication of articles from, 134–5
　properties, 133–5
　structure, 130–33
　uses, 129–30
metamorphic rocks, 59
methane
　bonding in, 24–5, 314
　in natural gas, 26, 27, 236, 237
　reaction with oxygen (equation), 17, 18–19
Meyer, Lothar, 289–90
mica, 95
milk, 166
milk of lime, 74, 187
minerals, 59–60
mining, 230–31, 238
mist, 165
model equations, 16, 33, 40–41
models, 88–90
molar masses, 108–9
　data, 323–4
mole, 107–9
molecular compounds, 90, 91–2, 136–7
　molar masses, 108–9
molecular formulae, 24, 25
molecular models, 89
molecules, 11, 12
　bonding in, 24–5, 90–92, 157, 312–14
　forces between, 91–2, 156–7, 172, 278
　of elements, 90, 91–2, 96
　of hydrocarbons, 23–5, 33, 39, 40, 41–2, 157
mordants, 61, 203, 204–5
Mossmorran, 36–7
　Fife Ethylene Plant, 32–5, 36
　gas separation plant, 28–30, 31, 32

names
　of carbon compounds, 147
　of inorganic compounds, 61, 188
National Parks, 75, 76
natural gas
　composition, 26–7, 237
　pipelines for, 27–8, 237
　sources in North Sea, 26, 236
　uses, 237
　　as a fuel, 226, 229, 236–7
　　as a source of hydrogen, 275–6
natural gasoline, 30
neutralization, ionic theory of, 219

neutralizing values, 266–7
neutrons, 298, 300
Newlands, John, 289
nickel–cadmium cell, 251
nitrates
    in food, 285
    in water, 199, 285
    see also nitrogen fertilizers
nitric acid
    in acid rain, 241
    production, 130, 277–8
nitrogen fertilizers
    composition, 269, 282
    fate in soil, 283–4
    history, 271–5
    issues related to use, 284–5
    manufacture, 275–82
    use, 268–70, 282
nitrogen fixation, 271–7
nitrogen oxides, 242
noble gases, 291, 306–7
noise (at oil terminal), 38
nucleic acids, 106, 145
nylon, 148

"octaves", 289
oil (petroleum), 20–23, 232–6
    fires in storage tanks, 245–7
    origins, 232
    pollution due to, 239
    refining of, 25–32, 234–6
    uses of fractions, 235
    see also petrochemicals
open cast mining, 238
optical brighteners, *see* brightening
        agents
ores, 77
organic chemistry, 22
organic compounds
    data, 323
    naming, 147
organic farming, 285
oxidation, 80, 141, 202, 226
oxides
    films on metals, 131
    metal oxides, 296
    non-metal oxides, 185, 296
oxidizing agent, 226
oxygen
    reactions
        with fuels, 224, 226
        with hydrogen, 12, 13, 16
        with metals, 141
        with methane (equation), 17,
            18–19
ozone, 242

paint, "emulsion", 165, 166–7
painting, 141
    electrostatic, 177
paper, 46
paracetamol, 215
Parkesine, 152
particles, 166
penicillin, 106
perfumes, 47
periodic law, 289
Periodic Table, 15, 288–97, 316–17
    exceptions, 291, 302
    groups, 15, 291, 293–5, 302
    periods, 15, 289, 293, 295–6, 306
Perkin, William, 66, 206–8
petrochemicals
    industry, 21, 27, 29–38, 42–5
    molecules in, 23–5
    processing of, 28–30, 32–4
    reactions, 38–42
    sources, 26
    transport of, 27–8, 31
    petrol, 30
pH of soil, 264–7
pharmacology, 214
phases, 168
phosphates, 182, 197, 199
phosphorus fertilizers
    production, 280–1

use, 269–70, 282
photographic film, 152, 153, 167
photosynthesis, 48, 49
pipelines, 27–8, 233
plants
    cell walls, 52, 53
    chemicals from, 46–56, 201–3,
        216–17
    elements needed by, 269
    preferred pH, 264–5
    roots, 262, 263
plastic materials, 124
plastics
    bonding in, 156–7
    fabrication of articles from,
        158–62
    properties, 154, 156–7
    structure, 41–3, 156–7
    uses, 45, 146–7, 153–5
    see also poly(chloroethene);
        polythene
plastics industry, 158–62
pollution
    air, prevention, 242
    sea, 239
    water, 185, 198–200, 241
poly(chloroethene), 42
polyester, 148
polyester–cotton blend, 212
polymerization, 41–2, 50–51,
    147–8, 150–51, 236
polymers, 42, 313
    natural, 149, 152
    properties, 156–7
    structure, 41–43, 150–51, 156–7
    synthetic, 154–5
    uses, 146–7, 154–5
    see also plastics
polystyrene, 146
polythene [poly(ethene)]
    manufacture, 42–4
    structure, 41–2, 147
polyunsaturated compounds, 40
polyvinyl chloride (pvc), *see*
    poly(chloroethene)
potassium, discovery, 71
potassium chloride, 280, 281
potassium fertilizers
    production, 280–1
    use, 269–70, 282–3
potato cells, 52, 53
pottery, 118, 124–6
power stations, 231, 240, 241–2,
    254
precipitation reactions, 189,
    195, 205
prills, 280
primary cells, 250, 251
printed circuits, 129
Procion dyes, 210–12
propane, 24, 25, 26–7, 29–30
proteins, 148, 166
    in the body, 145
    structure, 53–4
    see also enzymes; gelatine;
        insulin
protons, 298, 300
pumice, 165
pvc, *see* poly(chloroethene)

quarrying, 75–7
quartz (silicon dioxide), 94, 95, 138
quicklime (calcium oxide), 74
    neutralizing value, 266–7
    uses, 74, 75
quinine, 216

radioactive materials, 102, 303–6
radioisotopes, 303–4
Ramsay, William, 291–2
rate of reactions: effect of
    catalyst, 41, 55, 273
    concentration, 260
    pressure, 272–3
    surface area, 260, 262–3

temperature, 44, 243, 260, 272–3
Rattee, I. D., 209
RB211 engine, 130
reactants, 16
reactive dyes, 208, 209–11
reagents, purity, 220
recycling, 85, 126–8, 144
reduction, 80
refractory materials, 121
reinforced concrete, 134
reversible process, 197–8
Reye's syndrome, 215
"rock cycle", 58
rockets, 224
rocks, 57–9, 75
    particles in soil, 262
    weathering, 257–61
    see also limestone
rubber, 47, 146, 177
rusting, 141, 142, 143–4
Rutherford, Ernest, 102

sacrificial anode, 142
safety lamp, 71, 231
St Fergus gas separation plant, 27
salicylic acid, 216
salt, *see* sodium chloride
sandstone, 57, 58
saturated compounds, 40
scale, 195–7
sea, pollution, 239
seaweed, raw material, 64
secondary cells, 251–3
sedimentary rocks, 57, 58–9
silicon dioxide (silica)
    in glass, 114–16
    see also quartz
silk, 149, 204
silver bromide, 167
silver oxide cell, 250
slaked lime (calcium hydroxide), 74
    neutralizing value, 266–7
    uses, 74, 187, 265
slip casting, 125
smoke, 170, 231, 242
snow crystals, 259
soap, 180, 181–2, 194–5
soapless detergent, 180–81, 195
soda-lime glass, 114, 115, 116, 117
sodium
    discovery, 73
    reaction with chlorine, 96–7,
        306–7
sodium–sulphur cell, 253, 311–12
sodium carbonate, 67, 68, 69, 115
sodium chloride
    as raw material, 66–9
    electrolysis, 68, 73, 307, 310
    ions in, 96–7, 99, 306–7
    mining of, 66–7
    uses, 69
        in water softening, 197–8
sodium hydrogencarbonate,
    219, 220
sodium hydroxide, 67, 68, 69, 78
sodium phosphate, 182, 197
sodium stearate, 180, 194
soil, 269
    composition of, 262–3
    formation of, 256–61
    pH, 264–7
    profile, 262
sols, 167, 168, 170, 181
solder, 140
solution mining, 67
Solvay process, 68, 69
space craft, 253
space-filling models, 89
Spence, Peter, 65–6
stabilizers in food, 173
stainless steel, 139
starch, 49–51, 52, 171
    structure, 51
    test for, 52
state symbols, 13, 68
steel, 130, 136, 138–9

corrosion protection, 142–3
    grains in, 131
    strength of, 134, 135
Stephen, W. E., 209, 210
Stone, Edward, 216, 217
structures of substances, 90, 101,
    141, 318–20, 323–6
    key for determining, 101
    X-ray determination, 102–6
subsidence, 238
sucrose, 48
sugars, 48, 49, 55
sulphur, 96
sulphur dioxide, 187, 236, 240, 242
surface tension, 172–3
suspension bridge, 134
sweetness of sugars, 49
symbols
    for atoms, 11, 318–22
    for molecules, 12
        see also formulae
synthesis, 4, 147

tea, 214
tellurium, 291, 302
tensile strength of metals, 134
thermosetting plastics, 156–7, 161
thermosoftening plastics, 156–7
Thomson, J. J., 103
thorium, radioactive decay, 304
titanium dioxide sol, coagulation,
    176
Torrey Canyon, 239
tracer experiments, 303
Tyrian purple, 201, 203

ultra-violet light, 183
unsaturated compounds, 40
uranium, radioactive decay, 304
urine, source of ammonia, 64

vacuum forming, 158–9
vat dyes, 202, 208
vegetable oils, 9, 46
Visking tubing experiments,
    52, 171
vitamin B_{12}, 106
volcanic soil, 261
Volta, Alessandro, 249

washing powders, 182–3, 199
water
    as a product of burning, 12–13,
        16
    covalent bonding in, 314
    hard and soft, 192–8
    in soil, 262
    nitrate in, 285
    pollution of, 185, 198–200, 241
    reaction with ethene, 40–41
    rock weathering by, 258–9
    structure of ice, 259
    wetting power, 179–80
water cycle, 184–6
water treatment, 186–92, 214
Watson, James, 106
waxes, 47
weathering (rocks), 257–61
Whitby, alum industry, 61–4, 66
Wilkins, Maurice, 106
Wilson, C. T. R., 103
willow extract, 216, 217
woad, 201, 203
wool, 149
word equations, 12–13, 16
Wright, Jane C., 222

X-ray crystallography, 102–6

zinc
    crystals, 131
    used in galvanizing, 142
zinc–carbon cell, 250
zinc–titanium alloy, 129